Qualitative Estimates for Partial Differential Equations

An Introduction

James N. Flavin
Salvatore Rionero

CRC Press
Boca Raton New York London Tokyo

Library of Engineering Mathematics

Series Editor
Alan Jeffrey, University of Newcastle upon Tyne and University of Delaware

Linear Algebra and Ordinary Differential Equations
A. Jeffrey

Nonlinear Ordinary Differential Equations
R. Grimshaw (University of New South Wales)

Complex Analysis and Applications
A. Jeffrey

Perturbation Methods for Engineers and Scientists
A. W. Bush (University of Teesside)

Theory of Stability of Continuous Elastic Structures
M. Como and A. Grimaldi (University of Rome T.V.)

Qualitative Estimates for Partial Differential Equations: An Introduction
J.N. Flavin (University College Galway) and
S. Rionero (University of Naples, Federico II)

Asymptotic Methods in Reaction-Diffusion Equations
Z. Peradzýnski (Polish Academy of Sciences, Warsaw)

Continuum Mechanics with Applications
J. Ballmann (RWTH Aachen)

Numerical Methods for Differential Equations
J.R. Dormand (University of Teesside)

Qualitative Estimates for Partial Differential Equations

An Introduction

Library of Congress Cataloging-in-Publication Data

Flavin, James N.
Qualitative estimates for partial differential equations : an introduction / by James N. Flavin and Salvatore Rionero.
p. cm. — (Library of engineering mathematics)
Includes bibliographical references and index.
ISBN 0-8493-8512-1 (alk. paper)
1. Differential equations, Partial—Numerical solutions.
I. Rionero, Salvatore. II. Title. III. Series.
QA377.F55 1995
515'.353—dc20 95-40475
CIP

International Standard Book Number 0-8493-8512-1
Library of Congress Card Number 95-40475
Printed in the United States of America 1 2 3 4 5 6 7 8 9 0
Printed on acid-free paper

To our wives

Freda & Giuseppina

PREFACE

As partial differential equations (P.D.E.) are the principal subject of this book, let us first recall certain fundamental issues which arise in connection therewith. For any given partial differential equation subject to suitable supplementary conditions – boundary and/or initial conditions – the following issues arise:

i) *existence of solution* (i.e. can one find one function which satisfies all the conditions?);

ii) *uniqueness* (or *non-uniqueness*) *of solutions* (i.e. is there at most one solution?);

iii) *determination of the solution/s.*

Concerning (i) one must, of course, prescribe the class of functions in which the solution is to be found. A solution may exist in one class but not in another: it is often possible to prove the existence of weak solutions (i.e. solutions of limited smoothness) while the existence of "regular" solutions (i.e. ones with a considerable degree of smoothness) is more problematic.

With regard to (ii), we emphasize that uniqueness theorems have considerable importance. Generally, the P.D.E. and the associated supplementary conditions model a natural phenomenon: if uniqueness – in addition to existence – obtains, its behaviour can (in principle) be predicted exactly; stated otherwise, its behaviour evolves from the defining conditions in a deterministic manner.

Even if existence and uniqueness obtain, (iii) presents very considerable difficulties: one can only very rarely determine an explicit solution of a P.D.E. subject to suitable supplementary conditions. Excluding these rare cases, if one wishes to determine the characteristics of the solution one can either use approximate methods (e.g. numerical methods, asymptotic methods) or one may use the approach dealt with in this book. Generally speaking, this latter approach seeks to determine – by differential inequality techniques – the principal characteristics, or qualitative features, of the solution of a given *class* of partial differential equations subject to a given *class* of supplementary conditions; such methods address, ***inter alia***, the uniqueness issue mentioned above. In contradistinction to approximate methods, these methods and the conclusions to which they give rise are exact. Moreover – although qualitative methods rarely furnish accurate information on the solution at a point – they furnish very important information on the behaviour of the solution.

Specifically, this book deals with the determination of certain estimates for solutions of partial differential equations subject to boundary and/or initial conditions. Generally, the problems considered model phenomena arising in the natural sciences: fluid mechanics, solid mechanics, physics, chemistry, biology, etc.. The estimates considered are primarily of the following kind: a non-negative measure of the solution in the form of an integral (norm) is defined, which depends upon one of the independent variables, or, exceptionally, upon two. The objective is to obtain an ordinary – or, exceptionally, a partial – differential inequality for the measure, together with suitable supplementary conditions therefor; and to derive therefrom upper and/or lower bounds for the measure, in terms of the data of the problem. Very useful information for *classes* of problems is contained in these bounds; in fact, the properties which can be inferred therefrom may include: uniqueness of solution, continuous dependence, stability, non-existence of solution, pointwise bounds for the solution itself, asymptotic behaviour in time, etc..

Perhaps the most important aspect of the techniques outlined is the choice of the (non-negative) integrand of the integral measure. Sometimes the choice may be obvious and may emerge in a natural way; there may be more then one choice, and the best one may be dictated by the usefulness of the information which it gives or by analytical convenience, or sometimes by a compromise between the two. Also, indeed, the choice may be extremely elusive, and there are many problems which continue to defy a suitable choice.

In addition to the methods based upon differential inequalities which we have described, the book also treats a method based upon certain identities holding along solutions to partial differential equations (Lagrange identity method): in essence, this furnishes identities for integral measures of the solutions of certain P.D.E. in terms of data. Apart from its intrinsic interest, the principal reason for including it is that it provides alternative, complementary, means of dealing with some issues which may be resolved by differential inequality techniques – notably uniqueness and continuous dependence for certain ill-posed problems.

In addition to dealing with partial differential equations, other equations are occasionally treated in the same spirit: integral equations and integro-differential equations; apart from their intrinsic interest, these serve to enhance understanding of P.D.E. by setting them in a broader context.

In summary, ***differential inequality techniques for the estimation, in terms of***

data, of integral measures associated with problems for P.D.E. are the principal (though not exclusive) ***concern of this book.*** In the treatment of a wide range of such techniques, some repetition of treatments in existing monographs and expository articles is inevitable. Nevertheless, the material of the book and/or its treatment is, where possible, chosen for its novelty, and many of our results are appearing for the first time. Inevitably, the treatment is coloured by the interest of the authors.

The general aim is to introduce these techniques in a fairly elementary manner and to proceed systematically to show how they are used in contemporary research. Thus the book should be particularly suitable for senior undergraduates, first year postgraduates, and for others interested in, or contemplating research in the area, our general aim being to provide a rapid, efficient, and reasonably comprehensive introduction to some powerful techniques – and also to indicate suitable references for deeper and complementary studies. An integral part of the book is the large number of substantial exercises (of the order of one hundred and fifty) – a feature calculated to enhance its value to students, instructors, and other.

Consistent with the general aim outlined at the beginning of the last paragraph, the book is divided into two parts: in the main, Part I is devoted to *methodology*, Part II to *applications*. After an initial discussion of important fundamental issues, Part I, in the main, treats the basis for differential inequality techniques – as adumbrated above – and illustrates these by, on the whole, simple examples. Thus armed with a panoply of techniques, Part II treats more substantial examples which are at the forefront of contemporary research. We remark, however, that some of the techniques introduced in Part I are not used in Part II. For this and other reasons, elements of Part I may be regarded as self contained.

The book includes an Appendix on the fundamental inequalities used. This is done in order to furnish the reader with a compact compendium of these. The book does not treat the classification of P.D.E.s into elliptic, parabolic, hyperbolic types – assuming that the reader is already familiar with this.

Finally, we include an extensive bibliography entitled References. Our aim in providing this is two-fold: we wish to give a reasonably extensive indication of the literature – both seminal and current – relating to all topics dealt with. Secondly, we wish suitably to acknowledge various opera without which our work would not have been possible. The use of material drawn from references [136],[177] is by permission of Oxford University Press, and of material drawn from [231] by permission of Cambridge University Press.

Whereas it is not uncommon, in a preface, to describe, however briefly, the contents of each chapter, we prefer to refer the reader to the table of contents, the index, and the introduction to each chapter. Whereas it is the hope of the authors that many will read the book in its entirety, it is, of course, possible to read individual chapters and combinations of chapters. Indeed, a number of chapters are self-contained, and the extent to which any one is not should be clear from a quick perusal – particularly of the introduction to the relevant chapter.

We wish to make some remarks about mathematical and textual notations. In relation to the former, different notations sometimes occur in different context e.g. the absolute/bold faced notation and the indicial notation for tensor and vector quantities both occur – merely reflecting the diversity of notation occurring in the literature. Apart from a list of some notations at the beginning of the book, others are explained as they arise – with repetition sometimes in the interest of clarity. With regard to the numeration occurring throughout, the following practice is adopted. Each chapter is divided into sections, and these, sometimes, into subsections. In order to refer to these, the first number refers to the chapter, the second to the section, and – where relevant – the third to the subsection. Theorems, Lemmas and Remarks are numbered by two numbers, the first being the chapter number, the second referring to the sequence within the chapter. The same practice applies to equations and inequalities. Occasionally, however, to the last number are appended two or more suffices; this arises when two or more neighbouring equations/inequalities might reasonably be regarded as constituting a single entity.

The work for this book was jointly planned. Chapters 1,2,6,8 were written by author S.R. while Chapters 3,4,5,7 and Section 6.7, were written by author J.N.F. Chapter 9 was shared by both authors in about equal measure, Sections 9.2 and 9.7 being written by J.N.F., Sections 9.3, 9.4, 9.5, 9.6 by S.R.. There was considerable consultation throughout between both authors, and the work was carried out with the utmost harmony.

Naturally, every effort has been made to eliminate errors, but, of course, it is virtually impossible to succeed in this. The authors would appreciate if any such errors were brought to their attention.

Acknowledgements. We wish gratefully to acknowledge discussions with many people, over a considerable time, on various topics dealt with in the book. These include Professors S.Chiritá, R.J.Knops, G.Mulone, L.E.Payne, R.Russo, B.Straughan. Naturally, any defects in the book are attributable solely to the authors.

We gratefully acknowledge our indebtedness to Dr. Florinda Capone for her great care in preparing and processing the manuscript by electronic means. Our thanks are also due to Ms. Noelle Gannon for her help with the manuscript and for her assistance in other respects.

It is a pleasure to acknowledge Professor Alan Jeffrey. We also acknowledge the helpfulness and professionalism of the staff of CRC Press.

Galway and Naples, Spring 1995.

LIST OF FREQUENTLY USED SYMBOLS

$\mathbb{N}$	the set of natural numbers
$\mathbb{R}$	the real numbers $= (-\infty, \infty)$
$\mathbb{R}_-$	the nonpositive real numbers $= (-\infty, 0]$
$\mathbb{R}_+$	the nonnegative real numbers $= [0, \infty)$
$\bar{\mathcal{I}} = \mathrm{Cl}\, \mathcal{I}$	closure of the set $\mathcal{I}$
I	identity map
$\Vert \cdot \Vert = \Vert \cdot \Vert_X$	norm on the space X
X	normed space (in particular, Banach space)
$\emptyset$	empty set
$\nabla \times$	the curl operator
$\nabla \cdot$	the divergence operator
∇	the gradient operator
Δ (or ∇^2)	the Laplace operator
Δ_1 (or ∇_1^2)	the Laplace operator in two cartesian variables $\left(\Delta_1 = \frac{\partial^2}{\partial x^2} + \frac{\partial^2}{\partial y^2}\right)$
$L^p((0, T); X)$	the set of function u from $(0, T)$ into X such that $\int_0^T \vert u(\tau)\vert_x^p \, d\tau < \infty$
Lin	set of second order tensor
Sym	set of symmetric second-order tensor
$\mathrm{sym}\mathbf{T}, (\mathbf{T} \in \mathrm{Lin})$	symmetric part of $\mathbf{T}$

$f = O(g)$	$\left\|\frac{f}{g}\right\|$ bounded in appropriate limit
$f = O(1)$	f bounded in appropriate limit
$f = o(g)$	$\left\|\frac{f}{g}\right\| \to 0$ in appropriate limit
$f = o(1)$	$f \to 0$ in appropriate limit
$\frac{\partial(f_1, f_2, \ldots, f_n)}{\partial(x_1, x_2, \ldots, x_n)}$	jacobian of $f_1, f_2, \ldots, f_n$
$\mathbf{e}_i$	unit vector of x_i axes
$\mathbf{i}$	unit vector of x axis
$\mathbf{j}$	unit vector of y axis
$\mathbf{k}$	unit vector of z axis
$\mathbf{a} \cdot \mathbf{A}, \quad \mathbf{A} \in \mathrm{Lin}$	the vector $\sum_{i,j}^{1-3} a_i A_{ij} \mathbf{e}_j$
$\mathbf{a} \cdot \mathbf{A} \cdot \mathbf{b}, \quad \mathbf{A} \in \mathrm{Lin}$	the scalar $\sum_{i,j}^{1-3} a_i A_{ij} b_j$
$\nabla \mathbf{a}(\mathbf{x})$	second order tensor $\mathbf{A}$ such that $A_{ij} = \frac{\partial a_j}{\partial x_i}$
$\mathbf{A} : \mathbf{B}, \quad \mathbf{A}, \mathbf{B} \in \mathrm{Lin}$	the scalar $\sum_{i,j}^{1-3} A_{ij} B_{ij}$
$\mathbf{A}^2 = \mathbf{A} : \mathbf{A}$	$\forall\, \mathbf{A} \in \mathrm{Lin}$
$C_0(\Omega)$	linear space of continuous functions having compact support on Ω
$C^k(\Omega)$	linear space of functions having continuous partial derivatives up to the order k in Ω
$C_0^k(\Omega)$	$C_0(\Omega) \cap C^k(\Omega)$
$C^{2,1}\{(0,l) \times (0,T)\}$	linear space of functions of two variables (x,t) having continuous partial derivatives up to the order 2 in x and up to the order 1 in t
$L^p(\Omega)$	space of function f such that $\int_\Omega \|f\|^p d\Omega < \infty$
$W^{l,p}(\Omega)$	Sobolev space of functions belonging to $L^p(\Omega)$ along with their partial derivatives up to the order l

List of frequently used symbols

H^l	$W^{l,2}$
$W_l^p(\Omega)$	$W^{l,p}$
$H^0(\Omega)$	$L^2(\Omega)$
$H_0^s(\Omega)$	subset of H^1 constituted by the functions $u \in H^s$ having trace zero on $\partial\Omega$ together with their derivatives up to order $s-1, s = 1, 2...$

CONTENTS

Chapter one

PRELIMINARIES AND FUNDAMENTAL ISSUES

1.1 INTRODUCTION

The prediction of how real world phenomena evolve in time is of the greatest human interest. In order to describe a phenomenon, a mathematical model – called an *evolution equation* – is constructed, whose solutions are required to reflect the behaviour of that phenomenon. Consequently the problem of solving the evolution equation arises. But although one is able to obtain the explicit solution in some important cases, in general, this is not so.

The strategy of qualitative analysis is as follows: the determination of a priori estimates or properties of the solutions to the evolution equation, without solving it explicity. This book is principally devoted to the qualitative analysis of partial differential equations (P.D.E.s) modelling phenomena of the real world, and in this chapter relevant preliminaries and fundamental issues are discussed.

1.2 EVOLUTION EQUATION AND CHOICE OF THE STATE SPACE

Let $\mathcal{F}$ be a phenomenon taking place on a domain Ω of the physical three dimensional space $\mathbb{R}^3$ and let $u_i(\mathbf{x},t)$ – with $i = 1,2,...,n$ $(n<\infty)$, $\mathbf{x}$ a point of Ω and t an instant of time – represent the relevant quantities describing the *state* of $\mathcal{F}$. The vector $\mathbf{u}$, with components u_i, is the *state vector*. The phenomenon $\mathcal{F}$ is modelled by a P.D.E. if one can establish the existence of a function

$$\mathbf{F}(\mathbf{x},t,\mathbf{u},\frac{\partial u_i}{\partial x_r},\frac{\partial^2 u_j}{\partial x_r \partial x_s},...), \qquad i,j = 1,2..,n\,;\ r,s = 1,2,3$$

which governs the behaviour of the time derivative of $\mathbf{u}$, in the sense that, for any finite positive T,

$$\mathbf{u}_t = \mathbf{F}\,, \qquad \text{in } \Omega\times(0,T) \qquad (1.1)$$

subject to prescribed *initial data* $\mathbf{u}_0(x)$

$$\mathbf{u}(\mathbf{x},0) = \mathbf{u}_0(\mathbf{x}) \qquad \text{in } \Omega\,, \tag{1.2}$$

and to suitable ***boundary conditions***

$$A(\mathbf{u},\nabla\mathbf{u}) = \hat{\mathbf{u}} \qquad \text{on } \partial\Omega\times[0,T] \tag{1.3}$$

where $\hat{\mathbf{u}}(\mathbf{x},t)$ is prescribed and A is an assigned operator.

The initial-boundary value problem (I.B.V.P.) obtained is a mathematical model for the evolution of the state vector $\mathbf{u}$ of the phenomenon $\mathcal{F}$ and is the ***evolution equation*** of $\mathcal{F}$. The space of vector valued functions X defined on Ω which satisfy the prescribed boundary conditions, and which is endowed with a suitable topology, is called the ***state space***. Let us remark that the choice of topology is very important and may be said to be the core of the problem. In particular there may be more than one suitable choice. In order to emphasize this point, let us specify that the choice of functional topology has to be linked to the ***physics of the phenomenon*** – i.e. to meaningful aspects of the $\mathcal{F}$ state – and must not be made ***for the sake of mathematical convenience*** only.[1] For instance, let (1.1) be a mathematical model for the diffusion and growth of a population that occupies the bounded region Ω and let $\mathbf{u}(\mathbf{x},t)$ be the population density. Assuming that Ω is open, and bounded by a regular surface $\partial\Omega$, let us consider homogeneous boundary conditions

$$\mathbf{u}(\mathbf{x},t) = 0 \qquad \text{on } \partial\Omega\times[0,T]$$

which means that Ω is surrounded by a desert area. In this case X is the space of real valued functions defined on Ω, which satisfy the homogeneous boundary conditions. Concerning the choice of topology, let us consider two meaningful aspects of the phenomenon at hand:

i) behaviour in time of the total number of individuals of the given population;

ii) behaviour in time of the density maximum on Ω.

In case i) it is natural to choose the topology induced by the $L_1(\Omega)$-norm $\int_\Omega |f(x)|\,d\Omega < \infty$, while in case ii) the natural topology is that induced by the norm $\max_\Omega |f(x)| < \infty$. Consequently in case i) the state space is the real Banach space $L_1(\Omega)$, while in the case ii) it is $C_0(\Omega)$.

Remark 1.1 - *Let us remark, in passing, that in the state vector* $\mathbf{u}(\mathbf{x},t)$, $\mathbf{x}$ *can be any set of space-like variables and* t *any time-like parameter. Further, any P.D.E. can be written as an evolution equation. For instance, if* y *is a time-like parameter, the equation* $v_{xx} + v_{yy} = 0$ *can be written* $\mathbf{u}_t = \mathbf{F}$ *where* $y = t\,,\ \mathbf{u} = \begin{bmatrix} v \\ U \end{bmatrix}\,,\ \mathbf{F} = \begin{bmatrix} U \\ -v_{xx} \end{bmatrix}$.

[1] We refer to [1] for the question of *functional topology indifference properties*.

1.3 INITIAL AND BOUNDARY VALUE PROBLEMS

Two other very important problems arise for P.D.E. namely

i) ***initial value problems*** (I.V.P.)

ii) ***boundary value problems*** (B.V.P.).

Such problems might be envisaged as *degenerate* cases of I.B.V.P..

Concerning I.V.P., no boundary condition arises: the problem is that of solving (1.1) subject to the initial condition (1.2) only; this problem is often referred to as a *Cauchy problem*. Typically this arises in describing the propagation of wave-like disturbances in an unbounded medium.

Concerning boundary value problems (B.V.P.)

a) the time variable does not arise in the right hand side of the P.D.E. (1.1) and the left-hand side, therefore, is equated to zero, and

b) the initial condition does not arise.

That is to say, the problem is to determine a solution (solutions) to the P.D.E.

$$\mathbf{F}(\mathbf{x}, \mathbf{u}, \frac{\partial u_i}{\partial x_r}, \frac{\partial^2 u_j}{\partial x_r \partial x_s}, \ldots) = 0 \qquad \mathbf{x} \in \Omega$$

subject to the boundary condition $A(\mathbf{u}, \nabla\mathbf{u}) = \hat{\mathbf{u}}$ on $\partial\Omega$. When $A(\mathbf{u}, \nabla\mathbf{u}) = \mathbf{u}$ the problem is said to be a *Dirichlet problem*, while in the case $A(\mathbf{u}, \nabla\mathbf{u}) = \nabla\mathbf{u} \cdot \mathbf{n}$ where $\mathbf{n}$ is the unit outward normal to $\partial\Omega$, it is said to be a *Neumann problem*. If $A(\mathbf{u}, \nabla\mathbf{u}) = \lambda\mathbf{u} + \mu\nabla\mathbf{u} \cdot \mathbf{n}$ where λ, μ are scalars, the problem is said to be a *Robin problem*.

1.4 A FIRST REQUIREMENT OF THE MODEL

Let X be the state space of the evolution equation (1.1) endowed with a suitable topology, chosen in accordance with the physics of the phenomenon. As a first fundamental indication that the model is correct, one requires what is called *well posedness*. According to a classical definition of Hadamard [2] the problem is well-posed if in the space X:

i) *there exists a solution*;

ii) *it is unique*;

iii) *it depends continuously on the data.*

In this section the meaning of these three concepts are considered one after the other. Concerning the requirement i), let us state that, generally, it requires global (in time) existence, i.e. that the solution exists for every (finite) interval of time. Only very occasionally do we prove existence theorems, confining ourselves, in general, to recalling, where possible, the existence theorems for the equation at hand (supplemented by boundary and/or initial conditions), and the assumptions under which they have been obtained. Nevertheless, let us emphasize that solutions can be divided into two classes:

a) *strong (or smooth or regular or classical) solutions,*

b) *weak (or generalized or mild) solutions,*

and that each class admits different definitions.

Perhaps the most natural definition of a strong solution is one that requires solutions to (1.1) to exist in $X_1 = X \cap C^k[\Omega \times (0,T)]$ where k is the highest order of the $\mathbf{u}$ derivatives appearing in (1.1). But if one confines oneself to X_1 one can, in many cases, miss essential properties of the phenomenon at hand. For instance, this is so in the case of phenomena where shock waves can arise, and one is then constrained to weaken the notion of solution (see Sec.1.5).

Remark 1.2 - *Concerning the existence of classical solutions of the I.V.P., at least locally in time, we recall that it is ensured by the classical Cauchy-Kovalewski theorem under the assumptions that the data and* $\mathbf{F}$ *are analytic functions* [3]. *If this is not the case, there exist P.D.E.s for which even local existence of classical solution is not ensured. This is so in the case of Lewy's celebrated example*

$$u_x + iu_y - 2i(x+iy)u_t = f(x,y,t) \qquad (x,y,t) \in \mathbb{R}^3 \,, \tag{1.4}$$

where $i^2 = -1$. *Although the coefficients are all analytic functions, it is possible to show that* $\exists f \in C^\infty(\mathbb{R}^3)$ *such that* (1.4) *has no* C^1 *solution anywhere in* $\mathbb{R}^3$ [4,5].

Concerning the requirement ii) for well posedness, namely *uniqueness*, its meaning is obvious: there exists, at most, one solution to the problem (I.B.V. P., I.V.P., B.V.P., as appropriate).

Concerning the requirement iii), namely *continuous dependence*, it means the following: continuous dependence is said to obtain if, as a result of changes in the data (initial conditions and/or boundary conditions and/or coefficients in the P.D.E. etc.), the consequential change – in a suitable sense – in the solution, can be made arbitrarily small provided that the changes – in a suitable sense – in the data are made sufficiently small.

1.5 GENERALIZED SOLUTIONS

In order to fix ideas we consider the I.V.P.

$$\begin{cases} \mathbf{u}_t + \mathbf{F}_x(\mathbf{u}) = \mathbf{G}(\mathbf{u}) & (x,t) \in \mathbb{R} \times \mathbb{R}_+ \,, \\ \mathbf{u}(x,0) = \mathbf{u}_0(x) & x \in \mathbb{R} \end{cases} \tag{1.5}$$

where $\mathbf{u}=(u_1,...,u_n)\in \mathbb{R}^n$, $n\geq 1$, and $\mathbf{F}\in C^1(\mathbb{R}^n)$, $\mathbf{G}\in C(\mathbb{R}^n)$, $\mathbf{u}_0\in C(\mathbb{R})$ are prescribed real functions. Equation $(1.5)_1$ is commonly called a *balance law* and many meaningful phenomena are modelled by a P.D.E. of the type (1.5) [5,6]. For the sake of simplicity we consider a *conservation law* – i.e. $(1.5)_1$ in the absence of the source term $\mathbf{G}$ – in the case $n=1$

$$\begin{cases} u_t + F'(u)u_x = 0 & (x,t)\in \mathbb{R}\times \mathbb{R}_+ \\ u(x,0) = u_0(x) & x\in \mathbb{R} \end{cases} \tag{1.6}$$

Our aim is to show that (1.6) does not have *global classical* solutions, even if the data F and u_0 are analytic functions. Let u be a solution to (1.6) and let us consider the *characteristics*, i.e. the curves $\{x=x(s),\ t=t(s)\}$ along which u is constant

$$\frac{du}{ds} = u_t\frac{dt}{ds} + u_x\frac{dx}{ds} = 0 .$$

Then (1.6) implies that the characteristics are the curves such that

$$\frac{dt}{ds} = 1 \quad , \quad \frac{dx}{ds} = F'(u) \tag{1.7}$$

and hence – since they have constant slope $dt/dx = 1/F'(u)$ – are straight lines. For instance, the characteristic starting at $(\bar{x},0)$ has the slope $dt/dx= 1/F'[u_0(\bar{x})]$, and hence the equation $t=(x-\bar{x})/F'[u_0(\bar{x})]$. Let us consider the case $\{F'=u^p,\ p=\text{const.}(>0),\ u_0(x)\in C^1(\mathbb{R}),\ u_0'(x_0)<0,\ u_0(x_0)>0,\ x_0>0\}$.

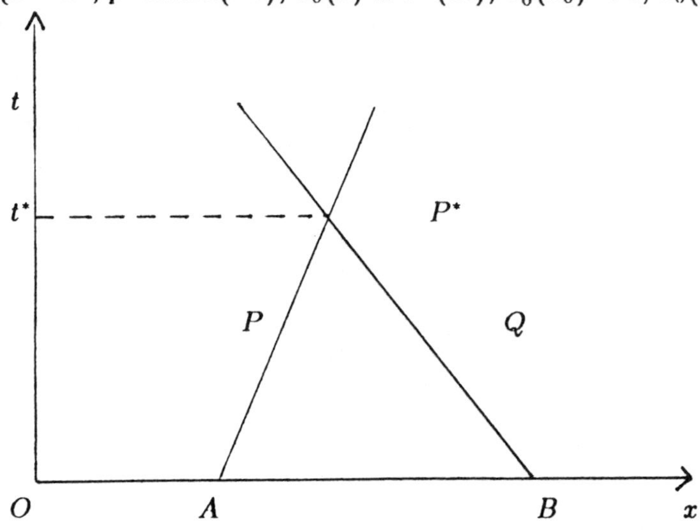

Figure 1.1

Then there exist two points x_1 , x_2 such that

$$0<x_1<x_2\ ,\ u_0(x_1)>u_0(x_2)\ ,\ \frac{1}{u_0^p(x_2)}>\frac{1}{u_0^p(x_1)} .$$

Then the characteristics r_1 and r_2, starting at the points $\{A=(x_1,0)\,,\ B=(x_2,0)\}$ will intersect at a point $P_*=(x_*,t_*)$ (See Figure 1.1). Therefore

$$u_0(x_1)=\lim_{\substack{P\to P_*\\ P\in r_1}} u[P(x,t)]\neq u_0(x_2)=\lim_{\substack{Q\to P_*\\ Q\in r_2}} u[Q(x,t)]$$

and hence the solution must be discontinuous at P_*. Consequently classical solutions can exist only *locally in time* i.e. in a subinterval of $[0,t^*)$.

Turning now to the introduction of weak solutions, let u be a smooth solution of (1.6) and let us multiply $(1.6)_1$ by a smooth test function i.e. a function $\varphi(x,t)\in C^1(\mathbb{R}\times\mathbb{R}_+)$, having compact support on $\mathbb{R}\times\mathbb{R}_+$. Precisely we assume that there exists a rectangle $I=[a,b]\times[0,T]$ such that φ is vanishing outside I and on the lines $x=a, x=b, t=T$, with $a<b$ and $T>0$. One has

$$(\varphi u)_t-\varphi_t u+(\varphi F)_x-\varphi_x F=0$$

and, therefore, integrating on $(a,b)\times(0,T)$, it follows that

$$\int_a^b[\varphi u]_{t=0}^{t=T}dx-\int_a^b\int_0^T\varphi_t u\,dx\,dt+\int_0^T[\varphi F]_{x=a}^{x=b}dt-\int_0^T\int_a^b\varphi_x F\,dt\,dx=0\,.$$

On letting $a\to-\infty$; $b,T\to\infty$ it turns out that

$$\int_{\mathbb{R}\times\mathbb{R}_+}(\varphi_t u+F\varphi_x)dx\,dt+\int_{\mathbb{R}}u_0\varphi(x,0)\,dx=0 \tag{1.8}$$

$\forall\varphi\in C_0^1(\mathbb{R}\times\mathbb{R}_+)$.

Conversely, if $v\in C^1(\mathbb{R}\times\mathbb{R}_+)$ and satisfies (1.8) $\forall\varphi\in C_0^1(\mathbb{R}\times\mathbb{R}_+)$, then, by integrating (1.8) by parts, it follows that

$$\int_{\mathbb{R}\times\mathbb{R}_+}[v_t+F'(v)v_x]\varphi\,dx\,dt=0\qquad\forall\varphi\in C_0^1(\mathbb{R}\times\mathbb{R}_+)\,.$$

Hence it turns out that

$$v_t+F'(v)v_x=0 \tag{1.9}$$

i.e. v satisfies $(1.6)_1$. Furthermore, multiplying (1.9) by $\varphi\in C_0^1(\mathbb{R}\times\mathbb{R}_+)$ and integrating by parts, one obtains

$$\int_{\mathbb{R}\times\mathbb{R}_+}(\varphi_t v+F(v)\varphi_x)dx\,dt+\int_{\mathbb{R}}v(x,0)\varphi(x,0)\,dx=0$$

Taking into account that v satisfies (1.9), it follows that

$$\int_{\mathbb{R}}[v(x,0)-u_0(x)]\varphi(x,0)dx=0$$

and hence, by the continuity of u_0 and the arbitrariness of $\varphi(x,0)$, one obtains $v(x,0)=u_0(x)$.

Formula (1.8) suggests a generalization of the notion of solution to (1.6) according to the following definition:

Definition 1.1 - *A weak solution to the I.V.P.* (1.6) – *where* u_0 *is bounded and measurable on* $\mathbb{R}$ – *is any function* $u(x,t)$, *bounded and measurable on* $\mathbb{R}\times\mathbb{R}_+$, *satisfying* (1.8) *for all* $\varphi\in C_0^1(\mathbb{R}\times\mathbb{R}_+)$.

Remark 1.3 - *Let us emphasize that, to the set of functions bounded and measurable on* $\mathbb{R}\times\mathbb{R}_+$, *belong functions which are discontinuous on a subset of zero measure i.e. functions having "jumps" on a curve* $\gamma\in\mathbb{R}\times\mathbb{R}_+$. *Such a curve represents a moving point on the* x *axis on which the "jump" takes place at instant* t *("shock wave"). We refer to* [5] *for a deeper discussion of this matter.*

Remark 1.4 - *Even in the context of generalized solutions, there are no general theorems of existence for P.D.E.s. Therefore the existence must be proved case by case. Furthermore, if existence has been proved on some functional space* X, *this does not imply that one is also able to prove uniqueness on* X. *In other words, for P.D.E.s, uniqueness must also be proved case by case and, possibly, in the existence space.*

1.6 DYNAMICAL SYSTEM GENERATED BY THE EVOLUTION EQUATION

Let $\mathbf{u}(\mathbf{u}_0,t)$, with $\mathbf{u}(\mathbf{u}_0,0)=\mathbf{u}_0$ be a global solution to the problem (1.1)-(1.3). Then $\mathbf{u}$ is a dynamical system according to the following definition [7,8]:

Definition 1.2 - *A dynamical system on a metric space* X *is a mapping*

$$\mathbf{v}:(\mathbf{v}_0,t)\in X\times\mathbb{R}\to\mathbf{v}(\mathbf{v}_0,t)\in X \tag{1.10}$$

such that

$$\mathbf{v}(\mathbf{v}_0,0)=\mathbf{v}_0\,. \tag{1.11}$$

For this reason, we recall some fundamental concepts of the theory of dynamical systems. For a dynamical system $\mathbf{v}$, the function

$$\mathbf{v}(\mathbf{v}_0,\cdot):t\in\mathbb{R}\to\mathbf{v}(\mathbf{v}_0,t)\in X$$

for a prescribed $\mathbf{v}_0\in X$, is called a *motion* associated with the *initial data* $\mathbf{v}_0$, and is denoted by $\mathbf{v}(\mathbf{v}_0,t)$ or $\mathbf{v}(t)$. If

$$\mathbf{v}(t)=\mathbf{v}_0\quad,\qquad\forall t\in\mathbb{R}\,, \tag{1.12}$$

the motion is *stationary* or *steady* and $\mathbf{v}_0$ is an *equilibrium* or *critical point*. Let $\mathbf{v}, \mathbf{w}$ be two motions and let

$$\mathbf{v}(0)=\mathbf{w}(0) \Longrightarrow \mathbf{v}(t)=\mathbf{w}(t) \qquad \forall t>0. \tag{1.13}$$

Then *the motion is unique,* **forward in time**, *with respect to the initial data.* If

$$\mathbf{v}(0)=\mathbf{w}(0) \Longrightarrow \mathbf{v}(t)=\mathbf{w}(t) \qquad \forall t<0,$$

then *the motion is unique,* **backward in time**. The set $\{t,\mathbf{v}(t)\}$ with t taking value on $\mathbb{R}_+$, is the *positive graph* of the motion $\mathbf{v}$, and its projection into X – i.e. the subset $\gamma=\{\mathbf{v}(t) : t\in\mathbb{R}_+\}$ – is the *positive orbit* or *trajectory* starting at $\mathbf{v}_0$. Similarly, when it exists, one can define the *negative graph* and the *negative orbit* or *trajectory* finishing at $\mathbf{v}_0$, on replacing t by $-t$. The union of the positive and negative orbits through $\mathbf{v}_0$, is the complete orbit going through $\mathbf{v}_0$. A subset $\mathcal{I} \subset X$, is *positively invariant* if $\mathbf{v}_0 \in \mathcal{I} \Rightarrow \mathbf{v}(t) \in \mathcal{I}$, $\forall t > 0$. If $\exists \mathcal{T} : \mathbf{v}(t+\mathcal{T})=\mathbf{v}(t), t\in\mathbb{R}$, the motion $\mathbf{v}$ is *periodic in time with period* $\mathcal{T}$.

Returning to Definition 1.2, the following additional property is usually required for a dynamical system

$$\mathbf{v}(\mathbf{v}_0,t+\tau)=\mathbf{v}(\mathbf{v}(\mathbf{v}_0,\tau),t) \quad , \quad (\mathbf{v}_0\in X,\, t,\tau\in\mathbb{R}_+) \tag{1.14}$$

Property (1.14) is ensured, *when it exists*, by the *forward uniqueness*. Properties (1.11) and (1.14) give to the one parameter family of operators $\mathbf{v}(\mathbf{v}_0,\cdot)$ the semigroup structure, in accordance with the following definition [8-10]:

Definition 1.3 - *A semigroup on a metric space X is a one parameter family $\{S(t)\}_{t\geq 0}$ of operators, $S(t) : X \to X$ such that, for all $t,\tau \in \mathbb{R}_+$, $x \in X$, one has*

i) $S(0) = I$

ii) $S(t+\tau) = S(t)S(\tau)$.

The equivalence between a semigroup $\{S(t)\}_{t\geq 0}$ and the dynamical system (1.10) with property (1.14) is immediately seen by setting

$$\mathbf{v}(\mathbf{v}_0,t) = S(t)\mathbf{v}_0 \qquad \mathbf{v}_0\in X,\, t\in\mathbb{R}_+ \tag{1.15}$$

Other basic properties may, in general, be needed in the study of a dynamical system. We recall here the following

$$\mathbf{v}(t,\cdot) : X \to X \qquad \text{is continuous} \quad \forall t\geq 0, \tag{1.16}$$

$$\mathbf{v}(\cdot,\mathbf{v}_0) : \mathbb{R}_+ \to X \qquad \text{is continuous} \quad \forall \mathbf{v}_0\in X, \tag{1.17}$$

$$\mathbf{v}(\cdot,\mathbf{v}_0) : \mathbb{R}_+ \to X \qquad \text{is injective}, \tag{1.18}$$

which, in terms of the semigroup $\{S(t)\}_{t\geq 0}$, are respectively

iii) $S(t) : X \to X$ is continuous $\forall t\geq 0$,

iv) $S(\cdot) : \mathbb{R}_+ \to X$ is continuous $\forall \mathbf{v}_0$,

v) $S(\cdot) : \mathbb{R}_+ \to X$ is injective.

When all the properties (1.11), (1.14), (1.16) and (1.17) hold, the dynamical system is a C_0-semigroup on the metric space X according to the following Definition [8-10]:

Definition 1.4 - *A C_0-semigroup on a metric space X is a one parameter family $\{S(t)\}_{t\geq 0}$ of operators, $S(t) : X \to X$ such that properties i)–iv) hold.*

In the sequel – because of the equivalence (1.15) – we often will refer to the dynamical system (1.10) with the properties (1.11), (1.14), (1.16), (1.17) as a C_0-semigroup.

Remark 1.5 - *Let us remark that the C_0-semigroup properties ensure "smooth" behaviours of the motions* [11] {*See Sec.* 1.7, *Theorem* 1.1}.

Remark 1.6 - *Property* (1.18) [v)] *ensures backward uniqueness.*

Remark 1.7 - *In the study of a dynamical system generated by a P.D.E., the existence of the operators $S(t)$ and their properties – consistent with Remark* 1.4 – *must be proved, case by case, in a preliminary way.*

1.7 CONTINUOUS DEPENDENCE

In modelling a real world phenomenon, it is inevitable that the mathematical model obtained will contain many errors. These arise in the measurements of the data (initial data, boundary data, forces, geometry of the domain on which the phenomenon takes place, parameters contained in the evolution equation,...), and on errors in formulating the model. The question arises therefore of how these errors may influence the solution. This is the concern of continuous dependence and, more generally, of stability. In this section we are concerned with continuous dependence on the initial data.

Let v be a dynamical system on a metric space X, $d : X \times X \to I\!R_+$ the metric and $S(x,r), r > 0$, the open ball centered at x and having radius r.

Essentially, the idea of continuous dependence of a particular (*basic*) motion $v(v_0,\cdot)$ is that any other motion $v(v_1,\cdot)$ – starting at the same initial instant from a position v_1 "sufficiently close" to v_0 – will remain "as close as desired" to the basic motion for all finite time $T>0$. We shall now specify this concept in a mathematically rigorous way.

Definition 1.5 - *A motion $v(v_0,\cdot)$ depends continuously on the initial data iff* $\forall T > 0, \forall \varepsilon > 0$

$$\exists \delta(\varepsilon,T) > 0 : v_1 \in S(v_0,\delta) \Rightarrow v(v_1,t) \in S(v(v_0,t),\varepsilon),\ \forall t \in [0,T] \tag{1.19}$$

Theorem 1.1 - *Let* $\mathbf{u} : X \times \mathbb{R}_+ \rightarrow X$ *be a dynamical system on a metric space* X*, continuous on* $X \times \mathbb{R}_+$ *and having the* C_0*-semigroup properties. Then any motion depends continuously on the initial data.*

Proof. - Assume that the Theorem does not hold. Then there exist two positive numbers T and ε such that:

$$\forall \delta_n > 0\,,\ \exists u_n \in S(u_0, \delta_n)\,,\ t_n \in [0, T] : d[u(u_0, t_n), u(u_n, t_n)] \geq \varepsilon$$

Letting $\delta_n \rightarrow 0$ as $n \rightarrow \infty$, then $\lim\limits_{n\rightarrow\infty} u_n = u_0$. Moreover – from the bounded sequence $\{t_n\}$ – one can choose a converging sequence $\{t_{n_k}\}$. Setting $t^* = \lim\limits_{n_k\rightarrow\infty} t_{n_k}$, one has $t^* \in [0, T]$. Consequently it follows that

$$\begin{cases} \lim\limits_{n_k\rightarrow\infty} u_{n_k} = u_0 \quad , \quad \lim\limits_{n_k\rightarrow\infty} t_{n_k} = t^*\,, \\ d\left[u(u_0, t_{n_k}), u(u_{n_k}, t_{n_k})\right] \geq \varepsilon\,. \end{cases}$$

But the triangle inequality and the continuity of u at the point (u_0, t^*) imply that:

$$\begin{aligned} \forall \sigma > 0, \exists \nu(\sigma) : n_k > \nu \Rightarrow d\left[u(u_0, t_{n_k}), u(u_{n_k}, t_{n_k})\right] \leq \\ d\left[u(u_0, t_{n_k}), u(u_0, t^*)\right] + d\left[u(u_0, t^*), u(u_{n_k}, t_{n_k})\right] \leq \sigma \end{aligned}$$

which for $\sigma < \varepsilon$, contradicts the previous inequality.

Remark 1.8 - *Continuous dependence with* $\{\delta = \varepsilon^{-\alpha}\,, 0 < \alpha \leq 1\}$ *in the formula* (1.19) *is called "Holder stability"* [3].

1.8 ILL-POSED PROBLEMS. COUNTEREXAMPLES TO UNIQUENESS

A problem which is not well posed is said to be *ill posed.* Lack of uniqueness implies ill posedness also, because it guarantees that continuous dependence cannot obtain. In fact, if X denotes a metric space, and u a dynamical system on X according to Definition 1.2, the following Theorem holds:

Theorem 1.2 - *A motion which is not unique cannot depend continuously on the initial data.*

Proof - Let u and v be two motions, and let

$$\begin{cases} u(0) = v(0) \quad , \quad u(t^*) \neq v(t^*); \qquad\qquad 0 < t^* < \infty \\ d[u(t^*), v(t^*)] = \varepsilon^* > 0\,. \end{cases}$$

Then $\forall T > t^*, \forall \varepsilon < \varepsilon^*$ (1.19) cannot hold.[2]

Because of their importance in a model, we shall present some counterexamples to uniqueness for some fundamental evolution equations.

a) Counterexamples to uniqueness in fluid mechanics. The Navier-Stokes equations

$$\begin{cases} \mathbf{v}_t = -\mathbf{v}\cdot\nabla\mathbf{v} - \nabla p + \nu\Delta\mathbf{v} + \mathbf{F}\,, \\ \nabla\cdot\mathbf{v} = 0\,, \end{cases} \qquad (\mathbf{x},t)\in\Omega\times\mathbb{R}_+ \qquad (1.20)$$

are a mathematical model describing the motion of an incompressible homogeneous viscous fluid occurring in a fixed region $\Omega\subset\mathbb{R}^3$. In (1.20), $\mathbf{x}=(x,y,z)\in\Omega$ is the space variable, $t\in\mathbb{R}_+$ the time, $\mathbf{v}(\mathbf{x},t)$ the velocity field, $p(\mathbf{x},t)$ the pressure field (divided by the constant density), $\nu(>0)$ the coefficient of kinematic viscosity and $\mathbf{F}(\mathbf{x},t)$ the body force acting on the fluid. When the fluid adheres completely to the boundary $\partial\Omega$ – which generally happens whenever $\partial\Omega$ is rigid – the initial-boundary conditions to append to (1.20) are:

$$\begin{cases} \mathbf{v}(\mathbf{x},0) = \mathbf{v}_0(\mathbf{x}) & \mathbf{x}\in\Omega\,, \\ \mathbf{v}(\mathbf{x},t) = \mathbf{v}^*(\mathbf{x},t) & (\mathbf{x},t)\in\partial\Omega\times\mathbb{R}_+\,, \end{cases} \qquad (1.21)$$

where $\mathbf{v}_0(\mathbf{x})$ (divergence free) and $\mathbf{v}^*(\mathbf{x},t)$ are prescribed vector functions.

Let us consider the case

$$\Omega\equiv\mathbb{R}^3 \quad , \quad \mathbf{v}_0\equiv\mathbf{F}\equiv 0\,.$$

Then one immediately obtains that the I.V.P. (1.20)–$(1.21)_1$, (there is no boundary when $\Omega\equiv\mathbb{R}^3$), is satisfied by the following three solutions [12,13], at least:

$$\mathbf{v}\equiv 0\,, \qquad p = p_0(t)\,; \qquad (1.22)$$
$$\mathbf{v}\equiv t^2(\mathbf{i}+\mathbf{j}+\mathbf{k})\,, \qquad p = -2t(x+y+z)+p_1(t)\,; \qquad (1.23)$$
$$\mathbf{v}\equiv (\sin\, t)\mathbf{i}\,, \qquad p = -x\,(\cos t) + p_2(t)\,; \qquad (1.24)$$

where $(\mathbf{x},t)\in\mathbb{R}^3\times\mathbb{R}_+$; $\mathbf{i}$, $\mathbf{j}$, $\mathbf{k}$ are unit vectors along the x, y and z axes respectively, and $p_i(t)$ $(i=0,1,2)$ are arbitrary functions.

b) Counterexamples to uniqueness in viscoelasticity. Beginning in 1979 G. Fichera has exhibited some celebrated counterexamples to uniqueness in viscoelasticity [14,15]. These counterexamples, because of their connection with the Coleman-Noll principle of fading memory [16], immediately attracted the attention of well known experts. For further details we refer the readers to [14,15,17]. An outline of the problem is also contained in [18]. Here we shall consider the counterexample presented in [1].

---

[2]The positive number ε^* depends, of course, on the topology, but not the proof. In this case, therefore, the lack of continuous dependence is independent of topology.

The equations governing the dynamics of a homogeneous viscoelastic solid in the linear case [18], are:

$$\begin{cases} \rho\ddot{\mathbf{u}} = \nabla\cdot\mathbf{T} + \rho\mathbf{F}\,, & \text{on } \Omega\times I\!R_+ \\ \mathbf{T}(\mathbf{x},t) = \mathbf{G}_0(\mathbf{x})\nabla\mathbf{u}(\mathbf{x},t) + \int_0^\infty \mathbf{G}'(\mathbf{x},s)\nabla\mathbf{u}(\mathbf{x},t-s)ds, & \end{cases} \tag{1.25}$$

$$\begin{cases} \mathbf{u}(\mathbf{x},t) = 0\quad, & \text{on } \partial\Omega\times I\!R_+ \\ \mathbf{u}(\mathbf{x},t) = \mathbf{u}_0(\mathbf{x},t)\quad, & \text{in } \Omega\times I\!R_- \end{cases} \tag{1.26}$$

where $\ddot{\mathbf{u}} = \mathbf{u}_{tt}$,$\Omega\subset I\!R^3$ is the domain occupied in the reference configuration, $\mathbf{T}\in$ Sym is the second order stress tensor $\mathbf{u}$ is the displacement, $\mathbf{G}$ is the fourth-order relaxation tensor related to the *Boltzmann function* $\mathcal{G}$ by $\{\mathcal{G}'(s) = \mathbf{G}(s),\ \mathcal{G}(0) = \mathbf{G}(0),\ s\in I\!R\}$, , ρ is the density, $\mathbf{F}$ the body force, $\mathbf{G}'$ the derivative with respect to s, and $\mathbf{u}_0$ a prescribed function. When the body is one-dimensional, when the quasi-static approximation obtains ($\ddot{\mathbf{u}} = 0$) and when $\mathbf{F} = 0$ – then the following is easily shown to hold:

$$\begin{cases} \dfrac{dT}{dx} = 0\,, \\ T(x,t) = G_0\varepsilon(x,t) + \int_{-\infty}^{t} G'(x,t-\tau)\varepsilon(x,\tau)d\tau. \end{cases} \tag{1.27}$$

Obviously $T\equiv\varepsilon\equiv0$ is a solution to (1.27) for any relaxation function G, and, therefore, nonuniqueness is obtained for any stress-relaxation function G such that the equation

$$G(0)\varepsilon(t) + \int_{-\infty}^{t} G'(t-\tau)\varepsilon(\tau)d\tau = 0 \tag{1.28}$$

has a nontrivial solution. Following [1], let us consider

$$G(s) = \frac{1}{2} + \frac{1}{2}e^{-s}. \tag{1.29}$$

Then, setting $\phi(t) = e^t\,\varepsilon(t)$, (1.28) becomes:

$$2\phi'(t) = \phi(t) \tag{1.30}$$

and hence $\varepsilon(t) = ce^{-\frac{1}{2}t}$, $c =$ const. Therefore nonuniqueness is obtained for any $c\neq 0$. We now introduce as ***influence function***

$$h(s) = e^{-\gamma s} \quad,\quad \gamma\in(0,1/2) \tag{1.31}$$

and immediately obtain:[3]

i) $h(s) > 0, \forall s \geq 0;$ ii) $\lim_{s \to \infty} s^p h(s) = 0, \forall p \geq 0;$

$$\text{iii)} \int_0^\infty \left(\frac{G'}{h}\right)^2 ds = \frac{1}{8(1-\gamma)} < \infty; \ \text{iv)} \, G(\infty) = \frac{1}{2} > 0;$$

$$\text{v)} \int_0^\infty G'(s) \sin \omega s \, ds = -\frac{1}{2} \int_0^\infty e^{-s} \sin \omega s \, ds \leq 0, \forall \omega \geq 0.$$

Properties i)-iii) are required by the principle of fading memory; v) has been derived by Graffi by requiring that energy be dissipated in any period of a sinusoidal strain function [19], and, finally, iv) is the inequality satisfied by the equilibrium elastic modulus, necessary to ensure $G(s) > 0$. (See [17], p.129). Fichera's counterexample satisfies all the requirements. For other kinds of counterexamples see [20,21].

c) Counterexamples to uniqueness in elastodynamics. The equations governing the dynamics of an elastic body are:

$$\rho \ddot{\mathbf{u}} = \nabla \cdot \mathbf{T} + \rho \mathbf{F} \qquad \text{on } \Omega \times I\!R_+ , \tag{1.32}$$

$$\mathbf{u} = \mathbf{u}_0 \quad , \quad \dot{\mathbf{u}} = \dot{\mathbf{u}}_0 \qquad \text{on } \Omega \times \{0\}, \tag{1.33}$$

$$\mathbf{u} = \hat{\mathbf{u}} \qquad \text{on } \partial_1\Omega \times I\!R_+ , \tag{1.34}$$

$$\mathbf{T} \cdot \mathbf{n} = \hat{\mathbf{s}} \qquad \text{on } \partial_2\Omega \times I\!R_+ , \tag{1.35}$$

where $\Omega, \mathbf{T}, \mathbf{u}, \rho, \mathbf{F}$ have the meaning introduced in b) supra; $\partial_1\Omega$ and $\partial_2\Omega$ are regular subsurfaces of $\partial\Omega$ such that $\partial_1\Omega \cup \partial_2\Omega = \partial\Omega, \overline{\partial_1\Omega} \cap \overline{\partial_2\Omega} = \emptyset$; $\hat{\mathbf{u}}$ (*surface displacement*), $\hat{\mathbf{s}}$ (*surface traction*), $\mathbf{u}_0$ (*initial displacement*), $\dot{\mathbf{u}}_0$ (*initial distribution of velocity*), are prescribed smooth functions; $\mathbf{n}$ denotes the outward unit normal to $\partial\Omega$ [22]. In the classical linear elasticity, the stress is related linearly to the ***strain tensor*** $\mathbf{E}$ (i.e. the symmetric part of $\nabla\mathbf{u}$) as follows:

$$T_{ij} = \frac{1}{2} C_{ijhk}(u_{h,k} + u_{k,h}) \Longleftrightarrow \mathbf{T} = \mathbf{C} \cdot \mathbf{E} \tag{1.36}$$

where $\mathbf{C} = \{C_{ijhk}\}$ is the fourth order *elasticity tensor*. If

$$C_{ijhk} = \lambda \delta_{ij} \delta_{hk} + \mu(\delta_{ih}\delta_{jk} + \delta_{ik}\delta_{jh}) \qquad \forall i,j,h,k \tag{1.37}$$

---

[3]For $\omega > 0$

$$\int_0^\infty e^{-s} \sin \omega s ds = \frac{1}{\omega} \sum_{n=0}^{\infty} \int_{2n\pi}^{2(n+1)\pi} e^{-\frac{x}{\omega}} \sin x dx ,$$

$$\int_{2n\pi}^{2(n+1)\pi} e^{-\frac{x}{\omega}} \sin x dx = \int_{2n\pi}^{(2n+1)\pi} e^{-\frac{x}{\omega}} \sin x dx + \int_{(2n+1)\pi}^{2(n+1)\pi} e^{-\frac{x}{\omega}} \sin x dx$$

$$= \int_{2n\pi}^{(2n+1)\pi} e^{-\frac{x}{\omega}} - e^{-\frac{x+\pi}{\omega}} \sin x dx = (1 - e^{-\frac{\pi}{\omega}}) \int_{2n\pi}^{(2n+1)\pi} e^{-\frac{x}{\omega}} \sin x dx \geq 0$$

where $\lambda(\mathbf{x})$ and $\mu(\mathbf{x})$ are scalar functions (known as *Lamé moduli*), then the elastic body is said to be *isotropic*.[4]
Following [23], let us consider the one-dimensional case $\Omega = [1,\infty), \partial_2\Omega = \emptyset$ (*displacement problem*) and $\mathbf{F} = \hat{\mathbf{u}} = \mathbf{u}_0 = \dot{\mathbf{u}}_0 \equiv 0$. Further let us assume that

$$\rho = p'(x) \quad , \quad C_{1111} = C(x) = \frac{1}{p'} \qquad x \in [1,\infty) \tag{1.38}$$

where $p : I\!R \to I\!R$ is a smooth function such that:[5]

$$p'(x) > 0\,, \, p(1) = 0\,, \, \lim_{x\to\infty} p(x) = l < \infty\,, \quad x \in [1,\infty). \tag{1.39}$$

Then (1.32)–(1.35) become

$$\begin{cases} p'\ddot{\mathbf{u}} = \partial_x \left(\frac{1}{p'} u_x\right) & \text{on} \quad [1,\infty) \times I\!R_+\,, \\ \mathbf{u} = \dot{\mathbf{u}} = 0 & \text{on} \quad [1,\infty) \times \{0\}\,, \\ \mathbf{u} = 0 & \text{on} \quad \{1\} \times I\!R_+\,. \end{cases} \tag{1.40}$$

Equation $(1.40)_1$ can be written

$$\left(\partial_t + \frac{1}{p'}\partial_x\right)\left(\partial_t - \frac{1}{p'}\partial_x\right) u = 0$$

and hence its general integral is given by the *d'Alembert integral*

$$u(x,t) = u_1[t - p(x)] + u_2\,[t + p(x)]\,. \tag{1.41}$$

In order to show nonuniqueness it is sufficient to prove the existence of nontrivial solutions to system (1.40). To this end, let $f \in C^2(I\!R_+)$ be such that:[6]

$$f(0) = f(l) = f(2l) = f'(l) = f'(2l) = f''(l) = f''(2l) = 0. \tag{1.42}$$

---

[4]Let us remark that, because of the symmetry of both the stress and the strain tensors, $\mathbf{C}$ has to satisfy the relations (*minor symmetries*)

$$C_{ijhk} = C_{jihk} = C_{ijkh} \qquad \forall i,j,h,k.$$

Further, in the isotropic case, $\mathbf{C}$ is symmetric (*major symmetry*)

$$C_{ijhk} = C_{hkij} \quad \forall i,j,h,k\,.$$

[5]For instance, $p(x) = l\left(1 - e^{1-x}\right)$.

[6]For instance, $f(t) = \left(\sin\frac{2\pi}{l}t\right)^n \quad , \quad n > 2$

Choosing

$$u_1(x,t)=\begin{cases}0, & \{x\in[1,\infty)\,,\,0\le t\le p(x)\}\,,\\ f[t-p(x)+l], & \{x\in[1,\infty)\,,\,t\ge p(x)\}\end{cases}\tag{1.43}$$

$$u_2(x,t)=\begin{cases}0, & \{x\in[1,\infty)\,,\,0\le t\le l-p(x)\}\,,\\ -f[t+p(x)+l], & \{x\in[1,\infty)\,,\,t\ge l-p(x)\}\end{cases}\tag{1.44}$$

it follows immediately that $u_i(i=1,2)$ is a solution to system $(1.40)_{1,2}$. Since $u_1(1,t)=-u_2(1,t)=f(t+l)$, on setting

$$u=u_1+u_2\tag{1.45}$$

one obtains a nontrivial solution to the whole system (1.40), for each function f.[7]

d) Counterexample to uniqueness in elastostatics. The equations governing elastostatics, which immediately follow from (1.32)–(1.35), are

$$\nabla\cdot\mathbf{T}+\rho\mathbf{F}=0 \qquad \text{on } \Omega\times I\!R_+\,,\tag{1.46}$$

$$\mathbf{u}=\hat{\mathbf{u}} \qquad \text{on } \partial_1\Omega\times I\!R_+\,,\tag{1.47}$$

$$\mathbf{T}\cdot\mathbf{n}=\hat{\mathbf{s}} \qquad \text{on } \partial_2\Omega\times I\!R_+\,.\tag{1.48}$$

The lack of uniqueness in elastostatics – because of its importance in practice – has attracted the attention, both in the past and currently, of many authors. For further treatments of this matter we refer the readers to [22,25] and to the references therein contained. We confine ourselves to considering a counterexample to uniqueness for the traction problem $\partial_2\Omega=\partial\Omega$ in homogeneous isotropic elastostatics. In view of the linearity of system (1.46)–(1.48), in order to prove nonuniqueness it is sufficient to show that the system

$$\begin{cases}\nabla\cdot\mathbf{T}=\mu\Delta\mathbf{u}+(\lambda+\mu)\nabla(\nabla\cdot\mathbf{u})=0 & \text{on } \Omega\,,\\ \mathbf{T}\cdot\mathbf{n}=0 & \text{on } \partial\Omega,\end{cases}\tag{1.49}$$

where $\Omega=I\!R^3/S(\mathbf{O},1)\overset{\text{def}}{=}S^{(c)}(\mathbf{O},1)$ admits a non trivial solution. Assuming

$$\lambda+\mu\neq 0\tag{1.50}$$

then it follows that

$$\mathbf{u}=(\alpha r+\beta r^{-2})\mathbf{e}_r \quad , \quad \mathbf{e}_r=\frac{P-O}{|OP|}\,,\tag{1.51}$$

[7]Nonuniqueness associated with (1.32)–(1.35) – which represent a purely mechanical model – may sometimes be eliminated by appealing to extra-mechanical laws i.e. thermodynamical laws e.g. [24].

satisfies $(1.49)_1$ $\forall\, \alpha, \beta \in I\!R$. But on $\partial\Omega$ is

$$\mathbf{T}(\mathbf{u}) \cdot \mathbf{e}_r = [(3\lambda + 2\mu)\alpha - 4\mu\beta]\mathbf{e}_r\,,$$

hence choosing α and β such that $\{\alpha\beta > 0,\ (3\lambda + 2\mu)\alpha = 4\mu\beta\}$ then (1.51) gives a nontrivial solution to system (1.49).

e) Counterexample to uniqueness for the heat equation. A celebrated counterexample due to Tikhonov [26,27] shows that the Cauchy problem for the one-dimensional heat equation

$$\theta_t = \theta_{xx} \qquad \text{on } I\!R \times [0,\infty)\,, \tag{1.52}$$

$$\theta = 0 \qquad \text{on } I\!R \times \{0\}, \tag{1.53}$$

(where θ is the temperature field) has, at least, a nontrivial solution in the class of the functions $\theta(x,t)$ smooth on $I\!R \times [0,\infty)$, and such that

$$|\theta(x,t)| < \text{const.}\, e^{x^2} + \sigma \tag{1.54}$$

for some $\sigma > 0$. The Tikhonov proof is too long to be reported here.

f) Counterexample to uniqueness for the Monge-Ampère equation. Let Ω be a bounded plane domain and consider the Dirichlet problem therein: $\varphi \in C^2(\Omega) \cap C(\overline{\Omega})$ satisfies the *Monge-Ampère* equation

$$\varphi_{xx}\varphi_{yy} - \varphi_{xy}^2 = f(x,y) \qquad \text{in } \Omega\,, \tag{1.55}$$

subject to

$$\varphi = g(x,y) \qquad \text{on } \partial\Omega\,. \tag{1.56}$$

Problems of this nature arise in fluid mechanics and its applications to geophysics: φ may represent the stream function, and f a quantity related to pressure [28,29]. It may be proved [30] that this problem has, at most, two solutions provided that $f > 0$. Moreover, a simple example suffices to establish that two solutions are, indeed, to be expected in general. Let Ω be the circle $x^2 + y^2 < 1$, (x,y) being rectangular cartesian coordinates and let $f = 4$, $g = 0$. Then it is clear that $\varphi_1 = 1 - x^2 - y^2$ and $\varphi_2 = -\varphi_1$ are both solutions.

g) Further counterexamples to uniqueness. In the exercises further counterexamples to uniqueness are presented. Let us end by presenting this very simple case concerning the parabolic I.B.V.P. [31]

$$\begin{cases} u_t = u_{xx} + u + (u \sin x)^{\frac{1}{2}} & \text{on } [0,\pi] \times I\!R_+\,, \\ u(x,0) = 0\,, & \\ u(0,t) = u(\pi,t) = 0 & \forall t \geq 0\,, \end{cases} \tag{1.57}$$

which has the nontrivial solution:

$$u = \frac{1}{4} t^2 \sin x. \tag{1.58}$$

Remark 1.9 - *The role of some counterexamples to uniqueness is to delimit the functional classes in which uniqueness can be obtained. For instance, let us consider the Cauchy problem for the Navier-Stokes equations in the whole space and the functional class of solutions* $\{\mathbf{v}, p\}$ *such that* $\mathbf{v}$ and $\nabla\mathbf{v}$ *are bounded or even grow at large spatial distances. Then, counterexample* (1.22)–(1.24) *shows that this class cannot be relaxed to one such that* ∇p *only is bounded, for otherwise one loses uniqueness.*

Remark 1.10 - *There exist many mathematical systems – modelling physical phenomena – which, although violating Hadamard's well posedness requirements, nevertheless give useful information. This happens, for example, in models concerning buckling where the knowledge of the load at which uniqueness ceases* (**critical load**), *is important. In some cases, therefore, Hadamard's well posedness conditions are too restrictive, and the problem of investigating how to recover continuous dependence, arises. For this question and, for treatment of ill posed problems, we refer to* [32-36].

1.9 LIAPUNOV STABILITY

The Liapunov stability of a basic motion $\mathbf{v}(\mathbf{v}_0, \cdot)$ of a dynamical system $\mathbf{v}$ extends the requirement of continuous dependence to the infinite interval of time $(0, \infty)$. Taking into account Theorem 1.2, one can assume that the dynamical system $\mathbf{v}$ is a semigroup.

Definition 1.6 - *A motion* $v(\cdot, v_0)$ *is Liapunov stable (with respect to perturbations in the initial data) iff:*

$$\begin{aligned} &\forall\varepsilon > 0, \exists\delta(\varepsilon) > 0 : v_1 \in S(v_0, \delta) \Rightarrow \\ &\qquad v(v_1, t) \in S(v(v_0, t), \varepsilon) \quad , \quad \forall t \in \mathbb{R}_+ \end{aligned} \tag{1.59}$$

A motion is *unstable* if it is not stable. Obviously (1.59) $\Longrightarrow$ (1.19)

Definition 1.7 - *A motion* $v(v_0, \cdot)$ *is said to be an attractor or attractive on a set* Y *if:*

$$v_1 \in Y \Rightarrow \lim_{t \to \infty} d[v(v_0, t), v(v_1, t)] = 0 . \tag{1.60}$$

The biggest set Y *on which* (1.60) *holds is called the basin (or domain) of attraction of* $v(v_0, \cdot)$.

Definition 1.8 - *The motion* $v(v_0, \cdot)$ *is asymptotically stable if it is stable, and if there exists* $\delta_1 > 0$ *such that* $v(v_0, \cdot)$ *is attractive on* $S(v_0, \delta_1)$. *In particular,* $v(v_0, \cdot)$ *is exponentially stable if there exist* $\delta_1 > 0, \lambda(\delta_1) > 0, M(\delta_1) > 0$ *such that:*

$$v_1 \in S(v_0, \delta_1) \Rightarrow d\left[v(v_1, t), v(v_0, t)\right] \le M e^{-\lambda t} d(v_1, v_0), \forall t \in \mathbb{R}_+ . \tag{1.61}$$

Liapunov stability of a set is of fundamental interest expecially in connection with the asymptotic behaviour of motions. Let us begin by recalling the definition of distance $\widehat{d}(A,B)$ between two subsets A, B of the metric space X:

$$\widehat{d}(A,B) = \inf_{x\in A, y\in B} d(x,y)\,. \tag{1.62}$$

We denote by $\widehat{S}(A,r), r > 0$, the open set $\{x \in X : \widehat{d}(x,A) < r\}$ where, according to (1.62), $\widehat{d} = \inf_{y\in A} d(x,y)$.

Definition 1.9 - *Let $\gamma(v_0)$ denote the positive orbit of the motion $v(v_0,\cdot)$. A set $A\subset X$ is Liapunov stable if*

$$\forall\varepsilon > 0, \exists\delta(\varepsilon) > 0 : v_0 \in \widehat{S}(A,\delta) \Rightarrow \gamma(v_0) \subset \widehat{S}(A,\varepsilon). \tag{1.63}$$

A set is *unstable* if it is not stable.

Definition 1.10 - *A set A is said to be an attractor or attractive on an open set $B \supset A$ if it is positive invariant and*

$$v_0 \in B \Rightarrow \lim_{t\to\infty} \widehat{d}[v(v_0,t), A] = 0\,. \tag{1.64}$$

The largest open set B on which (1.64) holds is the basin of attraction of A. If A is compact and attracts every bounded set of X, then A is a global attractor and the basin of attraction is the whole space X.

Definition 1.11 - *A set A is asymptotically stable if it is stable and if there exists $\delta_1 > 0$ such that A is attractive on $\widehat{S}(A,\delta_1)$. In particular, A is exponentially stable if there exist: $\delta_1 > 0, \lambda(\delta_1) > 0, M(\delta_1) > 0$ such that*

$$v_0 \in \widehat{S}(A,\delta_1) \Rightarrow \widehat{d}[v(v_0,t), A] \le Me^{-\lambda t}\widehat{d}(v_0,A) \quad , \quad \forall t \in I\!\!R_+\,. \tag{1.65}$$

If the basin of attraction is the whole space X, i.e. $\delta_1 = \infty$, then the asymptotic (or exponential) stability is said to be global.

Definition 1.12 - *A motion $v(v_0,\cdot)$ is said to be orbitally stable if its positive orbit $\gamma(v_0)$ is a stable set.*

Remark 1.11 - *From Definitions* 1.6–1.8, *it follows that an equilibrium point v_* (identified with the motion $v(v_*,t) = v_*, \forall t \in I\!\!R_+$) is*

i) *stable if*

$$\forall\varepsilon > 0, \exists\delta(\varepsilon) > 0 : v_0 \in S(v_*,\delta) \Rightarrow v(v_0,t) \in S(v_*,\varepsilon), \forall t \in I\!\!R_+\,; \tag{1.66}$$

ii) *an attractor on an open set $Y \supset v_*$ if*

$$v_0 \in Y \Rightarrow \lim_{t\to\infty} d[v(t,v_0), v_*] = 0\,; \tag{1.67}$$

iii) *asymptotically stable if it is stable and an attractor on a set* $S(v_*, \delta_1)$ *for* $\delta_1 > 0$. *In particular, it is exponentially stable if*

$$\exists\lambda(\delta_1) > 0\,, M(\delta_1) > 0 : v_0 \in S(v_*, \delta_1) \Rightarrow \\ d[v(v_0, t), v_*] \le M\, e^{-\lambda t} d(v_0, v_*)\,, \quad \forall t \in I\!R_+\,; \tag{1.68}$$

iv) *asymptotically (exponentially) globally stable if* $\delta_1 = \infty$.

Remark 1.12 - *Let X be a normed linear space,* $\|\cdot\|$ *the norm, d the metric induced by the norm and*

$$u(u_0, t) = v(v_1, t) - v(v_0, t) \qquad (v_1 = v_0 + u_0) \tag{1.69}$$

the perturbation at time t to the basic motion $v(v_0, \cdot)$. *Then* (1.59) *is equivalent to*

$$\forall\varepsilon > 0 \exists\delta(\varepsilon) > 0 : u_0 \in S(\mathbf{O}, \delta) \Rightarrow u(u_0, t) \in S(\mathbf{O}, \varepsilon)\,, \quad \forall t \ge 0 \tag{1.70}$$

where $\mathbf{O}$ *is the origin of X. Therefore it is always possible to express the stability of a given basic motion* $v(v_0, t)$ *through the stability of the zero solution of the perturbed dynamical system:*

$$u : (u_0, t) \in X \times I\!R_+ \rightarrow v(v_0 + u_0, t) - v(v_0, t) \tag{1.71}$$

Remark 1.13 - *If the dynamical system v is linear — i.e.* $v(\cdot, t)$ *is a linear operator of X on X,* $\forall t \in I\!R_+$ *— then the stability of every motion is determined by the stability of the zero solution. When v is nonlinear, the stability of the trivial solution does not determine the stability of every motion.*

Remark 1.14 - *For measuring the initial data, a metric* d^*, *different from d,is sometimes used* [37]. *In this case all the previous definitions continue to hold, taking into account the metric* d^* *for the initial data. For instance, denoting by* $S^*(\mathbf{x}, r)$, $r > 0$, *the open ball centered at* $\mathbf{x}$ *and having radius* r *in the* d^* *metric, Definitions* 1.5-1.6 *continue to hold on condition that* $S^*(\mathbf{v}_0, \delta)$ *takes the place of* $S(\mathbf{v}_0, \delta)$. *Concerning the relationship between* d *and* d^*, *it is, generally, merely required that* $d^* \to 0$ *implies* $d \to 0$.

Remark 1.15 - *On a set X with elements* $x_i \in X$, *a functional* $\rho : X \times X \to I\!R$ *is positive definite if*

$$\begin{cases} \rho(x_1, x_2) > 0 & & x_1 \ne x_2\,, \\ \rho(x_1, x_2) = 0 & \text{iff} & x_1 = x_2\,. \end{cases}$$

Sometime a positive definite function (which is not necessarily a metric) is chosen as a measure of the perturbations [7].

Remark 1.16 - *On a linear space X, normed by* $\|\cdot\| : X \to I\!R_+$, *many other norms can be introduced. For instance, for each positive constant a, setting*

$\|x\|_1 = \|ax\|$, a new norm is defined. Two norms $\|\cdot\|_1$ and $\|\cdot\|_2$ on X are equivalent, if there exist constants $c_2 \geq c_1 > 0$ such that $c_2\|x\|_2 \geq \|x\|_1 \geq c_1\|x\|_2$, $\forall x \in X$. Immediately then, it follows that stability (instability) with respect to a fixed norm implies stability (instability) with respect to an equivalent norm.

Remark 1.17 - *On a linear finite dimensional space X all possible norms are equivalent. Therefore the stability does not depend on the chosen norm. This is so in the case of the metric space $X = I\!R^n$ and consequently, in the case of phenomena with a finite number of degrees of freedom (modelled by O.D.E.s). When a phenomenon has an infinite number of degrees of freedom, its evolution is modelled by a P.D.E. embedded (generally) in a normed linear infinite dimensional space. Then it can turn out that a solution is stable with one choice of metric and unstable with another choice (Sec.1.10). This is a fundamental difference between O.D.E.s and P.D.E.s: for P.D.E.s stability depends on the topology. Another fundamental difference depends on the fact that only a normed linear space of finite dimension is locally compact.*[8] *Therefore, in the case of P.D.E.s, if one is also able to prove the boundedness of positive orbits, this does not imply their compactness (or precompactness). For a deep analysis of this point and how it influences the asymptotic behaviour of solutions, and, in particular, the existence of a global attractor, we refer to* [39].

1.10 TOPOLOGY DEPENDENT STABILITY

In order to show that for P.D.E.s stability is topology-dependent, let us consider the well-known example concerning the linear stability of *Couette flow* of an ideal incompressible fluid. The equations of motion of a perfect incompressible fluid (*Euler equations*) can be obtained from (1.20) on setting $\nu = 0$. When $\mathbf{F}$ is a conservative force ($\mathbf{F} = -\nabla U$), one has:

$$\begin{cases} \mathbf{v}_t + \mathbf{v}\cdot\nabla\mathbf{v} = -\nabla(p+U), \\ \nabla\cdot\mathbf{v} = 0, \end{cases} \qquad \text{in } S \times I\!R_+ \tag{1.72}$$

$$\begin{cases} \mathbf{v}(\mathbf{x},0) = \mathbf{v}_0(\mathbf{x}) & \text{in } S, \\ \mathbf{v}\cdot\mathbf{n} = 0 & \text{on } \partial S \times I\!R_+, \end{cases} \tag{1.73}$$

where $S \subset I\!R^3$ is a bounded domain filled with the fluid, and $(1.73)_2$ – in which $\mathbf{n}$ denotes the outward unit normal to ∂S – is the boundary condition used for

[8] A subset A of a metric space X is *precompact* if it is bounded, and every sequence $\{x_n\} \subset A$ contains a Cauchy sequence; is compact if it is precompact, and every sequence $\{x_n\} \subset A$ contains a subsequence converging to a point in A. A metric space X is *locally compact* if every closed and bounded subset is compact. The following Theorem holds [38]: a normed linear space is of finite dimension if and only if every bounded subset is compact.

a fixed boundary. Because of the vectorial identities

$$\begin{cases} \mathbf{a}\cdot\nabla\mathbf{a} = (\nabla\times\mathbf{a})\times\mathbf{a} + \frac{1}{2}\nabla a^2\,, \\ \nabla\times(\mathbf{a}\times\mathbf{b}) = \mathbf{b}\cdot\nabla\mathbf{a} - \mathbf{a}\cdot\nabla\mathbf{b} + (\nabla\cdot\mathbf{b})\mathbf{a} - (\nabla\cdot\mathbf{a})\mathbf{b}\,, \end{cases} \tag{1.74}$$

one obtains on taking the curl of both sides of $(1.72)_1$:

$$\Omega_t + \mathbf{v}\cdot\nabla\Omega = \Omega\cdot\nabla\mathbf{v} \tag{1.75}$$

where Ω is the *vorticity vector*:

$$\Omega = \nabla\times\mathbf{v}\,. \tag{1.76}$$

The Euler equations (1.75) for vorticity become simpler in the context of two dimensional motion

$$\mathbf{v} = (v_1, v_2, 0) \quad , \quad v_i = v_i(x_1, x_2, t)\,. \tag{1.77}$$

In fact, introducing the *stream function* $\Psi(x_1, x_2, t)$ and setting

$$\mathbf{v} = \nabla^{\perp}\Psi \quad , \quad \nabla^{\perp} = \left(\frac{\partial}{\partial x_2}, -\frac{\partial}{\partial x_1}, 0\right)\,, \tag{1.78}$$

one immediately obtains ($\mathbf{e}_i$ = unit vector along x_i axes)

$$\nabla\cdot\mathbf{v} \equiv 0 \quad , \quad \Omega = -\Delta\Psi\mathbf{e}_3 \quad , \quad \Omega\cdot\nabla\mathbf{v} = -\Delta\Psi\,\frac{\partial\mathbf{v}}{\partial x_3} = 0 \quad ; \tag{1.79}$$

and (1.75) becomes:

$$\frac{\partial\Delta\Psi}{\partial t} = \frac{\partial(\Psi, \Delta\Psi)}{\partial(x_1, x_2)}\,. \tag{1.80}$$

From $(1.73)_2$ it turns out that Ψ must be a constant on ∂S and therefore — because Ψ is defined modulo a constant — one can append to (1.80) the initial and boundary conditions

$$\Psi(x_1, x_2, 0) = \varphi(x_1, x_2) \qquad \text{in } S\,, \tag{1.81}$$

$$\Psi = c = \text{const.} \qquad \text{on } \partial S\,, \tag{1.82}$$

where φ is prescribed and c is arbitrary.

Let φ be a steady solution to (1.80), (1.82) and let us consider the stability of the *basic motion* $\mathbf{v}_* = \nabla^{\perp}\varphi$ with respect to planar perturbations $\mathbf{u} = \nabla^{\perp}\Phi(x_1, x_2, t)$. One immediately obtains the following initial boundary value problem

$$\frac{\partial\Delta\Phi}{\partial t} = \frac{\partial(\Phi, \Delta\Phi + \Delta\varphi)}{\partial(x_1, x_2)} + \frac{\partial(\varphi, \Delta\Phi)}{\partial(x_1, x_2)}\,, \tag{1.83}$$

$$\Phi(x_1, x_2, 0) = \Phi_0(x_1, x_2) \qquad \text{in } S\,, \tag{1.84}$$

$$\Phi = 0 \qquad \text{on } \partial S\,, \tag{1.85}$$

where Φ_0 is prescribed. *Linearizing* with respect to Φ, (1.83) gives

$$\frac{\partial \Delta\Phi}{\partial t} = \frac{\partial(\Phi, \Delta\varphi)}{\partial(x_1, x_2)} + \frac{\partial(\varphi, \Delta\Phi)}{\partial(x_1, x_2)} . \tag{1.86}$$

As basic motion $\mathbf{v}_*$, let us consider the *Couette flow* in the flat pipe $S = \{(x_1, x_2) : x_1 \in I\!R \,,\, |x_2| \leq 1\}$:

$$\mathbf{v}_* = \nabla^{\perp}\varphi \quad , \quad \varphi = \frac{1}{2}x_2^2 . \tag{1.87}$$

Then (1.83), (1.84) and (1.85) give

$$\begin{cases} \dfrac{\partial \Delta\Phi}{\partial t} + x_2 \dfrac{\partial \Delta\Phi}{\partial x_1} = 0 \,, \\ \\ \Phi(x_1, x_2, 0) = \Phi_0(x_1, x_2) \,, \\ \Phi(x_1, \pm 1, t) = 0 \qquad\qquad t \in I\!R_+ \,, \end{cases} \tag{1.88}$$

and hence

$$\begin{cases} \dfrac{\partial \omega}{\partial t} + x_2 \dfrac{\partial \omega}{\partial x_1} = 0 \,, \\ \\ \omega(x_1, x_2, 0) = \omega_0(x_1, x_2) \end{cases} \tag{1.89}$$

where $\omega = -\Delta\Phi$, $\omega_0 = -\Delta\Phi_0$. Assuming $\omega_0 \in C^1(S)$, immediately one has

$$\omega(x_1, x_2, t) = \omega_0(x_1 - x_2 t, x_2) .$$

Since

$$\sup_S |\omega(x_1, x_2, t)| = \sup_S |\omega_0(x_1, x_2)| \qquad \forall t \geq 0 \tag{1.90}$$

$$\begin{aligned} \sup_S |\omega_{x_2}| &= \sup_S |-t\omega_{0x_1}(x_1 - x_2 t, x_2) + \omega_{0x_2}(x_1 - x_2 t, x_1)| \\ &\geq t \sup_S |\omega_{0x_1}(x, y)| - \sup_S |\omega_{0x_2}(x, y)| \qquad \forall t \,, \end{aligned} \tag{1.91}$$

one has stability with respect the norm (1.90) but instability with respect the norm $\sup_S |\omega| + \sup_S |\omega_{x_2}|$.

For further examples of topology dependent stability we refer the readers to exercises. We end this section by considering a celebrated example of lack of stability i.e. the initial value problem for the *Laplace equation*:

$$\begin{cases} u_{tt} + u_{xx} = 0 & \{x \in [0,1], t \in I\!R_+\} \,, \\ \\ u(x,0) = u_0(x) \,,\, u_t(x,0) = u^*(x) & x \in [0,1] \end{cases} \tag{1.92}$$

where u_0 and u^* are prescribed functions.

Problem (1.92) – for $u_0 = u^* = 0$ – has the trivial solution. This solution is unstable with respect to the pointwise norm (L_∞-norm)

$$\|\cdot\|_\infty = \sup_{x\in[0,1]} |u(x,t)| . \tag{1.93}$$

In fact, choosing as perturbation to the initial data

$$u_0(x) = 0 \quad , \quad u^*(x) = \frac{a}{n} \sin n\pi x \quad , \quad (a \in I\!R,\, n \in I\!N_+) , \tag{1.94}$$

a solution $u_n \in C^2[0,1]\,, \forall t \in I\!R_+$, is

$$u_n(x,t) = \frac{u^*(x)}{n\,\pi} \sinh n\pi t . \tag{1.95}$$

Because of

$$\|u_n(t)\|_\infty = \sup_{x\in[0,1]} |u_n| = \frac{|a| \sinh n\pi t}{n^2\,\pi} , \qquad t \in I\!R_+ \tag{1.96}$$

it follows that, for $n \to \infty$, the data tends to zero while $\|u_n\|$ tends to $+\infty$.

Let us observe that, along the solutions to (1.92), one has

$$\frac{1}{2}\frac{d}{dt}\int_o^1 u_t^2\, dt = -\int_0^1 u_t\, u_{xx}\, dx = -\int_0^1 \left[\frac{\partial}{\partial x}(u_t\, u_x) - u_x\, u_{tx}\right] dx$$

i.e.

$$\frac{1}{2}\frac{d}{dt}\int_0^1 (u_t^2 - u_x^2)\, dx + [u_t\, u_x]_0^1 = 0 . \tag{1.97}$$

Because $\dfrac{\partial u_n}{\partial t} \in C_0[0,1]$, along the solutions (1.95), one has

$$\int_0^1 \left(\frac{\partial u_n^2}{\partial t} - \frac{\partial u_n^2}{\partial x}\right) dx = \text{const.} = \int_0^1 \left(\frac{\partial u_n^2}{\partial t} - \frac{\partial u_n^2}{\partial x}\right)_{t=0} dx$$

$$= \int_0^1 (u^*)^2\, dx .$$

Therefore, on the class of solutions (1.95), $1/2\int_0^1 (u_t^2 - u_x^2)\, dx$ can be assumed as measure of the perturbations {cfr. Remark 1.15} and the stability of the trivial solution is then recovered.

1.11 ERRORS IN FORMULATING THE MODEL

In Section 1.7 we considered the possible errors made in the measurement of the initial data. Here we take into account, also, errors made in formulating the model. Of course, this problem has, also, to be considered case by case and only few general results are known [40]. Here we study – by means of an example – the precise relationship between the solution to a nonlinear model and the solutions to its linearized version subject to proportional data.

Let us consider the Navier-Stokes equations governing the motion of an incompressible isothermal viscous fluid in a fixed bounded domain $\Omega \subset I\!R^3$, with rigid boundary $\partial\Omega$, smooth and at rest (See Sec. 1.8, a), under the action of a conservative force $\mathbf{F} = \nabla U$ (which is supposed to be absorbed into the pressure)

$$\begin{cases} \mathbf{v}_t + \mathbf{v} \cdot \nabla \mathbf{v} = -\nabla p + \nu \Delta \mathbf{v} \\ \\ \nabla \cdot \mathbf{v} = 0, \end{cases} \qquad \text{in } \Omega \tag{1.98}$$

$$\begin{cases} \mathbf{v}(\mathbf{x},0) = \mathbf{v}_0(\mathbf{x}) & \text{in } \Omega, \\ \mathbf{v}(\mathbf{x},t) = 0 & \text{on } \partial\Omega. \end{cases} \tag{1.99}$$

Their linearized version – called *Stokes equations* – is

$$\begin{cases} \mathbf{w}_t = -\nabla q + \nu \Delta \mathbf{w}, \\ \\ \nabla \cdot \mathbf{w} = 0, \end{cases} \qquad \text{in } \Omega \tag{1.100}$$

to which we append initial and boundary conditions analogous to (1.99)

$$\begin{cases} \mathbf{w}(\mathbf{x},0) = \mathbf{w}_0(\mathbf{x}) & \text{in } \Omega \\ \mathbf{w}(\mathbf{x},t) = 0 & \text{on } \partial\Omega. \end{cases} \tag{1.101}$$

Following Payne [41], we assume

$$\mathbf{v}_0 = \varepsilon \mathbf{w}_0 \qquad (\varepsilon = \text{const.} > 0). \tag{1.102}$$

Furthermore we assume $\mathbf{w}_0$, ν and Ω, are such that a solution $\mathbf{w}$ to the I.B.V.P. (1.100)–(1.101) and a solution $\mathbf{v}$ to the I.B.V.P. (1.98)–(1.99) exist and are smooth enough. The question is: under what conditions does $\varepsilon\mathbf{w}$ approximate $\mathbf{v}$ "sufficiently well"? Setting

$$\mathbf{v} = \varepsilon \mathbf{w} + \mathbf{u}, \tag{1.103}$$

in any norm $\|\cdot\|$, one has

$$\varepsilon\|\mathbf{w}\| - \|\mathbf{u}\| \le \|\mathbf{v}\| \le \varepsilon\|\mathbf{w}\| + \|\mathbf{u}\| \tag{1.104}$$

i.e. an upper and a lower bound for $\|\mathbf{v}\|$, if one is able to estimate $\|\mathbf{w}\|$ and $\|\mathbf{u}\|$ in terms of the data.

Let $||\cdot||$ be the L_2-norm. Then it follows that

$$\begin{aligned}\frac{d}{dt}||\mathbf{u}||^2 &= 2\int_\Omega \mathbf{u}\cdot\mathbf{u}_t d\Omega = 2\int_\Omega \mathbf{u}\cdot(\mathbf{v}_t-\varepsilon\mathbf{w}_t)d\Omega \\ &= 2\int_\Omega \mathbf{u}\cdot[\nu\Delta\mathbf{u}-\nabla(p-\varepsilon q)-\mathbf{v}\cdot\nabla\mathbf{v}]d\Omega\,. \end{aligned} \tag{1.105}$$

Noting that

$$\begin{cases} \mathbf{u}\cdot\Delta\mathbf{u} = \frac{1}{2}\nabla\cdot(\nabla\mathbf{u}^2)-(\nabla\mathbf{u})^2 \quad , \quad \nabla\cdot\mathbf{u}=0\,, \\ \mathbf{v}\cdot\nabla\mathbf{v}\cdot\mathbf{u} = \mathbf{v}\cdot(\varepsilon\nabla\mathbf{w}+\nabla\mathbf{u})\cdot\mathbf{u} = \varepsilon\mathbf{v}\cdot\nabla\mathbf{w}\cdot\mathbf{u}+\frac{1}{2}\mathbf{v}\cdot\nabla\mathbf{u}^2\,, \end{cases} \tag{1.106}$$

and that, for any scalar function $f\in C^1(\Omega)$ and any vector function $\mathbf{V}\in C^1(\Omega)$, one has

$$\nabla\cdot(f\mathbf{V}) = \nabla f\cdot\mathbf{V}+f\nabla\cdot\mathbf{V} \tag{1.107}$$

it follows from (1.103), (1.105) and the boundary conditions that

$$\frac{d}{dt}||\mathbf{u}||^2 = -2\nu||\nabla\mathbf{u}||^2 - 2\varepsilon^2\int_\Omega \mathbf{w}\cdot\nabla\mathbf{w}\cdot\mathbf{u}d\Omega - 2\varepsilon\int_\Omega \mathbf{u}\cdot\nabla\mathbf{w}\cdot\mathbf{u}\,d\Omega\,. \tag{1.108}$$

In view of

$$\begin{aligned}\int_\Omega \mathbf{w}\cdot\nabla\mathbf{w}\cdot\mathbf{u}d\Omega &= \int_\Omega \mathbf{w}\cdot\nabla(\mathbf{u}\cdot\mathbf{w})d\Omega - \int_\Omega \mathbf{w}\cdot\nabla\mathbf{u}\cdot\mathbf{w}d\Omega \\ &= -\int_\Omega \mathbf{w}\cdot\nabla\mathbf{u}\cdot\mathbf{w}d\Omega \end{aligned} \tag{1.109}$$

and Schwarz's inequality, one obtains from (1.108) that

$$\begin{aligned}\frac{d}{dt}||\mathbf{u}||^2 &\le -2\nu||\nabla\mathbf{u}||^2 + 2\varepsilon^2||\nabla\mathbf{u}||(\int_\Omega \mathbf{w}^4 d\Omega)^{1/2} \\ &+ 2\varepsilon||\nabla\mathbf{w}||(\int_\Omega \mathbf{u}^4 d\Omega)^{1/2}\,. \end{aligned} \tag{1.110}$$

Now for $\mathbf{V}\in\{\mathbf{u},\mathbf{w}\}$, the following Sobolev inequality holds (appendix, A.3)

$$\int_\Omega \mathbf{V}^4 d\Omega \le \alpha^2||\mathbf{V}||\cdot||\nabla\mathbf{V}||^3 \tag{1.111}$$

where α=const.>0. Therefore one obtains from (1.110):

$$\begin{aligned}\frac{d}{dt}||\mathbf{u}||^2 &\le -\,2\nu||\nabla\mathbf{u}||^2 + 2\varepsilon^2\alpha||\nabla\mathbf{u}||||\mathbf{w}||^{1/2}||\nabla\mathbf{w}||^{3/2} \\ &+ 2\varepsilon\alpha||\nabla\mathbf{w}||||\mathbf{u}||^{1/2}||\nabla\mathbf{u}||^{3/2}\,. \end{aligned} \tag{1.112}$$

We observe now that for the Poincaré inequality (appendix, A.4) is

$$||\mathbf{u}||^2 \le \lambda||\nabla\mathbf{u}||^2\,, \qquad \lambda = \text{const.} > 0\,. \tag{1.113}$$

Furthermore, for generalized Cauchy inequality (appendix, A.1), is

$$2\|\nabla \mathbf{u}\| \leq \frac{1}{\eta} + \eta\|\nabla \mathbf{u}\|^2 \qquad \eta = \text{const.} > 0\,. \tag{1.114}$$

Therefore (1.112) gives

$$\frac{d}{dt}\|\mathbf{u}\|^2 \leq -f(t)\|\nabla \mathbf{u}\|^2 + g(t) \tag{1.115}$$

where

$$\begin{cases} f(t) = 2\nu - \varepsilon^2\alpha\eta\|\mathbf{w}\|^{1/2}\|\nabla\mathbf{w}\|^{3/2} - 2\varepsilon\alpha\lambda^{1/4}\|\nabla\mathbf{w}\|\,, \\ g(t) = \dfrac{\varepsilon^2\alpha}{\eta}\|\mathbf{w}\|^{1/2}\|\nabla\mathbf{w}\|^{3/2}\,. \end{cases} \tag{1.116}$$

We now determine the conditions on the data and ε under which $f > 0$, $\forall t \geq 0$. To this end, let us obtain an upper bound for $\|\mathbf{w}\|$ and $\|\nabla\mathbf{w}\|$. One has

$$\begin{aligned} \frac{d}{dt}\|\nabla\mathbf{w}\|^2 &= 2\int_\Omega w_{i,j}w_{i,jt}\,d\Omega = 2\int_\Omega\left[\frac{\partial}{\partial x_j}\left(\frac{\partial\mathbf{w}}{\partial x_j}\cdot\mathbf{w}_t\right) - \Delta\mathbf{w}\cdot\mathbf{w}_t\right]d\Omega = \\ &= 2\int_{\partial\Omega}\frac{\partial\mathbf{w}}{\partial x_j}\cdot\mathbf{w}_t\mathbf{n}\cdot\mathbf{e}_j\,d\sigma - \frac{2}{\nu}\int_\Omega(\mathbf{w}_t + \nabla q)\cdot\mathbf{w}_t d\Omega\,, \end{aligned} \tag{1.117}$$

with $\mathbf{e}_j$ = unit vector in the x_j direction. Therefore, in view of

$$\begin{cases} \mathbf{w}_t = 0 & \text{on } \partial\Omega\,, \\ \nabla\cdot\mathbf{w}_t = 0 & \text{in } \Omega\,, \end{cases} \tag{1.118}$$

it follows that

$$\frac{d}{dt}\|\nabla\mathbf{w}\|^2 = -\frac{2}{\nu}\|\mathbf{w}_t\|^2\,. \tag{1.119}$$

Now – by Schwarz's inequality – one obtains

$$\int_\Omega \mathbf{w}\cdot\mathbf{w}_t d\Omega \leq \|\mathbf{w}\|\cdot\|\mathbf{w}_t\|\,,$$

and hence

$$\|\mathbf{w}_t\|^2 \geq \frac{\left(\displaystyle\int_\Omega \mathbf{w}\cdot\mathbf{w}_t d\Omega\right)^2}{\|\mathbf{w}\|^2}\,. \tag{1.120}$$

On the other hand, from (1.100)–(1.101), it turns out that

$$\int_\Omega \mathbf{w}\cdot\mathbf{w}_t d\Omega = -\nu\|\nabla\mathbf{w}\|^2 \tag{1.121}$$

and hence

$$\frac{d}{dt}\|\nabla\mathbf{w}\|^2 \leq -2\nu\frac{\|\nabla\mathbf{w}\|^4}{\|\mathbf{w}\|^2}\,. \tag{1.122}$$

Taking into account (1.113), it follows that

$$\frac{d}{dt}||\nabla \mathbf{w}||^2 \leq -\frac{2\nu}{\lambda}||\nabla \mathbf{w}||^2$$

which gives

$$||\nabla \mathbf{w}||^2 \leq ||\nabla \mathbf{w}_0||^2 e^{-2\nu\, t/\lambda}\,. \tag{1.123}$$

Similarly one obtains

$$\frac{d}{dt}||\mathbf{w}||^2 \leq -2\nu||\nabla \mathbf{w}||^2 \leq -\frac{2\nu}{\lambda}||\mathbf{w}||^2$$

and hence

$$||\mathbf{w}||^2 \leq ||\mathbf{w}_0||^2 e^{-2\nu\, t/\lambda}\,. \tag{1.124}$$

By means of (1.123)–(1.124) it turns out that

$$\begin{cases} f(t) \geq \left\{ \begin{array}{l} 2\nu - \varepsilon^2\alpha\eta||\mathbf{w}_0||^{1/2}\,||\nabla \mathbf{w}_0||^{3/2}\, e^{-2\nu\, t/\lambda} \\ -2\varepsilon\alpha\lambda^{1/4}||\nabla \mathbf{w}_0||\, e^{-\nu\, t/\lambda} \end{array} \right. , \\ \\ g(t) \leq \dfrac{\varepsilon^2\alpha}{\eta}||\mathbf{w}_0||^{1/2}\,||\nabla \mathbf{w}_0||^{3/2}\, e^{-2\nu\, t/\lambda}\,. \end{cases} \tag{1.125}$$

Hence, provided that

$$||\nabla \mathbf{w}_0|| < \frac{\nu}{\varepsilon\alpha\lambda^{1/4}}\,, \tag{1.126}$$

on choosing

$$0 < \eta < \frac{2(\nu - \varepsilon\alpha\lambda^{1/4}||\nabla \mathbf{w}_0||)}{\varepsilon^2\alpha||\mathbf{w}_0||^{1/2}\cdot||\nabla \mathbf{w}_0||^{3/2}}\,, \tag{1.127}$$

one obtains

$$f(t) \geq 2\alpha\nu \tag{1.128}$$

with

$$a = \frac{1}{2\nu}[2\nu - \alpha\varepsilon(\varepsilon\eta||\mathbf{w}_0||^{1/2}||\nabla \mathbf{w}_0||^{1/2} + 2\lambda^{1/4})||\nabla \mathbf{w}_0||] > 0\,. \tag{1.129}$$

Then, from (1.115), it follows that

$$\frac{d}{dt}||\mathbf{u}||^2 \leq -\frac{2a\nu}{\lambda}||\mathbf{u}||^2 + g(t) \tag{1.130}$$

and hence

$$\frac{d}{dt}[e^{2a\nu\, t/\lambda}||\mathbf{u}||^2] \leq \frac{\varepsilon^2\alpha}{\eta}||\mathbf{w}_0||^{1/2}\,||\nabla \mathbf{w}||^{3/2} e^{-2\nu(1-a)\, t/\lambda}\,. \tag{1.131}$$

Integrating, it turns out that

$$\|\mathbf{u}\|^2 \leq \|\mathbf{u}_0\|^2 e^{-2a\nu\, t/\lambda} +$$

$$+\begin{cases} \dfrac{\lambda\varepsilon^2\alpha}{\eta}\dfrac{\|\mathbf{w}_0\|^{1/2}\,\|\nabla\mathbf{w}_0\|^{3/2}}{2\nu(1-a)}\left(e^{-2\nu a/\lambda} - e^{-2\nu\, t/\lambda}\right), & a \neq 1 \\ \dfrac{\varepsilon^2\alpha}{\eta}\|\mathbf{w}_0\|^{1/2}\,\|\nabla\mathbf{w}_0\|^{3/2}\, te^{-2\nu a\, t/\lambda}, & a = 1 \end{cases} \tag{1.132}$$

Relation (1.132) shows that, for any $\varepsilon > 0$, in the $L^2(\Omega)$-space, the distance between $\mathbf{v}$ and $\varepsilon\mathbf{w}$ tends to zero exponentially as $t \to \infty$. In particular, for $\varepsilon = 1$, the solution to the problem for the Navier-Stokes equations, approaches exponentially, as $t \to \infty$, that of the corresponding problem for Stokes equations.

Remark 1.18 - *By handling in a different way the two integrals on the right hand side of* (1.108), *one may obtain results differing from but similar to* (1.132). *For this question and for continuous dependence on geometry, we refer to* [36,41-43].

Remark 1.19 - *Taking account of* (1.123)–(1.124) *and Definition* 1.10, *it follows that, for the Stokes equation, the critical point* $\{\mathbf{w} = 0,\ q = \text{const.}\}$ *is a global attractor in the* L_2*-norm of velocity, and in the* L_2*-norm of gradient of velocity.*

1.12 MATHEMATICAL MODELLING. REACTION DIFFUSION EQUATION

We consider now the question of obtaining the evolution equation of a phenomenon $\mathcal{F}$. Introduced the state vector $\mathbf{u}$, the problem consists of determining the function $\mathbf{F}$ appearing in (1.1). One appeals then to general laws governing $\mathcal{F}$. For mechanical, electromagnetic and other fundamental phenomena we refer to [44-46] where it is also shown how to obtain the evolution equation in the *local* form (1.1). One has to remember, in fact, that, normally, the laws governing phenomena are formulated in integral form.

For the sake of simplicity, let us consider here a phenomenon $\mathcal{F}$ whose state is described by a scalar function $u(\mathbf{x}, t)$. This is the case, for instance, for a biological or chemical phenomenon in which u represents the density of a population or the concentration of a substance. More generally, this is the case for any phenomenon $\mathcal{F}$ in which "a substance" or some "objects" move in a domain Ω with density or concentration $u(\mathbf{x}, t)$: cars on an autoroad, pollutants in a lake, heat along a bar,.... Denoting by S any regular subdomain of Ω and by $\mathbf{j}$ the *flux* of the substance, one requires that the *law of balance* holds: *the rate of change of the amount of "a substance" in S is equal to the*

rate at which the substance flows across ∂S into S plus the rate at which the substance is created in S. Therefore if $\mathbf{n}$ is the outward unit normal to ∂S and σ is the ***supply*** density of substance, it follows that

$$\frac{d}{dt}\int_S u\, dS = -\int_{\partial S} \mathbf{j}\cdot\mathbf{n}\, d\Sigma + \int_S \sigma\, dS\,. \tag{1.133}$$

Applying the divergence theorem, equation (1.133) becomes

$$\int_S (u_t + \nabla\cdot\mathbf{j} - \sigma)\, dS = 0\,.$$

Assuming the integrand to be continuous, and since S is arbitrary, the ***local balance*** or ***conservation equation*** follows:

$$u_t + \nabla\cdot\mathbf{j} = \sigma\,. \tag{1.134}$$

This equation holds for any flux transport $\mathbf{j}$ and any supply σ. Of course, σ and $\mathbf{j}$ can depend on u and its derivatives and can also depend explicitly on the position $\mathbf{x}$ and time t. The relationships of $\mathbf{j}$ and σ to u depend on the substance at hand and are called ***constitutive equations*** or ***response functions.*** Equation (1.134) models many phenomena and is referred to as a ***reaction diffusion*** equation.

a) *Heat or diffusion equation.* Let us consider the heat flow in a cylindrical rod extending from $x=0$ to $x=l$ and let Σ be the constant cross-sectional area (Figure 1.2), c_v the specific heat at constant volume of the material of which the bar is composed and T the temperature. We assume that T and any physical parameters are constant on Σ and that heat energy flows only in the direction orthogonal to Σ (bar is insulated laterally), under the boundary conditions $T(l,t)=T_2$ =const., $T(0,t)=T_1$ =const., with $T_1 > T_2$.

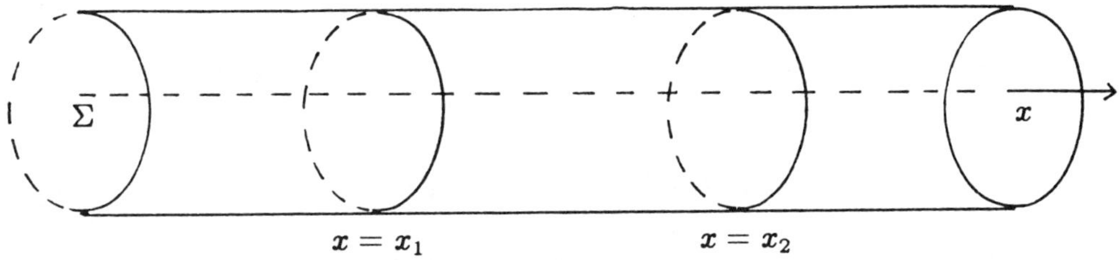

Figure 1.2

The heat energy density u is given by

$$u = \rho\, c_v T \tag{1.135}$$

where ρ is the mass density of the rod. Because ρ and c_v are assumed known, the true state parameter is T. Choosing as subdomain S of the rod that portion of it which is bounded by the cross-section at $x_1 \in [0,l[$ and by that at $x_2 \in]x_1, l]$, then it follows that

$$\frac{\partial}{\partial t}(\rho\, c_v\, T) + \nabla\cdot\mathbf{j} = \sigma\,. \tag{1.136}$$

As constitutive equation for the heat flux $\mathbf{j}$, let us assume the *Fourier's heat law*

$$\mathbf{j} = -K\,\nabla T \tag{1.137}$$

where $K > 0$ is the thermal conductivity. When ρ, c_v, K are independent of T and, in particular, constant, the linear *nonhomogeneous heat equation* follows immediately

$$T_t = kT_{xx} + \frac{1}{\rho\, c_v}\sigma \tag{1.138}$$

where $k = K/\rho c_v$ is the *thermal diffusivity*. The supply σ is called the *source term* if positive and the *sink term* if negative. The second case, for instance, happens when the bar is radiating heat to the external medium in a significant way. In the absence of supply, the full I.B.V.P. for the one-dimensional heat equation is

$$\begin{cases} T_t = kT_{xx}\,, \\ T(x,0) = T_0(x)\,, \\ T(0,t) = T_1(t)\,,\; T(l,t) = T_2(t)\,, \end{cases} \tag{1.139}$$

where T_0, T_1, and T_2 are prescribed. We end by remarking that other boundary conditions are possible. For instance, if one end of the rod, say $x = 0$, is insulated, then no heat flows through the cross-sectional area at $x = 0$ and (1.137) gives

$$T_x(0,t) = 0 \qquad\qquad \forall t\,. \tag{1.140}$$

b) *Diffusion equation of a biological population.* - Let $\mathcal{P}$ be a biological population living in a region Ω. In order to study the diffusion of $\mathcal{P}$ in Ω, we choose as state variable u the density $\rho(\mathbf{x},t)$ of $\mathcal{P}$. In this case $\sigma(\mathbf{x},t)$ represents, per unit volume, at $\mathbf{x}$ at time t, the difference between births and deaths. Denoting the diffusion velocity by $\mathbf{v}(\mathbf{x},t)$, one has $\mathbf{j} = \rho\mathbf{v}$ and hence (1.134) gives

$$\rho_t + \nabla\cdot(\rho\mathbf{v}) = \sigma\,. \tag{1.141}$$

As constitutive equations for σ and $\mathbf{v}$ are generally assumed:

$$\sigma = \sigma(\rho) \quad , \quad \mathbf{v} = \mathbf{v}(\rho, \nabla\rho)\,. \tag{1.142}$$

In particular, concerning the supply σ, the constitutive equations

$$\sigma = r\,\rho \qquad\qquad (r = \text{const.})\,, \tag{1.143}$$

$$\sigma = r\,\rho(1 - \frac{\rho}{s}) \qquad (r, s = \text{const.}) \tag{1.144}$$

are the *Malthusian* and *Verhulst* (or *logistic*) *laws* respectively. Turning to $(1.142)_2$, we consider the case in which the diffusion is caused by spatial variation in the population. Then a constitutive law for $\mathbf{v}$ which ensures that $\mathbf{v}$ is *isotropic*[9] – i.e. that there are no preferred directions for population flow – is[10]

$$\mathbf{v} = -K(\rho)\nabla\rho \tag{1.145}$$

[9]The function $\mathbf{v}(\rho, \nabla\rho)$ is said to be *isotropic* if, for every orthogonal linear transformation $\mathbf{Q}$ one has

$$\mathbf{Q}\mathbf{v}(\rho, \nabla\rho) = \mathbf{v}(\rho, \mathbf{Q}\nabla\rho)\,.$$

[10]According to the sense of $\nabla\rho$, $K(\rho) < 0$ for $\rho > 0$ means that, locally, $\mathcal{P}$ diffuses from a point of high density to one of low density.

and the corresponding flux is

$$\mathbf{j} = -K(\rho)\rho\nabla\rho\,. \tag{1.146}$$

When

$$K(\rho) = \frac{k}{\rho} \qquad (k = \text{const.} > 0)\,, \tag{1.147}$$

then

$$\mathbf{j} = -k\nabla\rho\,, \tag{1.148}$$

and consequently the diffusion equation of a biological population becomes

$$\rho_t = k\Delta\rho + \sigma\,. \tag{1.149}$$

When (1.144) obtains, this equation takes the form

$$\rho_t = r\rho(1 - \frac{\rho}{s}) + k\Delta\rho \tag{1.150}$$

and is now known as the ***Fisher equation***. This is so because the one dimensional version of (1.150) was proposed, in 1937, by R.A.Fisher [47-50] as a model for the spread of an advantageous gene in a population.

We remark that in [51] it has been pointed out that – at least when migration is due to crowding – (1.147) seems unreasonable. In this case, indeed, one would expect the diffusion velocity to decrease as ρ decreases. But (1.145), (1.147) give

$$\mathbf{v} = -\frac{k}{\rho}\nabla\rho \tag{1.151}$$

and hence the diffusion velocity can tend to infinity as $\rho \to 0$.

In order to circumvent this the following constitutive equation for the flux has been proposed in [51]

$$\mathbf{j} = -\nabla\varphi(\rho) \tag{1.152}$$

where

$$\varphi(\rho) = \int_0^\rho \xi K(\xi)d\xi\,. \tag{1.153}$$

This leads to the following ***non-linear*** P.D.E. for ρ

$$\rho_t = \Delta\varphi(\rho) + \sigma(\rho)\,. \tag{1.154}$$

c) ***Systems of reaction-diffusion equations.*** Let us consider the following generalization of (1.149)

$$\mathbf{u}_t = \nabla\cdot(D\nabla\mathbf{u}) + \mathbf{f} \qquad x \in \Omega\,,\, t > 0\,, \tag{1.155}$$

where $\mathbf{u}(\mathbf{x},t) = \{u_1(\mathbf{x},t), ..., u_n(\mathbf{x},t)\}$, D is a $n\times n$ matrix and $\mathbf{f}$ a vector supply term. Equation (1.155) models the diffusion through Ω of n interacting species or chemicals $\mathcal{P}_i$ $(i = 1, .., n)$, u_i being the density or concentration of $\mathcal{P}_i$ [47].

When there is no cross diffusion among the species, the diffusivity matrix D is simply a diagonal matrix:

$$D = \begin{Vmatrix} D_1 & 0 & \dots & 0 \\ 0 & D_2 & \dots & 0 \\ \dots & \dots & \dots & \dots \\ 0 & 0 & \dots & D_n \end{Vmatrix} \tag{1.156}$$

and then – for $D_i = \text{const.}$, $(i = 1, 2, .., n)$ – equation (1.155) becomes

$$\mathbf{u}_t = D\Delta\mathbf{u} + \mathbf{f}\,. \tag{1.157}$$

Equation (1.157) – with D a positive diagonal matrix (i.e. $D_i > 0\,, \forall i$) – can generate many interesting practical models [5,27,39,40,47,52-54]. We confine ourselves to mentioning the following ones:

$$\mathbf{u}_t = D\Delta\mathbf{u} + (1 - |\mathbf{u}|^2)\mathbf{u}\,; \tag{1.158}$$

$$\begin{cases} \dfrac{\partial u_1}{\partial t} = D_1 \Delta u_1 + u_1 f_1(u_1, u_2)\,, \\ \dfrac{\partial u_2}{\partial t} = D_2 \Delta u_2 + u_2 f_2(u_1, u_2)\,; \end{cases} \tag{1.159}$$

$$\begin{cases} \dfrac{\partial u_1}{\partial t} = D_1 \Delta u_1 + (u_2)^p g(u_1)\,, \\ \dfrac{\partial u_2}{\partial t} = D_2 \Delta u_2 + (u_2)^p g(u_1)\,; \end{cases} \tag{1.160}$$

$$\begin{cases} \dfrac{\partial u_1}{\partial t} = D_1 \Delta u_1 + \alpha(u_2 - u_1 u_2 + u_1 - \beta u_1^2)\,, \\ \dfrac{\partial u_2}{\partial t} = D_2 \Delta u_2 + \dfrac{1}{\alpha}(\gamma u_3 - u_2 - u_1 u_2)\,, \\ \dfrac{\partial u_3}{\partial t} = D_3 \Delta u_3 + \delta(u_1 - u_3)\,; \end{cases} \tag{1.161}$$

$$\begin{cases} \dfrac{\partial u_1}{\partial t} = D_1 \dfrac{\partial^2 u_1}{\partial x^2} + f_1(u_1, u_2, u_3, u_4)\,, \\ \dfrac{\partial u_2}{\partial t} = D_2 \dfrac{\partial^2 u_2}{\partial x^2} + g_1(u_1)[h(u_1) - u_2]\,, \\ \dfrac{\partial u_3}{\partial t} = D_3 \dfrac{\partial^2 u_3}{\partial x^2} + g_2(u_1)[h_2(u_1) - u_3]\,, \\ \dfrac{\partial u_4}{\partial t} = D_4 \dfrac{\partial^2 u_4}{\partial x^2} + g_3(u_1)[h_3(u_1) - u_4]\,; \end{cases} \tag{1.162}$$

$$\begin{cases} \dfrac{\partial u_1}{\partial t} = D_1 \dfrac{\partial^2 u_1}{\partial x^2} + f(u_1) - u_2\,, \\ \dfrac{\partial u_2}{\partial t} = D_2 \dfrac{\partial^2 u_2}{\partial x^2} + \alpha u_1 - \beta u_2\,. \end{cases} \tag{1.163}$$

Equations (1.158) arise in the study of superconductivity of liquids [5,40]; equations (1.159) describe the classical two-species interactions (Kolmogorov form) of Ecology when diffusion and spatial dependence are taken into account [5,47,52,53]. Equations (1.160) arise in combustion theory in the following circumstances: $u_1 (\geq 0)$ denotes temperature while $u_2 \in [0,1]$ is a concentration, and $g(u_1) = |u_1|^\gamma e^{-\alpha/s}$ where α, $\gamma(< 1)$ and p are positive constants [40]. Equations (1.161) – with $\alpha, \beta, \gamma, \delta$ positive constants – are the Feld-Noyes equations modelling the Belousov-Zhabotinsky reactions in chemical kinetics [5,47]. Finally, (1.162) where

$$\begin{cases} f_1 = \alpha u_2^3 u_3 (\beta_1 - u_1) + \alpha_2 u_4^2 (\beta_2 - u_1) + \alpha_3 (\beta_3 - u_1)\,, \\ g_i > 0\,,\ 1 > h_i > 0\,,\ \alpha_i = \text{const.}\,,\ i \in \{1,2,3\} \\ \beta_1 > \beta_3 > 0 > \beta_2\,, \end{cases} \tag{1.164}$$

are the Hodgkin-Huxley equations modelling nerve impulse transmission, while (1.163) correspond the Fitz-Hugh-Nagumo model of the same problem [5, 47, 55].

1.13 EXERCISES

Exercise 1.1 - Consider the Cauchy problem for the *Burgers' equation* (*without viscosity*)

$$\begin{cases} u_t + uu_x = 0 & (x,t) \in \mathbb{R} \times \mathbb{R}_+ \,, \\ u(x,0) = u_0(x) & x \in \mathbb{R} \,. \end{cases} \tag{1.165}$$

1) Determine the characteristics;
2) Verify that for

$$u_0(\overline{x}) = \begin{cases} 1\,, & \overline{x} < 0\,, \\ 1 - \overline{x}\,, & \overline{x} \in [0,1]\,, \\ 0\,, & \overline{x} \geq 1\,, \end{cases} \tag{1.166}$$

a continuous solution can exist only for $t<1$;
3) Verify that, for $t < 1$,

$$u(x,t) = \begin{cases} 1\,, & x < t\,, \\ \dfrac{1-x}{1-t}\,, & t \leq x \leq 1\,, \\ 0\,, & x > 1 \end{cases} \tag{1.167}$$

is a smooth solution corresponding to the initial data (1.166).

Hint. The Burgers equation can be put in the form

$$u_t + F'(u)u_x = 0\,, \qquad (F = \frac{u^2}{2}) \tag{1.168}$$

and is therefore a *conservative law.*

Answer.

1′) The characteristics are

$$t = \frac{(x - \overline{x})}{u_0(\overline{x})}\,. \tag{1.169}$$

2′) In the case (1.166) the characteristics are $x = \overline{x} + t\ \overline{x} < 0$; $x = \overline{x} + (1-\overline{x})t\ \overline{x} \in [0,1]$, $x = \overline{x}\ \overline{x} \geq 1$. The characteristics r_1 and r_2 – starting at the point $\{A = (0,0), B = (1,0)\}$ – are respectively $x = t$ and $x = 1$. Therefore r_1 and r_2 will intersect at the point $P_* = (1,1)$, and one has

$$1 = u_0(0) = \lim_{\substack{P \to P_* \\ P \in r_1}} u(P) \neq \lim_{\substack{P \to P_* \\ P \in r_2}} u(P) = u_0(1) = 0$$

3′) Since u is constant along the characteristics and (1.170) gives $\overline{x} = x - t$, $x < t$; $\overline{x} = \dfrac{x-t}{1-t}$, $t \leq x \leq 1$; $\overline{x} = x$, $x > 1$; then (1.167) follows from $u(x) = u_0[\overline{x}(x,t)]$, on taking (1.166) into account.

Exercise 1.2 - Consider a weak solution to the I.V.P. (1.6) and show that along a smooth curve γ across which u has a jump discontinuity, not every discontinuity is permissible but only those satisfying the *Rankine-Hugoniot jump condition*

$$[F(u)] = v[u] \tag{1.171}$$

where $v = dx/dt$ is the *speed of the discontinuity* and $[f]$ indicates the jump of f across γ.

Hint - By assumption u is smooth away from γ and has well-defined limits on both sides of γ. For any point $P \in \gamma$, consider a ball S centered at P and divided by γ in two parts S_i ($i = 1,2$), with S_1 to the left of γ. Then (1.8) – for $\varphi \in C_0^1(S)$ – gives

$$\sum_{1=1}^{2} \int_{S_i} (\varphi_t u + F\varphi_x) = 0\,.$$

Because u is smooth on S_i, ($i = 1,2$), and hence satisfies (1.6), it follows that

$$\sum_{i=1}^{2} \int_{S_i} \{(\varphi u)_t + (F\varphi)_x\} dx dt = 0\,. \tag{1.172}$$

But the divergence theorem gives

$$\int_{S_1} \{(\varphi u)_t + (F\varphi)_x\} dx dt = \int_{\partial S_1 \cap \gamma} \varphi(u_1 n_2 + F(u_1) n_1) dl$$

$$\int_{S_2} \{(\varphi u)_t + (F\varphi)_x\} dx dt = -\int_{\partial S_2 \cap \gamma} \varphi(u_2 n_2 + F(u_2) n_1) dl$$

where $\mathbf{n} = (n_1, n_2)$ is the unit normal to γ pointing from S_1 to S_2 and u_i, ($i = 1,2$), denotes the limit of u when the point approaches γ from S_i. From (1.172) it turns out that

$$\int_{S \cap \gamma} \varphi\{[u] n_2 + [F(u)] n_1\} dl = 0\,.$$

Assuming that $S \cap \gamma$ is given by $x = x(t)$, one can take $n_1 = \dfrac{1}{(1+\dot{x}^2)^{1/2}}$, $n_2 = -\dfrac{\dot{x}}{(1+\dot{x}^2)^{1/2}}$ and hence...

Exercise 1.3 - Consider the I.V.P. (1.165)–(1.166) and determine the speed of the discontinuity. By means of this relation show that an explicit non-classical solution of the I.V.P. (1.165)–(1.166) is given by (1.167) for $t < 1$, and by

$$u(x,t) = \begin{cases} 1\,, & x < \dfrac{1+t}{2} \\ 0\,, & x > \dfrac{1+t}{2} \end{cases} \tag{1.173}$$

for $t \geq 1$.

Hint. Use (1.171) with $u_2 = 1, u_1 = 0$ and recall that $F = u^2/2$.

Answer. $v = 1/2(u_1 + u_2)$ and hence for $u_2 = 1, u_1 = 0, v = 1/2$. Therefore $dx/dt = 1/2$, and $x(1) = 1$ implies $x = (1+t)/2$ and hence...

Exercise 1.4 - Consider the Burgers equation with the initial data

$$u_0(x) = \begin{cases} 0, & x < 0 \\ 1, & x > 0 \end{cases}$$

and verify that the functions

$$u^* = \begin{cases} 0, & x < \dfrac{t}{2} \\ 1, & x > \dfrac{t}{2} \end{cases}, \qquad u^{**} = \begin{cases} 0, & x < 0 \\ x/t, & 0 < x < t \\ 1, & x > t, \end{cases}$$

are both generalized solutions (cf. Remark 1.4).

Exercise 1.5 - Consider the Burgers equation with the initial data

$$u_0(x) = \begin{cases} -1, & x < 0 \\ 1, & x > 0 \end{cases}$$

and verify that

i) each function u_λ given by

$$u_\lambda = \begin{cases} -1, & x < \dfrac{(1-\lambda)t}{2}, \\ -\lambda, & \dfrac{(1-\lambda)t}{2} < x < 0, \\ \lambda, & 0 < x < \dfrac{(\lambda-1)t}{2}, \\ 1, & x > \dfrac{(\lambda-1)t}{2} \end{cases}$$

is a solution for each $\lambda \geq 1$

ii) only the solution u_1 satisfies the "*entropy*" condition

$$\frac{u(x+a,t) - u(x,t)}{a} \leq \frac{e}{t} \qquad a > 0, t > 0,$$

where e is independent of a, x and t.

Hint. See [5].

Exercise 1.6 - Consider the I.V.P.

$$\begin{cases} u_t + |u_x|^2 = 0 & \text{in } \mathbb{R} \times (0,\infty), \\ u = 0 & \text{on } \mathbb{R} \times \{0\}, \end{cases} \tag{1.174}$$

and show that for solutions $u(x,t)$ to (1.174) a.e.,[11] in the class of Lipschtz functions, there is non-uniqueness.

Hint. Verify that

$$u = \begin{cases} -(x+t)\,, & -t \le x < 0\,, \\ x - t\,, & 0 \le x \le t\,, \\ 0\,, & |x| > t\,, \end{cases}$$

is a Lipschitz function satisfying $(1.174)_2$ and solving $(1.174)_1$ a.e..

Exercise 1.7 - Verify that, for $\Omega \equiv I\!R^3$ and $\mathbf{F} \equiv \mathbf{v}_0 \equiv 0$, the Cauchy problem $(1.20) - (1.21)_1$ admits the following general class of potential-like solutions [13]:

$$\begin{cases} \mathbf{v}(\mathbf{x},t) = a(t)\,\nabla\Phi\,, \\ \\ p(\mathbf{x},t) = -\dfrac{da}{dt}\Phi + \dfrac{a^2}{2}\left(\nabla\Phi\right)^2 + p_0(t)\,, \end{cases}$$

where $a \in C^1(I\!R_+)$, $\Phi = \Phi(\mathbf{x}) \in C^2(I\!R^3)$, $\Delta_2\Phi = 0$ and $p_0(t)$ is an arbitrary function of t.

Exercise 1.8 - Show that a relaxation function G, and a positive constant $\overline{\omega}$ exist such that $\varepsilon(x,t) = \varepsilon_0(x)\,e^{i\overline{\omega}\,t}$ — where $\varepsilon_0(x)$ is differentiable but otherwise arbitrary — is a solution to (1.28) on $I\!R$.

Hint. Equation (1.28) can be written: $G(0)\varepsilon(x,t) + \int_0^t G'(s)\varepsilon(x,t-s)ds = 0.$

Answer. [20,21]

$$\begin{cases} G(s) = \dfrac{\alpha(\alpha-2)}{16\beta^3} - \displaystyle\int_0^S \left(z^2 - \frac{\alpha-1}{\beta}z + \frac{\alpha^2}{8\beta^2}\right)e^{-\beta\,z}dz\,, \\ \\ \overline{\omega} = \sqrt{\dfrac{8-\alpha}{\alpha}}\,\beta \quad , \quad \alpha \in (6,8)\,,\ \beta = \text{const.} > 0\,. \end{cases}$$

Exercise 1.9 - Denoting the elastic tensor by **C**, let us consider the functional $\varepsilon : \mathbf{L} \in \text{Lin} \to I\!R$:

$$\varepsilon(\mathbf{L}) = \mathbf{L}\cdot\mathbf{C}(\mathbf{L}) = C_{ijhk}\,L_{ij}\,L_{hk}\,.$$

By definitions, **C** satisfies the condition of

i) positive semidefiniteness	if $\varepsilon(\mathbf{L}) \ge 0,\ \forall\mathbf{L} \in \text{Lin}\,,$
ii) positive definiteness	if $\varepsilon(\mathbf{L}) > 0,\ \forall\mathbf{L} \in \text{Lin} - \{\mathbf{0}\}\,,$
iii) semi-strong ellipticity	if $\varepsilon(\mathbf{L}) \ge 0,\ \forall\mathbf{L} = \mathbf{a}\otimes\mathbf{b}\,,$
iv) strong ellipticity	if $\varepsilon(\mathbf{L}) > 0,\ \forall\mathbf{L} = \mathbf{a}\otimes\mathbf{b} \ne 0\,,$
v) uniform positive definiteness	if $\exists\alpha > 0 : \varepsilon(\mathbf{L}) \ge \alpha(\text{sym}\,\mathbf{L})^2\,, \mathbf{L} \in \text{Lin}\,,$
vi) uniform strong ellipticity	if $\exists\beta > 0 : \varepsilon(\mathbf{a}\otimes\mathbf{b}) \ge \beta\,a^2b^2\,, \forall\mathbf{a},\mathbf{b}\,.$

[11] a.e.=almost everywhere

Show that for a three dimensional, elastic isotropic body, the conditions i), ii), iii) and iv) are given respectively, in terms of Lamé moduli, by

$$\begin{array}{lll} \text{i)}' & \mu \geq 0\,, & 3\lambda + 2\mu \geq 0\,, \\ \text{ii)}' & \mu > 0\,, & 3\lambda + 2\mu > 0\,, \\ \text{iii)}' & \mu \geq 0\,, & \lambda + 2\mu \geq 0\,, \\ \text{iv)}' & \mu \geq 0\,, & \lambda + 2\mu > 0\,. \end{array}$$

Exercise 1.10 - Show that for a two dimensional elastic body, inequalities $\mu > 0, \lambda + 2\mu > 0$ guarantee the positive definiteness of the elastic tensor.

Exercise 1.11 - Show that when the body is isotropic and homogeneous (i.e. (1.37) holds and ρ, λ and μ are independent of $\mathbf{x}$)(1.32) and (1.35) become respectively:

$$\rho \ddot{\mathbf{u}} = (\lambda + 2\mu)\nabla(\nabla \cdot \mathbf{u}) - \mu \nabla \times \nabla \times \mathbf{u} + \rho \mathbf{F}\,, \tag{1.175}$$

$$2\mu \mathbf{E} \cdot \mathbf{n} + \lambda(\nabla \cdot \mathbf{u})\mathbf{n} = \widehat{\mathbf{s}}\,. \tag{1.176}$$

Hint. Obtain $\rho \ddot{\mathbf{u}} = \mu \Delta \mathbf{u} + (\lambda + \mu)\nabla(\nabla \cdot \mathbf{u}) + \rho \mathbf{F}$. Then take into account that $\Delta \mathbf{u} = \nabla(\nabla \cdot \mathbf{u}) - \nabla \times \nabla \times \mathbf{u}$.

Exercise 1.12 - Verify that the function $u(x,t) = u_1 - u_2$ with u_i $(i = 1, 2)$ given by (1.43)–(1.44) is a counterexample to the traction problem i.e. the problem (1.40), when $(1.40)_3$ is replaced by $u_x = 0$ on $\{1\} \times \mathbb{R}_+$.

Exercise 1.13 - Let Ω be the exterior of the sphere $S(\mathbf{O}, 1)$ and consider the inhomogeneous, isotropic elastic case with Lamé moduli depending only on $r = |OP|$ and such that $\mu > 0$, $\lambda + \mu > 0$. Taking into account Section 8c, and the result of the previous exercise, show that the initial boundary value problem (1.32)–(1.35), where $\mathbf{F} \equiv \mathbf{u}_0 \equiv \dot{\mathbf{u}}_0 \equiv \hat{\mathbf{u}} \equiv \hat{\mathbf{s}} \equiv 0$, admits nontrivial solutions.

Hint. Consider a solution of the type $\mathbf{u} = u(r,t)\mathbf{e}_r$, $\mathbf{e}_r = \dfrac{P - O}{r}$, and assume

$$\rho = \frac{p'(r)}{r^{n-1}}\,, \; \mu = \frac{1}{r^{n-1}p'}\,, \; p'(r) > 0\,, \; p(1) = 0\,, \; \lim_{r \to \infty} p(r) = l < \infty\,.$$

In this case $(1.32)_1$ is equivalent to

$$p'(r)\ddot{u} = \partial_r \left(\frac{1}{p'} \partial_r u \right)\,.$$

Exercise 1.14 - Let (r, θ, z) be a cylindrical, polar coordinate system, Ω the exterior domain $\{r \geq 1;\ \theta, z \in \mathbb{R}\}$ and let us consider the inhomogeneous, isotropic elastic case with Lamé moduli $\lambda > 0$, $\mu = \mu(r) > 0$. Show that, for $\mathbf{F} = \mathbf{u}_0 = \dot{\mathbf{u}}_0 = \hat{\mathbf{u}} = \hat{\mathbf{s}} = 0$, the initial value problem (1.32)–(1.35) admits nontrivial solutions.

Hint. Consider solution of the type $\mathbf{u} = u(r,t)\mathbf{k}$, $\mathbf{k}$ being the unit vector along the z- axis, and assume

$$\rho = \frac{p'(r)}{r}, \; \mu = \frac{1}{r\,p'}, \; p'(r) > 0, \; p(1) = 0, \; \lim_{r\to\infty} p(r) = l < \infty .$$

Exercise 1.15 - Consider the quasi-linear parabolic initial-boundary value problem:

$$\begin{cases} \dfrac{\partial u}{\partial t} = \displaystyle\sum_{i,j=1}^{3} \frac{\partial}{\partial x_i}\left(a_{ij}(x)\frac{\partial u}{\partial x_j}\right) + \lambda u + \phi^{1-\alpha} u^{\alpha} & \text{on } \Omega \times I\!R\,, \\ u = 0 & \text{on } \Omega \times \{0\}\,, \\ u = 0 & \text{on } \partial\Omega \times I\!R\,, \end{cases} \tag{1.177}$$

where $\Omega \subset I\!R^3$ is a regular bounded domain, α = const. $\in (0,1)$, and $\{\lambda, \phi > 0\}$ is a solution of the eigenvalue problem:

$$\begin{cases} \displaystyle\sum_{i,j=1} \frac{\partial}{\partial x_i}\left(a_{ij}\frac{\partial \phi}{\partial x_j}\right) + \lambda\phi = 0 & \text{on } \Omega\,, \\ \phi = 0 & \text{on } \partial\Omega\,. \end{cases} \tag{1.178}$$

Show that (1.177) admits nontrivial solutions.

Hint. Consider solution of the type $u = \rho(t)\phi(x)$.

Answer. [31]: $\rho(t) = [(1-\alpha)t]^{1/(1-\alpha)}$

Exercise 1.16 - Show that there exist nontrivial solutions to the homogeneous linear integral equation

$$f(t) = 4t^2 \int_0^1 s\, f(s)\, ds\,.$$

Hint. Set $c = \int_0^1 s\, f(s)\, ds$ *Answer.* $f = 4ct^2, \quad c = \text{const.} \neq 0\,.$

Exercise 1.17 - Show that there exists a nontrivial solution to the homogeneous nonlinear integral equation (of the Hammerstein type [56])

$$f(t) = \lambda t^2 \int_0^1 s\, f^2(s)\, ds\,, \qquad \lambda = \text{const.} \neq 0$$

Hint. Set $c = \int_0^1 s f^2(s)\,ds$. *Answer.* $f(t) = \frac{6}{\lambda} t^2$.

Exercise 1.18 - Show that there exist two solutions to the non-homogeneous integral equation:

$$f(t) = 1 + \lambda \int_0^1 f^2(s)\,ds, \qquad 0 \neq \lambda < \frac{1}{4}$$

Hint. Set $c = \int_0^1 f^2(s)\,ds$. *Answer.* $f(t) = \frac{1 \pm \sqrt{1-4\lambda}}{2\lambda}$.

Exercise 1.19 - Consider the B.V.P.: $u(x,y)$ is a classical solution in the rectangular domain $\Omega = \{0 < x < a, 0 < y < b\}$ of

$$\begin{aligned} &\Delta u + \omega^2 u = 0 && \text{in } \Omega, \\ &u = 0 && \text{on } \partial\Omega \end{aligned}$$

where ω is a constant. Show that $u = 0$ is not a unique solution for *all* ω.

Hint. Choose $u = \sin\frac{n\pi x}{a} \sin\frac{m\pi x}{b}$; $n, m = 1, 2$. *Answer.* $\omega^2 = \pi^2(\frac{n^2}{a^2} + \frac{m^2}{b^2})$.

Exercise 1.20 - Let Ω be the *interior* of an ellipsoid $x^2 + a^2y^2 + b^2z^2 < 1$ where a and b are constants. Verify that, provided that the (constant) Lamé moduli λ, μ satisfy $\mu(\lambda + 2\mu) = -\mu^2(a^2 + b^2) < 0$, then the B.V.P.

$$\begin{cases} \mu\Delta\mathbf{u} + (\lambda + \mu)\nabla(\nabla \cdot \mathbf{u}) = 0 & \text{in } \Omega, \\ \mathbf{u} = 0 & \text{on } \partial\Omega, \end{cases}$$

admits the Cosserat classical solution $u_1 = x^2 + a^2y^2 + b^2z^2 - 1$, $u_2 = u_3 = 0$.

Hint. See [22,57].

Exercise 1.21 - Let Ω be a (bounded) plane domain in the portion $x > 0$ of the xy plane. Consider the B.V.P.:$u(x,y) \in C^2(\Omega) \cap C(\overline{\Omega})$ and satisfies

$$\begin{aligned} &\left(\frac{u_x}{x}\right)_x u_{yy} - \frac{1}{x} u_{xy}^2 + \frac{1}{2}\left(\frac{u_y^2}{x^2}\right)_x = \pi(x,y) && \text{in } \Omega, \\ &u = g(x,y) && \text{on } \partial\Omega. \end{aligned}$$

Problems of this nature are analogous to the Dirichlet problem for the Monge-Ampére equation [See Section 8,f)]. They are of some relevance to axisymmetric fluid flow. It may be proved [28] that the above problem has, at most, two solutions *provided* that $\pi > 0$. Prove that two solutions are indeed to be expected in general.

Hint. Let Ω be the circular domain $(x - x_0)^2 + y^2 < a^2$ where x_0, a are positive constants such that $x_0 > \frac{a}{2}(1 + \sqrt{5})$ and let $g = x^2$.

Answer. $u = x^2 \pm [(x - x_0)^2 + y^2 - a^2]$, $\pi = 4x^{-2}(x_0 - x^{-1}y^2)$.

Exercise 1.22 - Consider the Cauchy problem for the P.D.E.:

$$\begin{cases} \dfrac{\partial u}{\partial t} - z\dfrac{\partial u}{\partial z} = 0 & (z,t) \in \mathbb{R} \times \mathbb{R}_+ \,, \\ u(z,0) = f(z) & z \in \mathbb{R} \,, \end{cases}$$

and show that the zero solution ($f = 0$) is
i) Liapunov stable in the $L^\infty(u)$-norm,
ii) unstable in the $L^\infty(\nabla u)$-norm, if $f \neq$ const.

Hint. The solution has the form: $u(t,z) = f(e^t z)$. *Answer.* Cfr. [58] page 101.

Exercise 1.23 - Consider the Cauchy problem for the P.D.E.:

$$\begin{cases} \dfrac{\partial u}{\partial t} = \left(\dfrac{2}{t} - 6t^5 x^2\right) u(x,t) & (x,t) \in [-1,1] \times \mathbb{R}_+ \,, \\ u(x,1) = f(x) & \end{cases}$$

and its zero solution ($f \equiv 0$). Consider the class of initial perturbations: $f(x) = ae^{-x^2}$, $a =$ positive constant, and show that the zero solution is
i) asymptotically stable in the $L^1[-1,1]$-norm,
ii) unstable in the $L^\infty[-1,1]$-norm.

Hint. The solution has the form [1] $u(x,t) = at^2 e^{-t^6 x^2}$.

Exercise 1.24 - Consider the I.B.V.P. for the diffusion equation

$$\begin{aligned} & u_t = u_{xx} \,, && (x,t) \in (0,1) \times \mathbb{R}_+ \,, \\ & u(x,0) = u_0(x) \,, && x \in [0,1] \,, \\ & u(0,t) = u(1,t) = 0 \,, && t \in \mathbb{R}_+ \,. \end{aligned}$$

Show that the critical point $u = 0$, is a global attractor in the L^2-norm.

Hint. Determine the time derivative of the L^2-norm and take into account the Poincaré inequality.

Answer. $\|u\|^2 \leq \|u_0\|^2 e^{-2\pi^2 t}$.

Exercise 1.25 - Consider the I.V.P. for the linear diffusion equation with a linear source term

$$u_t = u_{xx} + au\,, \qquad (x,t) \in \mathbb{R} \times \mathbb{R}_+\,,$$
$$u(x,0) = u_0(x)\,, \qquad x \in \mathbb{R}\,,$$

and show, provided $a > 0$, that the zero solution is unstable with respect to the L^∞-norm.

Hint. Choose $u_0(x) = \varepsilon \cos kx$, $u(x,t) = u_0(x)\,e^{\lambda t}$, where ε, λ, k are constants.

Answer. For $k^2 < a$, $\lambda = a - k^2$, there exists the solution $u = \varepsilon \cos kx\, e^{(a-k^2)t}$ which is unbounded in the L^∞-norm.

Exercise 1.26 - Consider the I.B.V.P.

$$\begin{cases} u_t = u_{xx} + au & (x,t) \in (0,1) \times \mathbb{R}_+ \\ u(x,0) = u_0(x) & x \in [0,1]\,, \\ u(0,t) = u(1,t) = 0 & t \in \mathbb{R}_+\,, \end{cases}$$

and verify that the zero solution is stable in the L^2-norm for $a < \pi^2$, while it is unstable in any L^p-norm for $a > \pi^2$ and $p \geq 1$.

Hint. We refer to [59] for the L^2-stability when $a < \pi^2$. For L^p instability, when $a > \pi^2$, consider the initial data $u_0 = \varepsilon \sin \pi x$, ($\varepsilon =$ const.), and the perturbations $u = u_0 e^{\lambda t}$ with $\lambda =$ const.

Answer. For $a > \pi^2$, there exists the perturbation $u = u_0 e^{(a-\pi^2)t}$.

Exercise 1.27 - Consider the I.B.V.P. of Exercise 1.26, for $x \in [0,l]$ with $l =$ const. > 0, under homogeneous boundary conditions, and verify that $a = \pi^2/l^2$ is the stability-instability boundary.

Exercise 1.28 -Consider the I.B.V.P. for the Burgers equation (with viscosity):

$$\begin{cases} u_t + uu_x = u_{xx} & (x,t) \in (0,1) \times \mathbb{R}_+\,, \\ u(x,0) = u_0(x) & x \in (0,1)\,, \\ u(0,t) = u(1,t) = 0 & t \in \mathbb{R}_+\,. \end{cases}$$

Verify that $u = 0$ is a global attractor in the L^2-norm, at least for strong solutions.

Hint. See Exercise 24 and consider that $\int_0^1 u^2 u_x\, dx = 0$.

Exercise 1.29 - Consider the I.B.V.P. for the Navier-Stokes equation (1.98)–(1.99) with $\Omega \subset \mathbb{R}^3$, bounded. Show that the critical point ($\mathbf{v}=0, p=$const.) is a global attractor in the L^2-norm, at least for strong solutions.

Hint. Follow Section 11 and take into account that

$$\int_\Omega \mathbf{v}\cdot\nabla\mathbf{v}\cdot\mathbf{v}d\Omega = \frac{1}{2}\int_\Omega \mathbf{v}\cdot\nabla v^2 d\Omega = \frac{1}{2}\int_\Omega v^2\mathbf{v}\cdot\mathbf{n}d\Omega = 0\,.$$

Exercise 1.30 - Estimate – in the L^2-norm – the error made in approximating the I.B.V.P. of Exercise 1.28 with its linearized version (Exercise 1.24).

Hint. Follow Section 11.

Exercise 1.31 - Consider the *backward heat equation*

$$\begin{cases} u_t = u_{xx} & x \in I\!R\,,\ t < 0 \\ u(x,0) = f(x) & x \in I\!R \end{cases}$$

and show that the I.V.P. is not well-posed.

Hint. Consider the lack of stability, with respect to initial data, of the zero solution.

Answer. For $f = f_n = \dfrac{1}{n}\sin nx$, $(n = 1,2,\ldots)$, exists the solution $u_n = \dfrac{e^{-n^2t}}{n}\sin nx$. Hence $|u_n| = \dfrac{e^{n^2|t|}}{n}|\sin nx|$, $\lim_{n\to\infty}|f_n| = 0$, $\lim_{n\to\infty}|u_n| = \infty$ for $x \neq k\pi$.

Exercise 1.32 - Estimate – in the pointwise norm – the error made in approximating the B.V.P.

$$\begin{cases} u_t = u_{xx} + \varepsilon[F(u)]_{xx}\,, & x > 0\,,\ |t| < T \\ u = h_1(t)\,,\ u_x = g_1(t) & x = 0\,,\ |t| \le T \end{cases}$$

with ε and T positive constant and $F(u)$ differentiable nonlinear function, with its linearized version

$$\begin{aligned} &v_t = v_{xx}\,, && x > 0\,,\ |t| < T \\ &v = h_1\,,\ v_x = g_1\,, && x = 0\,,\ |t| \le T \end{aligned}$$

Hint. See section 3 of [60].

Chapter two

ESTIMATES BASED ON FIRST ORDER INEQUALITIES 1: LIAPUNOV DIRECT METHOD

2.1 INTRODUCTION

As discussed in the introduction to chapter one, the strategy of qualitative analysis is to obtain estimates and properties of the state vector $\mathbf{u}$ of a phenomenon $\mathcal{F}$ without solving explicitly the P.D.E. modelling $\mathcal{F}$. In this strategy a central role is played by the inequalities that one is able to obtain from the P.D.E. at hand. In order to be more precise, let U be a scalar function of a time like variable t – linked in some definite manner to the state vector $\mathbf{u}$ (for instance, a suitable norm of $\mathbf{u}$). Let us assume that one is able to prove that, along the solutions to the P.D.E. modelling $\mathcal{F}$, U satisfies one of the following inequalities:

$$\dot{U} \leq f(t)U + g(t) \qquad t \geq t_0 \tag{2.1}$$

$$\ddot{U} \leq a(t)\dot{U} + b(t)U + c(t)\,, \qquad t \geq t_0 \tag{2.2}$$

where f, g, a, b, c are known functions of t. Then the following question arises: what estimates for, or properties of, the function U can one obtain from (2.1) or (2.2) (or similar inequality) and what can one deduce for the state vector $\mathbf{u}$. Issues such as these – which form the central theme of the book – are dealt with in this and subsequent chapters. In the present chapter we consider the case of the first order inequality (2.1) and, first of all, we recall (Sec.2.2) a well known general estimate together with a generalization holding for any U satisfying (2.1). Subsequently we will concentrate on the simple special case $\dot{U} \leq 0$ in order to show, through the Liapunov direct method, how one can obtain much important information on the behaviour of the state vector $\mathbf{u}$ and hence on the phenomenon at hand.

2.2 GENERAL ESTIMATES BASED ON FIRST ORDER INEQUALITIES

We begin by recalling a very well known general estimate obtained from the first order inequality (2.1), i.e. the so called Gronwall lemma and its generalization.

The following is the Gronwall lemma.

Lemma 2.1 - *Let* (2.1) *hold and let* $U, \dot{U}, f$ *and* g *belong to* $L^1_{loc}(]t_0, \infty[)$, *i.e. are locally integrable. Then the following estimate holds:*

$$U(t) \leq U(t_0)\exp(\int_{t_0}^{t} f(\tau)d\tau) + \int_{t_0}^{t} g(\tau)\exp(\int_{\tau}^{t} f(s)ds)d\tau, \qquad t \geq t_0. \tag{2.3}$$

Proof - Setting

$$w(t) = \exp(-\int_{t_0}^{t} f(\tau)d\tau) \ ,$$

it follows that

$$\frac{d}{dt}(wU) \leq wg$$

and hence, integrating on (t_1, t_2) with $t_2 \geq t_1 \geq t_0$, one has

$$w(t_2)U(t_2) \leq w(t_1)U(t_1) + \int_{t_1}^{t_2} w(\tau)g(\tau)d\tau \, . \tag{2.4}$$

The usual Gronwall estimate (2.3) then immediately follows for $t_1 = t_0$, $t_2 = t$.

Let us remark now that if f, g, U are positive then $w(t) \leq 1$, $\forall t \geq t_0$, and therefore (2.4) – for $t_1 = t_0, t_2 = t + \xi, \xi \geq 0$ – implies that

$$U(t+\xi) \leq [U(t_1) + \int_{t_1}^{t+\xi} g(\tau)d\tau]\exp\int_{t_1}^{t+\xi} f(\tau)d\tau \, , \tag{2.5}$$

and the following lemma holds

Lemma 2.2 - *Let the assumptions of Lemma* 2.1 *hold and let* U, f *and* g *be non-negative. If there exist three positive constants* α, β, δ *such that* $(\forall t \geq t_0)$

$$\int_{t}^{t+\delta} f(\tau)d\tau \leq \beta \quad , \quad \int_{t}^{t+\delta} g(\tau)d\tau \leq \alpha \, ; \tag{2.6}$$

then, for $\xi \in]0, \delta[$ *and* $t \geq t_0$, *the following estimate holds:*

$$U(t+\xi) \leq \left[\frac{1}{\xi}\int_{t}^{t+\xi} U(\tau)d\tau + \alpha\right] e^{\beta} \, . \tag{2.7}$$

Proof - Inequality (2.7) is an immediate consequence of (2.5). In fact for $t_1 \in [t, t+\xi]$, (2.5) gives

$$U(t+\xi) \leq [U(t_1) + \alpha]e^{\beta} \, ,$$

and hence, integrating with respect to t_1 on $(t, t+\xi)$, (2.7) follows.

Let us emphasize the importance of (2.7) considering the case $f=\lambda=\text{const.}(>0), g=t_0=0$. Then (2.3) gives

$$U(t) \leq U(0)e^{\lambda t}$$

i.e., a bound for U growing (*exponentially*) with t. In contrast to this result, but ***under a stronger assumption on*** U, (2.7) allows us to obtain an ***uniform bound*** with respect to t. In fact, if

$$\exists \gamma = \text{const.} > 0 : \int_t^{t+\delta} U(\tau)d\tau \leq \gamma \tag{2.8}$$

then (2.7) gives the estimate ***independent of*** t

$$U(t+\xi) \leq (\alpha + \frac{\gamma}{\xi})e^{\beta} \quad , \quad \forall t \geq t_0 \, . \tag{2.9}$$

We end this section by recalling that

i) Lemma 2.2 with the uniform estimates (2.9) has been obtained by Foias and Prodi [61].

ii) further versions of the Gronwall Lemma can be found in [62].

2.3 LIAPUNOV FUNCTIONS

In 1893 A.M.Liapunov – in order to establish conditions ensuring stability of solutions of O.D.E.s – introduced a method which is called *the direct or second method* [63]. This method – based on knowing the sign of the time derivative, along the solutions, of an auxiliary function, but without any recourse to them – has been recognized to be very general and powerful, and has been used for over 65 years in the qualitative theory of O.D.E.s [64 – 67]. The first generalizations of the Liapunov direct method to P.D.E.s and, in general, to evolution equations other than O.D.E.s, appeared only in the years 1957-59 [68,69,37]. To some extent, the Liapunov direct method may be considered to have been rediscovered in the western world in the years 1955-1960. For instance, this is so in the case of the so-called "*energy method*" for stability in fluid mechanics. Although its origin can be found at least as far back as the work of Kelvin on perfect fluids (1887) [70], the modern version of the energy method (1959) [59,71,72] can be considered to be a particular case of the Liapunov direct method. Our aim is to introduce the fundamental ideas and problems of the Liapunov direct method in the light of its applications to phenomena which are modelled (essentially) by P.D.E.s For further analysis in this field we refer the readers to [7,8,39].

Definition 2.1 - *Let v be a dynamical system on a metric space X. A functional $V : X \to \mathbb{R}$ is a Liapunov function on a subset $I \subset X$ if V is continuous on I, and a nonincreasing function of time along the solutions having the* *initial data on I.*

In order to ensure that $V[v(x,\cdot)]$ is a nonincreasing function of time, *in the sequel we assume that V is differentiable with respect to time and that the derivative is non-positive.* However, it is standard in the literature to ensure that V is non-increasing by requiring that the generalized time derivative

$$\overset{\bullet}{V} \stackrel{\text{def}}{=} \liminf_{t\to O^+} \frac{1}{t}\{V[v(x,t)] - V(x)\}\ , \quad x \in I\,, \tag{2.10}$$

(coinciding with the ordinary derivative when V is differentiable) is non-positive (see exercise 2.1).

In the sequel – for some $\alpha \leq \infty$ – we denote by $\Sigma_\alpha \subset X$ a subset of the set $\Sigma(X,\alpha) \stackrel{\text{def}}{=} \{x \in X : V(x) < \alpha\}$ and by $\Sigma_{(\alpha,\beta)}$ the intersection $\Sigma_\alpha \cap \Sigma(X,\beta)$ for $\beta \leq \alpha$. The following Theorem holds:

Theorem 2.1 - *Let v be a dynamical system on a metric space X and let V be a Liapunov function on Σ_α, having a non-positive time derivative. Then*

i) $\Sigma_{(\alpha,\beta)}$ and $\overline{\Sigma}_{(\alpha,\beta)}$, $\forall \beta \leq \alpha$, *are positive invariant,*

ii) $V[v(x,\cdot)]$, $\forall x \in \overline{\Sigma}_\alpha$, *is a non-increasing function of time,*

iii) $V[v(x,\cdot)]$ *is differentiable a.e. with*

$$V[v(x,t)] \leq V(x) + \int_0^t \overset{\bullet}{V}[v(x,\tau)]d\tau \quad , \quad (x,t) \in \Sigma_\alpha \times I\!R_+\,. \tag{2.11}$$

The proof is immediate when $\overset{\bullet}{V}$ is the ordinary derivative. In the case of the generalized derivative (2.10) we refer to { [8], pag. 142, Corollary 1.1 }.

Remark 2.1 - *Theorem 2.1 shows that the Liapunov functions can be used for the determination of some positive invariant sets. We emphasize the importance of this role because if a bounded (or precompact) set $S \subset X$ can be shown to be positive invariant, then $x \in S \to \gamma(x) \in S$ and hence the positive orbit $\gamma(x)$ is bounded (or precompact). For this reason we notice that Theorem 2.1 continues to hold under weaker conditions on V. In fact, instead of the continuity of V, it is enough to require its* **lower semicontinuity** [8], *i.e.* $\{\forall x \in X, \forall \alpha \in I\!R\}$ *the set $\{x \in X, V(x) \leq \alpha\}$ is closed. However the continuity of V is needed in the Liapunov Direct method and for this reason we use Definition 2.1 for V.*

2.4 DIRECT METHOD

As observed in Remark 1.12 the stability of a given motion can be expressed — for X a normed linear space — through the stability of the zero solution of the perturbed dynamical system. For this reason, one can introduce the direct method for investigating the stability of an equilibrium position only. Assuming — for the sake of simplicity — that X *is a normed linear space*, and denoting by $\mathcal{F}_r$, $r =$ const. > 0, the set of the function $\varphi : [0,r) \to R_+$ continuous, strictly increasing and satisfying $\varphi(0) = 0$, then the Liapunov direct method can be summarized by the following two Theorems.

Theorem 2.2 - *Let u be a dynamical system on X and let* $\mathbf{O}$ *be an equilibrium point. If V is a Liapunov function on the open ball $S(\mathbf{O},r)$, for some $r > 0$, such that:*

i) $V(\mathbf{O}) = 0$,

ii) $\exists f \in \mathcal{F}_r : V(u) \geq f(\|u\|)\,, \forall u \in S(\mathbf{O},r)\,,$

then $\mathbf{O}$ *is stable. If, in addition,*

iii) $\exists g \in \mathcal{F}_r : \overset{\bullet}{V}(u) \leq -g(\|u\|)\,, \forall u \in S(\mathbf{O},r),$

then $\mathbf{O}$ *is asymptotically stable.*

Proof. Let us assume $\varepsilon < r$ and introduce

$$\alpha = \inf_{\|u\|=\varepsilon} V(u) \geq f(\varepsilon) > 0 \quad , \qquad\qquad (\varepsilon \neq 0)\,.$$

In view of i), ii) and Theorem 2.1, it follows that $S(\mathbf{O},\varepsilon)$ contains a positive invariant component Σ_α of $\Sigma(X,\alpha)$. The stability is then immediately obtained observing that — by i) and the V continuity — there exists $\delta(\varepsilon) > 0$ such that $S(\mathbf{O},\delta) \subset \Sigma_\alpha$ and therefore $u_0 \in S(\mathbf{O},\delta) \Rightarrow \gamma(u_0) \subset \Sigma_\alpha \subset S(\mathbf{O},\varepsilon)$.

Turning now to the asymptotic stability, by (2.11) and ii)-iii), it follows that:

$$0 \leq f\,[\||u(u_0,t)\||] \leq V[u(u_0,t)] \leq V(u_0) - \int_0^t g(\|u(u_0,\tau)\|)d\tau \qquad (2.12)$$

$\forall u_0 \in S(\mathbf{O},\delta)$. Because $V[u(u_0,t)] : I\!R_+ \to I\!R_+$ is a bounded nonincreasing function, then there exists a $\beta \in I\!R_+$ such that:

$$0 \leq \inf_{t \in I\!R_+} V[u(u_0,t)] = \beta \leq V(u_0) \leq \alpha\,.$$

But $\beta > 0$ implies $\gamma(u_0) \cap \Sigma(X,\beta) = \emptyset$ and — by the V continuity — the existence of $r^* > 0$ such that:

$$\gamma(u_0) \cap S(\mathbf{O},r^*) = \emptyset;\ \|u(u_0,t)\| > r^*\ ;\ g(r^*) \leq g\left(\|u(u_0,t)\|\right), \forall t \in I\!R_+\,.$$

Consequently (2.12) gives:

$$0 < V[u(u_0,t)] \leq V(u_0) - \int_0^t g(r^*)d\tau \leq V(u_0) - tg(r^*) < 0,\ t > \frac{V(u_0)}{g(r^*)}$$

which is impossible. Therefore $\beta=0$ and the asymptotic stability then follows.

Theorem 2.3 - *Let u be a dynamical system on $X \times I\!R_+$ and let* **O** *be an equilibrium point. If V is a Liapunov function on the open set $A_r = S(\mathbf{O},r) \cap \Sigma(X,0)$, for some $r > 0$, and*

i) $V(\mathbf{O}) = 0,$

ii) $\exists g \in \mathcal{F}_r : \overset{\bullet}{V}(u) \leq -g[-V(u)], u \in A_r$

iii) $A_\varepsilon \neq \emptyset \quad , \quad \forall \varepsilon > 0 \, ,$

then **O** *is unstable.*

Proof. Because of i) and the V continuity, there exists $0 < \varepsilon < r$ such that $u \in S(\mathbf{O},\varepsilon) \Rightarrow V(u) > -1$. The point **O** cannot be stable for otherwise one could find a $\delta(\varepsilon) > 0$ such that $u_0 \in S(\mathbf{O},\delta) \Rightarrow \gamma(u_0) \in S(\mathbf{O},\varepsilon)$ and hence, by ii), $u_0 \in A_\delta \Rightarrow \gamma(u_0) \in A_\varepsilon$ and $V[u(u_0,t)] \leq V(u_0) < 0, \forall t \in I\!R_+$. Consequently $g[-V(u)] \geq g[-V(u_0)]$ on $\gamma(u_0)$ and (2.11) gives:

$$-1 < V[u(u_0,t)] \leq V(u_0) - \int_0^t g[-V(u_0)]d\tau \leq V(u_0) - tg[-V(u_0)] < -1$$

for $t > \dfrac{V(u_0)}{g[-V(u_0)]}$, which is impossible. Therefore **O** is unstable.

Remark 2.2 - *In the case $X = I\!R^n$, assumption* ii) *of Theorem* 2.2 *is simply replaced by $V(u) > 0$ for $u \neq 0$. This is because in $I\!R^n$ the surface $\|u\| = \varepsilon$ is compact and hence V – being continuous – has a positive minimum there. When X is an infinite dimensional space, the surface $\|u\| = \varepsilon$ is no longer compact and therefore V may not have a minimum. This difficulty is overcome by assumption* ii).

Remark 2.3 - *Let u be a dynamical system on X and let* **O** *be an equilibrium point. If V is a Liapunov function on the open ball $S(\mathbf{O},r)$ and is positive definite, i.e.*

$$V(\mathbf{O}) = 0 \quad , \quad V(u) > 0 \qquad u \neq \mathbf{O},$$

then the stability with respect to the measure V of the perturbation immediately follows [*see Remark* 1.15]. *If, moreover, there exists a positive constant c such that along the motions*

$$\overset{\bullet}{V} \leq -cV$$

then one has

$$V \leq V(u_0)e^{-ct} \, ,$$

i.e. asymptotic exponential stability in the measure V.

2.5 THE NORM AS LIAPUNOV FUNCTION

Let us consider the most frequent case – at least for the stability of fluid motion – that $V = ||u||$. Then, denoting by λ, α and β positive constants, it follows immediately that:

$$\dot{||u||} \leq 0 \qquad \Rightarrow \quad \text{stability of } \mathbf{O}\,, \tag{2.13}$$

$$\dot{||u||} \leq -\lambda ||u|| \qquad \Rightarrow \quad ||u|| \leq ||u_0|| e^{-\lambda t}\,, \tag{2.14}$$

$$\begin{cases} \dot{||u||} \leq -\lambda||u|| + \beta||u||^{1+\alpha} \\ \\ ||u_0|| < \left(\dfrac{\lambda}{\beta}\right)^{1/\alpha} \end{cases} \Rightarrow \begin{cases} ||u|| \leq ||u_0|| e^{-\delta t} \\ \\ \delta = \lambda - \beta ||u_0||^{\alpha} \end{cases} \tag{2.15}$$

Let us remark that, for the stability of fluid motion at least [59], one usually obtains – instead of $(2.15)_1$ – a more involved inequality. In fact, in this case, denoting by $||\cdot||$ the L_2-norm, usually one obtains:

$$\dot{||\mathbf{u}||} \leq R\mathcal{I} - \mathcal{D} + \beta ||\mathbf{u}||^{\alpha} \mathcal{D} \tag{2.16}$$

where R is a non-dimensional number related to the basic motion, α and β are positive constants and $\mathcal{I}$ and $\mathcal{D}$ are quadratic integral functionals (i.e. integrals) of $\mathbf{u}$ and $\nabla \mathbf{u}$, with $\mathcal{D}$ positive definite. The stability is then linked to the variational problem (along the motions)

$$\frac{1}{R^*} = \max \frac{\mathcal{I}}{\mathcal{D}} \tag{2.17}$$

In fact, if a Poincaré inequality holds, i.e.

$$\exists \gamma = \text{const.} > 0 \; : \; ||\mathbf{u}|| \leq \gamma \mathcal{D} \tag{2.18}$$

then (exercise 2.2):

$$\begin{cases} R < R^* \\ \\ ||\mathbf{u}_0||^{\alpha} < \dfrac{R^* - R}{\beta R^*} \end{cases} \Rightarrow ||\mathbf{u}|| \leq ||\mathbf{u}_0|| \exp\left[-\frac{1}{\gamma}\left(\frac{R^* - R}{R^*} - \beta ||\mathbf{u}_0||^{\alpha}\right) t\right] \tag{2.19}$$

2.6 ASYMPTOTIC BEHAVIOUR IN TIME OF THE STATE VECTOR

The knowledge of the asymptotic behaviour of motions as $t \to \infty$ is of primary importance in order to predict the state of the system after a transient period.

In this matter a fundamental role is played by the so-called positive limit set.

Definition 2.2 - *Let v be a dynamical system on a metric space X and let $x \in X$. A set $\Omega(x) \subset X$ is the positive limit set of the motion $v(x,t)$ if, $\forall y \in \Omega(x)$, there exists a sequence $\{t_n(y)\}$, $t_n \in I\!R_+$, such that:*

$$\begin{cases} \lim_{n\to\infty} t_n = \infty \\ \\ \lim_{n\to\infty} d[v(x,t_n),y] = 0 \end{cases} \tag{2.20}$$

Of course, if x is an equilibrium point, one has $\Omega(x) = x$; if $v(x,t)$ is periodic in time, then $\Omega(x) = \gamma(x)$, where $\gamma(x)$ is the orbit of $v(x,t)$. In general, obviously, $\Omega(x)$ belongs to the closure of $\gamma(x)$.

It is easy to see that

$$\Omega(x) = \bigcap_{\tau\geq 0} \overline{(\bigcup_{t\geq\tau} v(x,t))} = \bigcap_{\tau\geq 0} \overline{\gamma(v(x,\tau))}. \tag{2.21}$$

Therefore $\Omega(x) \subset \overline{\gamma(x)}$ and hence $\gamma(x)$ bounded $\Rightarrow \Omega(x)$ bounded, $\gamma(x)$ precompact $\Rightarrow \Omega(x)$ precompact.

Theorem 2.4 - *Let v be a dynamical system on a metric space X with the properties of C_0-semigroup. Then $\Omega(x)$ is closed and positive invariant.*

Proof. - Let $y \in \Omega(x)$ and let $\{t_n\}$ satisfy (2.20). By (1.16) and (1.14) it follows that:

$$\forall \bar{t} \in I\!R_+ \,,\; v(y,\bar{t}) = v(\lim_{n\to\infty}, \bar{t} v(x,t_n)) = \lim_{n\to\infty} v(x,\bar{t}+t_n).$$

But by definition of $\Omega(x)$ $\lim_{n\to\infty} v(x,\bar{t}+t_n) \in \Omega(x)$, and hence $v(y,\bar{t}) \in \Omega(x)$, $\forall \bar{t} \in I\!R_+$.

Let now $\bar{y} \in X$ be an accumulation point of $\Omega(x)$ and let $\{y_k\}$, for $k \in N_+$ and $y_k \in \Omega(x)$, be a sequence such that $d(y_k,\bar{y}) < 1/k$ for $k = 1,2,....$ Then $y_k \in \Omega(x)$ implies the existence of a sequence $\{t_n(y_k)\}$ such that

$$\lim_{n\to\infty} v(x,t_n(y_k)) = y_k$$

and hence

$$\exists \bar{n} \in N \,:\, n > \bar{n} \Rightarrow d[v(x,t_n(y_k)),y_k] < \frac{1}{k}.$$

Therefore it turns out that

$$d[v(x,t_n(y_k)),\bar{y}] \leq d[v(x,t_n(y_k)),y_k] + d(\bar{y},y_k) \leq \frac{2}{k}.$$

Letting $k \to \infty$, one has $\bar{y} \in \Omega(x)$ and hence $\Omega(x)$ is closed.

Theorem 2.5 - *Let v be a dynamical system on a metric space X with the C_0-semigroup properties and let X be complete and $\gamma(x)$ precompact. Then $\Omega(x)$ is: i) nonempty, compact, connected, invariant*[1] *and $d[v(x,t),\Omega(x)] \to 0$ as $t \to \infty$; ii) the smallest closed set approached by $v(x,t)$ as $t \to \infty$.*

Proof. - We refer to [8] page 167.

For further properties and analysis of $\Omega(x)$ and of global attractors we refer the readers to [8,39,40]. We end this section by mentioning the La Salle Invariance Principle [64,8,73,74] whose goal is to enlarge the role of the Liapunov functions. In fact, in order to obtain information on the asymptotic behaviour of motions, one may (Theorem 2.6) use Liapunov functions that need not satisfy the assumptions of Theorem 2.2 or 2.3.

Theorem 2.6 - *Let v be a dynamical system on a metric space X, with the C_0-semigroup properties and let V be a Liapunov function on a set $A \subset X$. If*

$$i)\quad V(x) > -\infty \qquad \forall x \in \bar{A},$$
$$ii)\quad \gamma(x) \subset A,$$

then $\Omega(x)$ belongs to the largest positive invariant subset $\mathcal{M}^+$ of $\Omega^* = \{x \in \bar{A} : \dot{V}(x) = 0\}$. Further – if X is complete and $\gamma(x)$ is precompact – then $\lim_{t\to\infty} d[v(t,x), \mathcal{M}^+] = 0$.

Proof. - We refer to $\{[8], pp. 168 - 170\}$ where a complete proof is given allowing V to be only lower semicontinuous. We confine ourselves here to giving the essential lines of the proof in the case where V is continuous and $\Omega(x) \neq \emptyset$. By Theorem 2.1, $V[v(x,t)]$ is a nonincreasing finite valued function on $\mathbb{R}_+$. Therefore $\lim_{t\to\infty} V[v(x,t)] = \alpha = \inf_{t\in\mathbb{R}_+} V[v(x,t)]$. But the continuity of V implies that:

$$V\left[\lim_{n\to\infty} v(x,t_n)\right] = \lim_{n\to\infty} V[v(x,t_n)] = \alpha$$

on each sequence $\{t_n\}$ such that $\lim_{n\to\infty} t_n = \infty$. Therefore, on noting (2.20), it turns out that

$$V(z) = \alpha \qquad \forall z \in \Omega(x). \tag{2.22}$$

But $\{\Omega(x) \subset \bar{\gamma}, \gamma \subset A\} \Rightarrow \Omega(x) \subset \bar{A}$, hence i) implies $\alpha > -\infty$. Furthermore – because $\Omega(x)$ is positive invariant – (2.22) implies that

$$\dot{V}(z) = 0 \qquad \forall z \in \Omega(x)$$

---

[1] A set $A \subset X$ is invariant under the dynamical system v, if there exists a function $S : A \times \mathbb{R} \to A$:

$$S(x,0) = x\,,\ S(x,t+\tau) = v[S(x,\tau),t]$$

for $x \in A$, $\tau \in \mathbb{R}$, $t \in \mathbb{R}_+$.

and hence $\Omega(x) \subset \mathcal{M}^+$.

Let us consider now X complete and $\gamma(x)$ precompact. Each diverging sequence $\{t_n\}$, generates a sequence $\{v(x,t_n)\} \subset \gamma(x)$ that – because $\gamma(x)$ is precompact – contains a Cauchy sequence $\{v(x,t_k)\}$. But X is complete, therefore:

$$\exists y \in X : \lim_{k\to\infty} v(x,t_k) = y \Rightarrow y \in \Omega(x) \to \lim_{t\to\infty} d[v(x,t), \mathcal{M}^+] = 0$$

Remark 2.4 - *Let us emphasize that the La Salle Invariance Principle works very well when $X \equiv \mathbb{R}^n$ but generally it needs help when X is infinite dimensional. This depends on the requirement of precompactness of positive orbits contained in Theorem 2.6. In fact – although the Liapunov functions (at least the lower semicontinuous Liapunov functions) can be used, sometimes, to establish precompactness of positive orbits – generally they allow one merely to obtain the boundedness of such orbits and only if X is locally compact does boundedness imply precompactness. But for X a Banach space, X is locally compact iff it is finite dimensional (see Remark 1.17). Then in using Theorem 2.6 one needs also conditions ensuring precompactness of positive orbits for X infinite dimensional.*

Remark 2.5 - *Concerning the relevant literature on the Invariance Principle, we mention the papers of Hale [74], Dafermos [75,76], Dafermos and Slemrod [77] and Haraux [78]. Further we recall that, in order to circumvent the requirement of precompactness, many efforts, in different directions, have been made. We mention the use of weak topologies [79,80,81], and the use of the theory of one parameter families of Liapunov functions [82,83] introduced in [84,85] for finite dimensional dynamical systems. Finally, we recall that, there exist P.D.E.s for which a Poincaré-Bendixon theorem has been obtained, i.e. a theorem ensuring that the ω-limit set of any bounded solution consists in precisely one-periodic solution or consists of solutions tending to equilibrium as $t \to \pm\infty$. This happens, for instance, in the case of the scalar reaction-diffusion equation [86].*

2.7 LIAPUNOV FUNCTIONS FOR THE REACTION-DIFFUSION EQUATIONS

In this section we will introduce and collect some Liapunov functions useful in the application of the Direct Method. To be precise, we will try to furnish examples of positive definite functionals (see Remark 1.15) which – at least along strong solutions of the equations at hand – have non-positive time derivative. We emphasize that our goal is merely to introduce possible Liapunov functions for some equations of general form. Of course, we do not claim that they are the best possible. In fact – for any given equation – different Liapunov functions can be used depending, among other things, on the topology.

For the sake of simplicity we consider the I.B.V.P. for the parabolic equations with one space variable. Let us consider the I.B.V.P.

$$\begin{cases} u_t = u_{xx} + f(x,t,u,u_x,u_{xx}) & x \in (0,1)\,,\, t > 0 \\ u(0,t) = u(1,t) = 0 & \forall t \geq 0 \\ u(x,0) = u_0(x) & x \in [0,1]\,. \end{cases} \tag{2.23}$$

This problem – at least when f is independent of u_{xx} – is well known and conditions on the data and on f ensuring global existence and uniqueness of strong solutions have been obtained { see for instance, [27] pp. 201- 206; [87] pp. 560-570, [88,89] }. But problems in which f depends on u_{xx} are also well known in the literature as models of real world phenomena [47,60,90]. For instance this is so in the case of heat diffusion when the thermal conductivity depends on the temperature. Then, instead of (1.137), one has

$$\mathbf{j} = -[k + g(T, \nabla T)]\nabla T \qquad k = \text{const.}$$

and, instead of (1.138), one obtains an equation like $(2.23)_1$ with f depending on u_{xx}. This happens, for instance, in heat diffusion in "cold ice" – which is commonly found in glaciers [60].

Setting $\{w = u_x, D_T = (0,1) \times (0,T)$ with $T = \text{const.}(> 0)$, $[0,1] \times [0,T] \times \mathbb{R}^2 = \overline{D}_T \times \mathbb{R}^2\}$, in the case

$$f = -g(x,t,u,w) \qquad (x,t,u,w) \in \overline{D}_T \times \mathbb{R}^2$$

the following theorems of existence and uniqueness hold[2]

Theorem 2.7 *Assume that* $u_0 \in C^2([0,1])$ *and*

i) g *is Hölder continuous in bounded subsets of* $\overline{D}_T \times \mathbb{R}^2$,

ii) for some positive constants A_i $(i = 1,2)$

$$uf(x,t,u,0) \leq A_1 u^2 + A_2\,, \qquad (x,t) \subset D_T,\, u \in \mathbb{R}\,,$$

---

[2]Let X be a metric space and $d(x_1,x_2)$ the distance between two points $x_1, x_2 \in X$. A function f defined on a bounded closed set S (with values on $\mathbb{R}$) is said to be *Hölder continuous of exponent* α $(0 < \alpha < 1)$ in S if there exists a constant k such that

$$|f(x_2) - f(x_1)| \leq k[d(x_2,x_1)]^\alpha\,, \qquad \forall x_1, x_2 \in S\,.$$

If S is an open set and f is Hölder continuous of exponent α for every bounded closed set $B \subset S$ with some k which may depend on B, then f is said to be *locally* continuous (of exponent α) in S. If k can be taken independently of B, then f is *uniformly* Hölder continuous (of exponent α). We specify that the metric introduced in D_T, is generally supposed to be {[27] p. 61, [87]}

$$d(P_1,P_2) = [|x_2 - x_1|^2 + |t_2 - t_1|]^{1/2}\,, \qquad \forall P_i = (x_i,t_i) \in D_T.$$

iii) for some positive monotone increasing function $A(|u|)$ and for some sufficiently small $\mu = const. > 0$ (depending on D_T)

$$|f(x,t,u,w)| \le A(|u|) + \mu|w|, \quad (u,v) \in \mathbb{R}^2 .$$

If u_0, u_0', u_0'' are Hölder continuous for some $\delta < 1$ and

$$u_0''(x) = g(x,0,0,u_0'), \qquad x \in \{0,1\}$$

then there exists a solution u of (2.23) bounded in D_T and such that u, u_x, u_t are uniformly Hölder continuous of exponent δ on D_T, while u_{xx} is uniformly Hölder continuous of exponent γ, for some $0 < \gamma < 1$.

Theorem 2.8 *If $g(x,t,u,w)$ is monotone nondecreasing in u (or is Lipschitz continuous in u, uniformly with respect to (x,t,u,w) in bounded sets of $\overline{D}_T \times \mathbb{R}^2$) then there exists at most one solution of (2.23).*

For further details and for the proofs of theorems 2.7-2.8 we refer to pp.201-206 of [27]. Similar theorems can be found in [87,88,89]. In the sequel we assume $u = 0 \Rightarrow f = 0$ and we concentrate on the stability of the solution $u \equiv 0$, with respect to smooth global perturbations.

a) *The diffusion equation* - The homogeneous case ($f = 0$) is the simplest. In this case each of the following functionals

$$\begin{cases} U = \dfrac{1}{2}\|u(t)\|^2 = \dfrac{1}{2}\displaystyle\int_0^1 u^2(x,t)dx \\[2ex] V = \dfrac{1}{2}\|u_x\|^2 = \dfrac{1}{2}\displaystyle\int_0^1 u_x^2 dx \\[2ex] W = U + V \end{cases} \tag{2.24}$$

is a Liapunov function. In fact, along the solutions, taking account of the boundary conditions, it turns out that

$$\begin{cases} \dot{U} = \displaystyle\int_0^1 uu_t dx = \int_0^1 uu_{xx} dx = [uu_x]_0^1 - \int_0^1 u_x^2 dx = -\int_0^1 u_x^2 dx \\[2ex] \dot{V} = \displaystyle\int_0^1 u_x u_{xt} dx = [u_x u_t]_0^1 - \int_0^1 u_{xx} u_t dx = -\int_0^1 u_{xx}^2 dx \\[2ex] \dot{W} = -\displaystyle\int_0^1 (u_x^2 + u_{xx}^2) dx . \end{cases} \tag{2.25}$$

Because of the Poincaré inequality (exercises 2.4 - 2.6)

$$\int_0^1 u^2 dx \le \frac{2}{\gamma}\int_0^1 u_x^2 dx, \qquad (\gamma = \frac{1}{2\pi^2}) \tag{2.26}$$

it immediately follows that U, V, W are Liapunov functions and that the solution $u = 0$ is asymptotically exponentially stable in accordance with

$$\begin{cases} U \le U_0 e^{-\gamma t}, \; V \le V_0 e^{-\gamma t}, \\ W \le W_0 e^{-\gamma t} . \end{cases} \tag{2.27}$$

b) *Two nonhomogeneous linear cases* - We will consider now the following two cases

$$f = (g(x)u)_x \tag{2.28}$$

$$f = g(x)u_x \tag{2.29}$$

with $g \in C^1([0,1])$ in the case (2.28) and $g \in C([0,1])$ in the case (2.29). Our aim is to show that the zero solution remains asymptotically exponentially stable in the L^2-norm. In both cases we choose as Liapunov function [90]

$$\mathcal{V} = \frac{1}{2}\int_0^1 e^{\varphi(x)} u^2 dx \tag{2.30}$$

where

$$\varphi(x) = \int_0^x g(\xi) d\xi . \tag{2.31}$$

Because

$$\begin{cases} \inf\limits_{x \in [0,1]} \varphi \equiv \alpha \leq \varphi(x) \leq \beta \equiv \sup\limits_{x \in [0,1]} \varphi , \\ e^{\alpha} \int_0^1 u^2 dx \leq \int_0^1 e^{\varphi(x)} u^2 dx \leq e^{\beta} \int_0^1 u^2 dx , \end{cases} \tag{2.32}$$

the weighted L^2-norm $\|e^{\varphi/2}u\|$ is equivalent to the L^2-norm of u.

In the case (2.28) it turns out that

$$\begin{aligned} \dot{\mathcal{V}} &= \int_0^1 e^{\varphi} u u_t dx = \int_0^1 e^{\varphi} u[u_{xx} + (gu)_x] dx = -\int_0^1 e^{\varphi} (gu + u_x)^2 dx \\ &= -\int_0^1 e^{-\varphi} [(e^{\varphi} u)_x]^2 dx \leq -e^{-\beta} \int_0^1 [(e^{\varphi} u)_x]^2 dx \leq \\ &- \frac{e^{-\beta}\gamma}{2} \int_0^1 e^{2\varphi} u^2 dx \leq -\frac{e^{\alpha-\beta}\gamma}{2} \int_0^1 e^{\varphi} u^2 dx \end{aligned} \tag{2.33}$$

i.e.

$$\begin{cases} \dot{\mathcal{V}} \leq -c\mathcal{V} \\ c = \gamma e^{\alpha-\beta} . \end{cases} \tag{2.34}$$

From (2.32)–(2.33) it follows that

$$\|u\|^2 \leq 2e^{-\alpha}\mathcal{V} \leq 2e^{-\alpha}\mathcal{V}_0 e^{-ct} \leq e^{\beta-\alpha}\|u_0\|^2 e^{-ct} . \tag{2.35}$$

Turning now to the case (2.29), it turns out that

$$\begin{aligned} \dot{\mathcal{V}} &= \int_0^1 e^{\varphi} u u_t dx = \int_0^1 e^{\varphi} u(u_{xx} + g u_x) dx \\ &= [e^{\varphi} u u_x]_0^1 - \int_0^1 e^{\varphi} (gu + u_x) u_x dx + \int_0^1 e^{\varphi} g u u_x dx = -\int_0^1 e^{\varphi} u_x^2 dx \\ &\leq -\int_0^1 e^{-\varphi} (e^{\varphi} u)_x^2 dx \end{aligned}$$

i.e. one again one obtains relation (2.33) and (2.35) therefore follows.

c) *The nonlinear cases $f = g(u_x)u_{xx}$ and $f = g(u_x)$* - Considering the first case, equation $(2.23)_1$ then becomes

$$u_t = [1 + g(u_x)]u_{xx} \tag{2.36}$$

and one has a diffusion equation with the diffusion coefficient $k = 1 + g$. It is quite natural to require that $1 + g \geq 0$. In this case, stability with respect to the L_2-norm of u_x immediately follows. In fact, considering the function $(2.24)_2$ it turns out that

$$\overset{\bullet}{V} = -\int_0^1 (1+g)u_{xx}^2\,dx\,. \tag{2.37}$$

Further the asymptotic stability is ensured by the assumption :

$$\exists\varepsilon = \text{const.} > 0 : g(\xi) > \varepsilon - 1, \quad \forall\,\xi \in I\!R\,. \tag{2.38}$$

In fact then it follows that

$$\overset{\bullet}{V} \leq -\varepsilon\int_0^1 u_{xx}^2\,dx \tag{2.39}$$

and hence (exercise 2.6)

$$\overset{\bullet}{V} \leq -\varepsilon\pi^2 V\,. \tag{2.40}$$

Let us now consider the case

$$u_t = u_{xx} + g(u_x) \tag{2.41}$$

with $g(0) = 0$, subject to the boundary conditions

$$u_x = 0 \quad , \quad x \in \{0,1\}, t \geq 0 \tag{2.42}$$

and show that in this case *the condition $g \in L^1_{\text{loc}}(I\!R)$ is enough to ensure that the zero solution is asymptotically exponentially stable in the L_2-norm of u_x.* In this case too we consider the positive definite functional $(2.24)_2$. It follows that

$$\overset{\bullet}{V} = -\int_0^1 u_{xx}u_t\,dx = -\int_0^1 u_{xx}^2\,dx - \int_0^1 u_{xx}g(u_x)\,dx$$

i.e.

$$\overset{\bullet}{V} = -\int_0^1 u_{xx}^2\,dx - \int_0^1 \frac{\partial}{\partial x}\int_0^{u_x} g(\xi)\,d\xi\,.$$

But, in view of (2.42),

$$\int_0^1 \frac{\partial}{\partial x}\int_0^{u_x} g(\xi)\,d\xi = \int_{u_x(0,t)}^{u_x(1,t)} g(\xi)\,d\xi = 0$$

and hence, also taking into account (2.26) for $u = u_x$, it follows that

$$\overset{\bullet}{V} \leq -\int_0^1 u_{xx}^2\, dx \leq -\gamma V$$

i.e.

$$V \leq V_0\, e^{-\gamma t}, \qquad\qquad (\gamma = \frac{1}{2\pi^2}).$$

d) *The nonlinear case* $f = -\varphi(u)$ - In the case at hand, equation $(2.23)_1$ becomes

$$u_t = u_{xx} - \varphi(u), \qquad \varphi(0) = 0, \tag{2.43}$$

and is a reaction-diffusion equation containing as a particular case the Fisher equation (1.150). One immediately obtains

$$\overset{\bullet}{V} = -\int_0^1 u_{xx} u_t\, dx = \begin{cases} -\displaystyle\int_0^1 u_{xx}^2 dx - \int_0^1 \varphi'(u) u_x^2 dx, & \varphi \in C^1(I\!R) \\ \text{or} & \\ -\displaystyle\int_0^1 u_t^2 dx - \frac{d}{dt}\int_0^1 F(u) dx, & \varphi \in L^1_{\text{loc}}(I\!R) \end{cases} \tag{2.44}$$

with

$$F(u) = \int_0^u \varphi(\xi) d\xi. \tag{2.45}$$

From $(2.44)_1$ it follows that

$$\overset{\bullet}{V} \leq -\int_0^1 (\varphi' + \pi^2) u_x^2\, dx$$

and *the condition*

$$\exists \varepsilon = const. > 0 : \varphi'(\xi) > \varepsilon - \pi^2, \qquad \forall \xi \in I\!R$$

ensures the asymptotic exponential stability with respect to the norm of H^1. We remark that the simple stability is guaranteed by the condition $\{\xi\varphi(\xi) \geq 0, \forall \xi \in I\!R\}$. In fact in this case $\{F(\xi) \geq 0, \forall \xi \in I\!R\}$ (see exercise 2.12) and choosing as Liapunov function

$$V^* = V + \int_0^1 F(u) dx, \tag{2.46}$$

from $(2.44)_2$ it follows that

$$\overset{\bullet}{V^*} = -\int_0^1 u_t^2 dx. \tag{2.47}$$

e) *The nonlinear case* $f(u, u_x) = g(u) - uu_x$ - Now equation $(2.23)_1$ becomes

$$u_t + uu_x = u_{xx} + g(u) \tag{2.48}$$

i.e. one has the *Burgers equation* with viscosity and the source term $g(u)$ (see exercise 1.28). For a good and simple analysis of the case

$$g(u) = au^2 \qquad a = \text{const.} > 0 \tag{2.49}$$

we refer the reader to [59] (see also exercises). We confine ourselves to proving the following theorem:

Theorem 2.9 - *For*

$$a < \frac{2\pi}{2+q}\sqrt{q} \tag{2.50}$$

the zero solution is a global exponential attractor with respect to the norm

$$\int_0^1 (u^{3m+1})^{1/m}\, dx = \int_0^1 u^{1+q}\, dx \tag{2.51}$$

where m is an odd integer and $q = 2 + \dfrac{1}{m}$.

Proof. Let us introduce the weighted norm

$$W = \int_0^1 e^{-px} u^{1+q}\, dx \tag{2.52}$$

with

$$p = (2+q)a\,. \tag{2.53}$$

It follows immediately that – along the solutions – one has

$$\dot{W} = (1+q)\int_0^1 e^{-px} u^q u_{xx}\, dx\,. \tag{2.54}$$

But since

$$\int_0^1 e^{-px} u^q u_{xx} dx = p\int_0^1 e^{-px} u^q u_x dx - q\int_0^1 e^{-px}(u^{(q-1)/2}u_x)^2\, dx$$
$$= \frac{p^2}{q+1}\int_0^1 e^{-px} u^{q+1} dx - \frac{4q}{(q+1)^2}\int_0^1 e^{-px}\left(\frac{\partial}{\partial x}u^{(1+q)/2}\right)^2 dx\,,$$

and taking into account the *weighted Poincaré inequality*

$$\int_0^1 e^{-kx} f^2\, dx \le \frac{4}{k^2+4\pi^2}\int_0^1 e^{-kx} f_x^2\, dx \tag{2.55}$$

which holds for any $f \in H_0^1(0,1)$ and for any positive k (see [59] and exercises 2.10-2.11), one finds that for $k = p$ one has

$$\dot{W} \le -\lambda\int_0^1 e^{-kx}(\frac{\partial}{\partial x}u^{(1+q)/2})^2\, dx \tag{2.56}$$

where

$$\lambda = \frac{4(4q\pi^2 - p^2)}{(1+q)(p^2+4\pi^2)} \cdot \tag{2.57}$$

Now (2.50) ensures $p^2 < 4q\pi^2$ i.e. $\lambda > 0$, it therefore follows from (2.55)–(2.56) that

$$\overset{\bullet}{W} \le -\frac{\lambda(p^2+4\pi^2)}{4} W , \tag{2.58}$$

i.e.

$$\begin{cases} W \le W_0 e^{-\gamma t} \\ \gamma = \dfrac{\lambda}{4}(p^2+4\pi^2) , \end{cases} \tag{2.59}$$

and hence

$$\int_0^1 u^{1+q}\, dx \le W_0 e^{(p-\gamma t)} . \tag{2.60}$$

Remark 2.6 - *The function $F(y) = 2\pi\sqrt{y}/(2+y)$, $y \in \mathbb{R}_+$ achieves its maximum value $\pi/\sqrt{2}$ for $y = 2$. Therefore the largest bound for a (i.e. $a < \pi/\sqrt{2}$) appears as an asymptotic value, for $m \to \infty (\equiv q \to 2)$. This value – following the argument used in the proof of Theorem 2.9 – can be reached only for positive solutions. In fact $q = 2 \Rightarrow u^{q+1} = u^3$ and hence $u^{q+1} < 0$ for $u < 0$.*

f) *Diffusion with variable conductivity* - We consider the I.B.V.P.

$$\begin{cases} u_t = \dfrac{\partial}{\partial x}[(k+g(u))u_x] & x \in (0,1),\ t > 0 \\ u(0,t) = u(1,t) = 0 & t \ge 0 \\ u(x,0) = u_0(x) & x \in [0,1] \end{cases} \tag{2.61}$$

with $u = u(x,t)$, $k = \text{const.}(>0)$ and $g \in C^1(\mathbb{R})$. In the case of heat diffusion, this happens when the thermal conductivity depends on the temperature. For instance, this is so in the case of heat diffusion in the "cold ice" of glaciers for which one can assume [60]

$$\begin{cases} k = 1 \quad , & g = \varepsilon_1 u + \varepsilon_2 u^2 \\ \varepsilon_1 = 0.261 \quad , & \varepsilon_2 = 1.163 . \end{cases} \tag{2.62}$$

Let us add that $(2.61)_1$, more generally, can be written:

$$u_t = \Delta\varphi(u) \tag{2.63}$$

with

$$\varphi(u) = \int_0^u [k+g(s)]ds , \tag{2.64}$$

and models many other phenomena like diffusion of: i) biological populations [47,91], ii) fluids through porous media [92], iii) heat in the Stefan problem [93]. For a deep analysis of this matter we refer to [94,95,96] for the case $\varphi(u) = u^m$ $(m \neq 1)$, and to [97] for the general case. Setting

$$F(u) = \int_0^u g(\xi)d\xi \tag{2.65}$$

it follows that

$$u = 0 \Rightarrow F = 0 \tag{2.66}$$

$$u_t = ku_{xx} + \frac{\partial^2}{\partial x^2}F(u). \tag{2.67}$$

Hence it turns out that

$$\begin{aligned} \overset{\bullet}{U} &= -k\int_0^1 u_x^2 dx - \int_0^1 u_x \frac{\partial F}{\partial x} = \\ &= -\int_0^1 (k+g)u_x^2 dx \end{aligned}$$

i.e. *the global asymptotic exponential stability of the zero solution is ensured if*

$$\exists \varepsilon = \text{const.} > 0 : \; k + g(\xi) \geq \varepsilon, \; \forall \xi \in I\!R.$$

In the case (2.62), immediately it follows that $\varepsilon > 3/4$.

g) *The case* $u_t = [F(x,u)]_{xx}$ - We end this section by considering the equation generalizing $(2.61)_1$[3],

$$u_t = [F(x,u)]_{xx}, \qquad (x,t) \in (0,1) \times I\!R_+ \tag{2.68}$$

with $u = u(x,t)$, under the conditions $(2.61)_2$–$(2.61)_3$. This equation may also model many phenomena of the real world [90,97]. We assume, $F \in C^2([(0,1) \times I\!R])$

$$\begin{cases} F(x,0) = F_{xx}(x,0) = 0, & \forall x \in (0,1), \\ G(x,u) = \displaystyle\int_0^u F(x,\xi)d\xi > 0, \end{cases} \tag{2.69}$$

and – following [90] – we consider the functional

$$\mathcal{E} = \int_0^1 G[x,u(x,t)]dx. \tag{2.70}$$

Then it turns out that

$$\overset{\bullet}{\mathcal{E}} = \int_0^1 u_t F(x,u)dx = \int_0^1 F(x,u)[F(x,u)]_{xx}dx = -\int_0^1 ([F]_x)^2 dx \tag{2.71}$$

---

[3] i.e. $u_t = \dfrac{\partial F}{\partial u}u_{xx} + \dfrac{\partial^2 F}{\partial u^2}u_x^2 + 2\dfrac{\partial^2 F}{\partial u \partial x}u_x + \dfrac{\partial^2 F}{\partial x^2}$

and the stability of the zero solution in the $\mathcal{E}$-measure is ensured by $(2.69)_1 - (2.69)_2$ and (exercise 2.12)

$$\xi F(x,\xi) \geq 0\,, \qquad \{\forall\, \xi \neq 0\,, \quad \forall\, x \in [0,1]\}. \tag{2.72}$$

Remark 2.7 *Let* $u \in H_0^1([0,1])$. *Then by the Schwarz inequality immediately it follows that*

$$|u| = |\int_0^x u_x(s)ds| \leq \left(\int_0^x ds\right)^{1/2}\left(\int_0^x u_x^2 ds\right)^{1/2} \leq \|u_x\|_2\,, \qquad \forall\, x \in [0,1]$$

and hence $\sup\limits_{x\in[0,1]} |u| \leq \|u_x\|_2$. *Therefore in the equations at hand, under homogeneous boundary conditions, the stability in the* L_2*-norm of the gradient implies the same type of stability in the pointwise norm.*

Remark 2.8 *Let* $I_m \subset H_0^1([0,1])$ *be such that* $\|u_x\|_2 \leq m, \forall\, u \in I_m$ *for some positive constant* m. *Then immediately it follows*

$$u^2 = \int_0^x du^2 \leq 2\|u_x\|_2\|u\|_2\,, \qquad \forall\, x \in [0,1]$$

and hence $\sup\limits_{x\in[0,1]} |u| \leq \sqrt{2}m^{1/4}\|u\|_2^{1/2}$. *Therefore the stability in the* L_2*-norm with respect to the perturbations belonging to* I_m *implies the same type of stability in the pointwise norm.*

2.8 EXERCISES

Exercise 2.1 - Show that a Liapunov function, having the generalized derivative (2.10) non-positive, is nonincreasing along the solutions.

Hint - The generalized derivative (2.10) is the right lower derivative of the function $f(t) = V[v(t,x)]$. Let a be a positive constant and let $f \in C([0,a))$ with right derivative non-positive on $[0,a)$. Consider then the function $\varphi = f - \varepsilon t$, some $\varepsilon > 0$.

Answer - Assume – on the contrary – that $\exists t_i \in [0,a)$, $(i = 1,2)$, such that

$$\begin{cases} t_1 < t_2 \\ f(t_2) > f(t_1) \,. \end{cases}$$

Then for

$$0 < \varepsilon < \frac{f(t_2) - f(t_1)}{t_2 - t_1}$$

it turns out that $\varphi(t_2) > \varphi(t_1)$. Consequently, because of the continuity of φ, $\exists t^* > t_1 : \varphi(t) > \varphi(t_1)\,, \forall t \in (t_1, t^*)$. Hence denoting by D_+ the right lower derivative, it follows that $D_+\varphi(t_1) \geq 0$. But this is impossible because

$$\{D_+ f(t_1) \leq 0\,, \varphi = f - \varepsilon t\} \to D_+\varphi(t_1) = D_+ f(t_1) - \varepsilon \leq -\varepsilon < 0\,.$$

Exercise 2.2 - Consider the inequality (2.16) and assume (2.17). Show that

$$\begin{cases} R < R^* \\ \\ \|\mathbf{u}_0\|^\alpha < \dfrac{R^* - R}{\beta R^*} \end{cases} \Rightarrow \|\mathbf{u}\| \leq \|\mathbf{u}_0\| \exp\left[-\frac{1}{\gamma}(\frac{R^* - R}{R^*} - \beta\|\mathbf{u}_0\|^\alpha)t\right]$$

Hint - By (2.17)–(2.18) it follows that

$$\|\dot{u}\| \leq (\beta\|u\|^\alpha - \frac{R^* - R}{R^*})\mathcal{D}$$

and then (by assumptions)

$$\|\dot{u}\|_{t=o} \leq -(\frac{R^* - R}{R^*} - \beta\|u_0\|^\alpha)\mathcal{D}_0 \leq -\frac{1}{\gamma}(\frac{R^* - R}{R^*} - \beta\|u_0\|^\alpha)\|u_0\| < 0\,.$$

Use recursive arguments.

Exercise 2.3 - Consider the function space

$$\mathcal{H} = \{u \in C^2(0,1) \cap C_0(0,1)\}$$

and assume the validity of the Poincaré inequality

$$\exists \gamma = \text{const.} > 0 \, : \, ||f'||^2 \geq \gamma ||f||^2 \qquad \forall f \in \mathcal{H} \, ,$$

with $||\cdot|| = L_2$-norm. Show that γ cannot exceed π^2.

Hint - Consider the exercise 1.26 and obtain

$$\frac{1}{2}\frac{d}{dt}||u||^2 = -||u_x||^2 + a||u||^2 \leq (a-\gamma)||u||^2$$

Then take into account that the I.B.V.P. of exercise 1.26 admits the solution $u = e^{(a-\pi^2)t} \sin \pi x \in \mathcal{H} \times I\!R_+$ for $a > \pi^2$ also.

Exercise 2.4 - In the space $\mathcal{H}$ introduced in exercise 2.3 show the validity of the Poincaré inequality

$$||f'||^2 \geq \pi^2 ||f||^2 \qquad \forall f \in \mathcal{H} .$$

Hint - Assume the existence of $\dfrac{1}{R^*} = \max\limits_{I} \dfrac{||f||^2}{||f'||^2}$ on a space $I \supset \mathcal{H}$ (see exercise 2.5).

Answer - Let $g \in I$ be the function on which $\dfrac{||f||^2}{||f'||^2}$ attains the maximum. Then, letting ε be a positive parameter, for any $f \in I$, the function

$$F(\varepsilon) = \frac{||g+\varepsilon f||^2}{||g'+\varepsilon f'||^2}$$

attains its maximum for $\varepsilon = 0$ and therefre one has

$$0 = \left[\frac{d}{d\varepsilon}F(\varepsilon)\right]_{\varepsilon=0} =$$

$$= \left[\frac{||g'+\varepsilon f'||^2 \frac{d}{d\varepsilon}||g+\varepsilon f||^2 - ||g+\varepsilon f||^2 \frac{d}{d\varepsilon}||g'+\varepsilon f'||^2}{||g'+\varepsilon f'||^4}\right]_{\varepsilon=0} =$$

$$= \frac{1}{||g'||^2}\left[\left(\frac{d}{d\varepsilon}||g+\varepsilon f||^2\right)_{\varepsilon=0} - \frac{||g||^2}{||g'||^2}\left(\frac{d}{d\varepsilon}||g'+\varepsilon f'||^2\right)_{\varepsilon=0}\right] =$$

$$= \frac{2}{||g'||^2}\int_0^1 (gf - \frac{1}{R^*}g'f')dx = \frac{2}{||g'||^2}\int_0^1 g(f + \frac{1}{R^*}f'')dx .$$

Since g is arbitrary in I, f has to satisfy the (Euler-Lagrange) equation

$$f'' + R^* f = 0 \qquad f \in I\,.$$

It follows that

$$f = c \sin \sqrt{R^*}\, x$$

where

$$\sqrt{R^*} = n\pi \qquad n = \pm 1, \pm 2, ..$$

and c=const. Therefore $R^* = \pi^2$.

Exercise 2.5 - By using the Direct Method of Variational Calculus [98], show that on the function space $H_0^1(0,1) = H^1(0,1) \cap L_0^2(0,1)$

i) the ratio $\int_0^1 f^2 dx / \int_0^1 f'^2 dx$, is bounded;

ii) there exists $\overline{f} \in H_0^1(0,1)$ such that $\max\limits_{f \in H_0^1} \dfrac{\int_0^1 f^2 dx}{\int_0^1 f'^2 dx} = \dfrac{\int_0^1 \overline{f}^2 dx}{\int_0^1 (\overline{f}')^2 dx}$.

Hint - For any $\varphi \in C^1(0,1)$ it follows that

$$\int_0^1 \varphi' f^2 dx = -2 \int_0^1 \varphi f f' dx \leq \varepsilon \int_0^1 \varphi f^2 dx + \frac{1}{\varepsilon} \int_0^1 \varphi f'^2 dx$$

with ε = const. > 0. Choosing $\{\varphi = x,\ \varepsilon = 1/2\}$ it turns out that

$$\int_0^1 f^2 dx \leq 4 \int_0^1 f'^2 dx$$

and therefore $\exists l$ = const. > 0 such that

$$\sup_{f \in H_0^1} \frac{\displaystyle\int_0^1 f^2 dx}{\displaystyle\int_0^1 f'^2 dx} = l < \infty\,.$$

Using the normalization $f \Rightarrow \dfrac{f}{(\int_0^1 f'^2 dx)^{1/2}}$, one can proceed to the point ii) showing the existence of $\overline{f} \in H_0^1$ with $\displaystyle\int_0^1 (\overline{f}')^2 dx = 1$ such that $\displaystyle\int_0^1 \overline{f}^2 dx = l$. Because $l < \infty$, there exists a sequence $\{f_n\}$, $f_n \in H_0^1\ \forall n \in \mathbb{N}$, such that

$$\int_0^1 (f_n')^2 dx = 1\,, \qquad \lim_{n \to \infty} \int_0^1 f_n^2 dx = l \quad n = 1, 2, ...$$

By the weak compactness of the (Hilbert) space $H_0^1(0,1)$...

Answer - See [98].

Exercise 2.6 - Consider the function space $\mathcal{H}$ of exercise 2.3 and show that the following Poincaré inequality holds

$$\int_0^1 u_x^2 dx \leq \frac{1}{\pi^2} \int_0^1 u_{xx}^2 dx.$$

Hint - Let $u \in \mathcal{H}$. Because $u \in C_0([0,1])$, using the Schwarz inequality, it follows

$$\int_0^1 u_x^2 dx = -\int_0^1 u u_{xx} dx \leq \|u\|_2 \|u_{xx}\|_2 .$$

Hence by the Poincaré inequality (2.26), it follows $\|u_x\|_2^2 \leq \dfrac{1}{\pi}\|u_x\|\|u_{xx}\|_2$ and hence...

Exercise 2.7 - Consider the I.B.V.P. of exercise 1.26. Show that $a = \pi^2$ is a **bifurcation point**. Precisely, show that a steady solution, other than zero, appears for $a = \pi^2$.

Hint - Solve the equation

$$u_{xx} + \pi^2 u = 0 \qquad\qquad u = 0 \quad x = 0, 1 .$$

Answer - For $a = \pi^2$, $u = \sin \pi x$ is a critical point.

Exercise 2.8 - Consider the *Burgers equation with variable viscosity*

$$u_t + u u_x = \frac{\partial}{\partial x}[k + g(u) u_x] \qquad\qquad x \in (0,1),\, t > 0$$

with $k =$ const. > 0 and $g \in L^1_{\mathrm{loc}}(\mathbb{R})$. Consider the boundary and initial conditions $(2.23)_2$–$(2.23)_3$ and show that the zero solution is asymptotically exponentially stable in the L^2-norm.

Hint - See subsection f) of Section 2.7.

Exercise 2.9 - Consider the I.B.V.P.

$$\begin{cases} u_t + u u_x = u_{xx} + \beta u^2 & \in (0,1),\, t > 0 \\ u(x,0) = u_0(x) & x \in [0,1] \\ u(0,t) = u(1,t) = 0 & t \geq 0 \end{cases}$$

and show that the zero solution is asymptotically exponentially stable in the L^2-norm under restriction on the initial data.

Hint - Find an inequality like (2.16). *Answer* - See [59] pp. 10-12.

Exercise 2.10 - Verify that the weighted Poincaré inequality

$$\int_0^1 e^{-kx} f^2 dx \leq \gamma \int_0^1 e^{-kx} f'^2 dx, \qquad \gamma = \text{const.} > 0$$

($k = \text{const.} > 0$, $f \in C_0^1(0,1)$) immediately follows on integrating $\frac{d}{dx}(e^{-kx} f^2)$ and show that the lowest value of γ that one can obtain in this way is $4/\pi^2$. Is this value enough in the proof of Theorem 2.9?

Hint - Recall the inequality $2ab \leq \varepsilon a^2 + \frac{1}{\varepsilon} b^2$ with $\varepsilon = \text{const.} > 0$.

Answer - For $p = (2+q)a$, $k = p$ and $\gamma = \frac{4}{p^2}$, from (2.54) it follows that

$$\dot{W} = p^2 \int_0^1 e^{-px} u^{q+1} dx - \frac{4q}{q+1} \int_0^1 e^{-px} \left(\frac{\partial}{\partial x} u^{(q+1)/2} \right)^2 dx \leq$$
$$\leq -\lambda \int_0^1 e^{-px} \left(\frac{\partial}{\partial x} u^{(q+1)/2} \right)^2 dx$$

with $\lambda = 4\left(\frac{q}{q+1} - 1 \right)$. *But now* $\lambda < 0, \forall q \in \mathbb{R}_+$!

Exercise 2.11 - Find the value $\frac{4}{k^2 + 4\pi^2}$ for the constant γ in the inequality of the exercise 2.10.

Hint - Determine – via variational methods – the maximum

$$\frac{1}{R^*} = \max_{H_0^1(0,1)} \frac{\int_0^1 e^{-kx} f^2 dx}{\int_0^1 e^{-kx} f'^2 dx}.$$

Follow the answer of exercise 2.4 and establish that the Euler-Lagrange equation of the variational problem is $f'' - kf' + R^* f = 0$.

Exercise 2.12 - Let $g \in L^1_{\text{loc}}(\mathbb{R})$. Show that $x g(x) \geq 0, \quad \forall x \in \mathbb{R}$ implies $\int_0^x g(s) ds \geq 0, \quad \forall x \in \mathbb{R}$.

Hint - Consider, per absurdum, that $\exists a \in \mathbb{R}$: $\int_0^a g(s) ds < 0$. If $a > 0$, then $\exists \bar{x} \in (0, a) : g(\bar{x}) < 0$ and hence ...
If $a = -|a| < 0$ then

$$\int_0^a g(s) ds = -\int_{-|a|}^0 g(s) ds < 0 \Leftrightarrow \int_{-|a|}^0 g(s) ds > 0$$

and hence $\exists \bar{x} \in (-|a|, 0) : g(\bar{x}) > 0$ and hence...

Exercise 2.13 - Consider the I.B.V.P.

$$\begin{cases} u_t = u_{xx} + \sin u_x & x \in (0,1), t > 0\,, \\ u(0,x) = u_0(x) & x \in [0,1]\,, \\ u_x(0,t) = u_x(1,t) = 0 & t > 0\,. \end{cases}$$

Determine the stability of the zero solution with respect to the L^2-norm of u_x.

Hint - See Sec.2.7, subsection c).

Exercise 2.14 - Consider the I.B.V.P.

$$\begin{cases} u_t = u_{xx} - \sin 2u & x \in (0,1), t > 0\,, \\ u(0,x) = u_0(x) & x \in [0,1]\,, \\ u_x(0,t) = u_x(1,t) = 0 & t > 0\,. \end{cases}$$

Determine the stability of the zero solution with respect to the L^2-norm of u_x.

Hint - $f = -\varphi$, $\varphi = \sin 2u$, $F(u) = \sin^2 u$. See Section 2.7, subsection d).

Exercise 2.15 - Consider the I.B.V.P.

$$\begin{cases} u_t = [1 + g(u_x)]u_{xx} & x \in (0,1), t > 0 \\ u(0,x) = u_0(x) & x \in [0,1] \\ u(0,t) = u(1,t) = 0 & t > 0\,. \end{cases}$$

Assume $g \in L^1_{\mathrm{loc}}(I\!R)$ and verify that the function

$$E = V + \int_0^1 dx \int_0^{u_x} G(s)ds$$

with V and G given respectively by $(2.24)_2$ and (2.38) is a Liapunov function at least along the smooth solutions. Is E non-negative?

Answer - One has $\dot{E} = -\int_0^1 u_t^2 dx$. The condition (see exercise 2.12) $\xi G(\xi) \geq 0, \quad \forall \xi \in I\!R$ assures that E is non-negative. This condition is certainly satisfied if $g(\xi) \geq 0 \, \forall \xi \in I\!R$.

Exercise 2.16 - Consider the I.B.V.P.

$$\begin{cases} u_t = \dfrac{\partial^2}{\partial x^2}\varphi(u) & x \in (0,1)\,, \quad t > 0\,, \\ u(0,t) = u(1,t) = 0 & t \geq 0\,, \\ u(x,0) = u_0(x) & x \in [0,1]\,, \end{cases}$$

with $\{\varphi(0) = 0, \varphi \in C^2(\mathbb{R})\}$. Show that $\{\varphi'(\xi) > 0, \forall \xi \in \mathbb{R}\}$ ensures the stability of the zero solution with respect to the measure

$$E = \frac{1}{2}\int_0^1 (\varphi^2 + \varphi_x^2)dx\,,$$

while $\{\varphi'(\xi) \geq \varepsilon = \text{const.} > 0, \forall \xi \in \mathbb{R}\}$ ensures asymptotic exponential stability. Consider, in particular, the case (2.64) with k and g given by (2.62).

Hint. Set $\mathcal{V} = \frac{1}{2}\int_0^1 \varphi_x^2 dx$. Then it follows that

$$\dot{\mathcal{V}} = \int_0^1 \varphi_x \varphi_{xt} dx = -\int_0^1 \varphi_{xx}\varphi_t dx = -\int_0^1 \varphi_{xx}\varphi' u_t dx = -\int_0^1 \varphi'(\varphi_{xx})^2 dx\,.$$

Further take into account that $x \in \{0,1\} \Rightarrow \varphi = 0$ hence by the Poincaré inequality (exercises 2.4,2.6)...

Answer. $E \leq \left(1 + \frac{1}{\pi^2}\right)\mathcal{V}, \forall t \in \mathbb{R}_+$. Hence

$$\begin{cases} \varphi' \geq 0 \quad \Rightarrow & \mathcal{V}(t) \leq \mathcal{V}(0), \quad \forall t \geq 0 \\ \varphi' \geq \varepsilon = \text{const.} > 0 \quad \Rightarrow & \dot{\mathcal{V}} \leq -2\pi^2\varepsilon\mathcal{V} \leftrightarrow \mathcal{V} \leq \mathcal{V}_0 e^{-2\pi^2\varepsilon t}\,. \end{cases}$$

In the case (2.64), with k and g given by (2.62), one has

$$\begin{cases} \varphi' = 1 + \varepsilon_1 u + \varepsilon_2 u^2 > 0, & \forall u \in \mathbb{R} \\ \varphi'' = 2\varepsilon_2 u + \varepsilon_1 \quad , & \varphi''' = 2\varepsilon_2 \end{cases}$$

and hence

$$\varphi' \geq 1 + \varepsilon_1\left(-\frac{\varepsilon_1}{2\varepsilon_2}\right) + \varepsilon_2\frac{\varepsilon_1^2}{4\varepsilon_2^2} = 1 - \frac{1}{4}\frac{\varepsilon_1^2}{\varepsilon_2} > 1 - \frac{1}{4} = \frac{3}{4}.$$

Exercise 2.17 - Consider the I.B.V.P. of the exercise 2.16 with $\{\varphi(0) = 0, \varphi \in C^2(\mathbb{R})\}$ and show that $\{\varphi'(\xi) \geq 0, \forall \xi \in \mathbb{R}\}$ ensures the stability of the zero solution with respect to the L^2-norm while $\{\varphi'(\xi) \geq \varepsilon = \text{const.} > 0, \forall \xi \in \mathbb{R}\}$ ensures asymptotic exponential stability.

Answer. $\dot{U} \leq -\int_0^1 \varphi'(u)u_x^2 dx$. If $\varphi' \geq \varepsilon$, $\dot{U} \leq -2\varepsilon\pi^2 U$

Exercise 2.18 - Consider the I.B.V.P. for Fisher's equation

$$\begin{cases} u_t = 2ku_{xx} + u(2a - 3bu) & (x,t) \in (0,1) \times \mathbb{R}_+\,, \\ u(x,0) = u_0(x) & x \in [0,1]\,, \\ u(0,t) = u(1,t) = 0 & t \geq 0\,, \end{cases}$$

with a, b, k positive constants and $u_0 \in C^1([0,1])$. Show that, for any positive constant α, the condition

$$||u_0'||^2 = \int_0^1 (u_0')^2 dx < r$$

with r given by

$$r = \frac{-k + \sqrt{k^2 + 4\alpha\gamma}}{2\gamma}, \quad \gamma = \frac{b^2}{16a}$$

implies

$$V^* = k \int_0^1 u_x^2 dx + \int_0^1 (bu - a)u^2 dx \leq \alpha, \quad \forall t \geq 0$$

at least along smooth solutions.

Hint. V^* is a Liapunov function [section 2.7, d)]. By the Schwarz and the arithmetic-geometric inequalities, it follows that

$$V^*(0) < k \int_0^1 (u_0')^2 dx + \frac{b\varepsilon}{2} \int_0^1 u_0^4 dx +$$
$$+ \left(\frac{b}{2\varepsilon} - a\right) \int_0^1 u_0^2 dx \qquad \varepsilon = \text{const.}(> 0).$$

Choose $\varepsilon = 1/2a$ and use the Sobolev inequality (see Appendix): $\int_0^1 u_0^4 dx \leq \frac{1}{4}[\int_0^1 (u_0')^2 dx]^2$

Answer. $V^*(0) \leq k||u_0'||^2 + \gamma||u_0'||^4$; $||u_0'||^2 \in (0, r) \Rightarrow V^*(0) \leq \alpha$

Exercise 2.19 - Consider the I.B.V.P. of exercise 2.18. Show that, at least along the smooth solutions,

$$\begin{cases} a < k\pi^2 \\ ||u_o|| < \dfrac{4(k\pi^2 - a)}{3b} \end{cases} \Rightarrow \frac{d||u||^2}{dt} \leq -2(2k\pi^2 - 2a - 3b||u_0||) \cdot ||u||^2$$

and hence, $||u||^2 \leq ||u_0||^2 e^{-\lambda t}$ with $\lambda = 4(k\pi^2 - a) - 3b||u_0||$.

Hint. Let $|| \cdot ||$ be the L_2-norm. Then it follows that

$$\frac{1}{2}\frac{d}{dt}||u||^2 = -2k||u_x||^2 + 2a||u||^2 - 3b \int_0^1 u^3 dx.$$

Use the Sobolev inequality (see Appendix) and follow section 2.5.

Exercise 2.20 - Consider the I.V.P. for Fisher's generalized one-dimensional equation

$$\begin{cases} u_t = u_{xx} + f(u) & (x,t) \in I\!R \times I\!R_+ , \\ u(x,0) = u_0(x) & x \in I\!R \end{cases} \tag{2.73}$$

and the corresponding (*kinetic*) equation

$$\frac{du}{dt} = f(u). \tag{2.74}$$

Show that a zero of f is a stable solution of (2.73) with respect to the L_∞-norm iff it is a stable solution of (2.74).

Hint. The following comparison theorem holds {[50], Theorem 4.1, p. 85 and [27], p. 52 }: let u_i ($i = 1,2,3$) be global solutions of (2.73) with $u_0 = \varphi_i$ ($i = 1,2,3$). Then $\{\varphi_1(x) \le \varphi_2(x) \le \varphi_3(x), \forall\, x \in I\!R\} \Rightarrow \{u_1 \le u_2 \le u_3, \forall\,(x,t) \in I\!R \times I\!R_+\}$.

Answer. Solutions of (2.74) are also solutions of $(2.73)_1$. Therefore an unstable critical point of (2.74) is an unstable solution of (2.73) in the L_∞-norm. Vice versa, let $\overline{u}$ be a stable critical point of (2.74). Then $\forall\, \varepsilon > 0$, $\exists \delta(\varepsilon) > 0$ such that the solutions v, w of (2.74) corresponding respectively to the initial data $\{v(0) = \overline{u} - \delta,\ w(0) = \overline{u} + \delta\}$ satisfy

$$|v(t) - \overline{u}| < \varepsilon\ ,\ |w(t) - \overline{u}| < \varepsilon, \qquad \forall t > 0.$$

Let $u(x,t)$ be a solution of (2.73) with

$$|u_0(x) - \overline{u}| < \delta \Longleftrightarrow v(0) = \overline{u} - \delta \le u_0(x) \le \overline{u} + \delta = w(0).$$

Then, by the comparison theorem, it follows that

$$v(t) \le u(x,t) \le w(t), \qquad \forall t > 0$$

and hence

$$\overline{u} - \varepsilon \le u(x,t) \le \overline{u} + \varepsilon, \qquad \forall t > 0.$$

Exercise 2.21 - Let Ω be a smooth bounded domain in $I\!R^n$. Consider the I.B.V.P.

$$\begin{cases} u_t = \Delta u - g(\mathbf{x}, u) & (\mathbf{x}, t) \in \Omega \times I\!R_+ \\ u(\mathbf{x}, 0) = u_0(\mathbf{x}) & \mathbf{x} \in \Omega \\ u(\mathbf{x}, t) = 0 & (\mathbf{x}, t) \in \partial\Omega \times I\!R_+ \end{cases}$$

with $g \in C([\Omega \times I\!R])$. Verify that $E = \dfrac{1}{2}\|\nabla u\|_2^2 + G(u)$, with

$$G(u) = \int_\Omega d\Omega \int_0^{u(\mathbf{x},t)} g(\mathbf{x}, \xi)\, d\xi \tag{2.75}$$

is a Liapunov function, at least along smooth solutions.

Hint. See section 2, d).

Exercise 2.22 - Let Ω be a smooth bounded domain in $\mathbb{R}^n$. Consider the I.B.V.P.

$$\begin{cases} u_t = \sum_{i,j=1}^{n} \frac{\partial}{\partial x_i}[a_{ij}(\mathbf{x})\frac{\partial u}{\partial x_j}] - g(\mathbf{x}, u) & (\mathbf{x}, t) \in \Omega \times \mathbb{R}_+ \\ u(\mathbf{x}, 0) = u_0(\mathbf{x}) & \mathbf{x} \in \Omega \\ u(\mathbf{x}, t) = 0 & (\mathbf{x}, t) \in \partial\Omega \times \mathbb{R}_+ \end{cases}$$

with $g \in C([\Omega \times \mathbb{R}])$, $a_{ij} = a_{ji}\, \forall \mathbf{x} \in \Omega$, $a_{ij} \in C(\Omega)$. Verify that

$$E = \frac{1}{2}\sum_{ij}\int_{\Omega} a_{ij}\frac{\partial u}{\partial x_j}\frac{\partial u}{\partial x_i} d\Omega + G(u)$$

with $G(u)$ given by (2.75), is a Liapunov function at least along the smooth solutions. Further assume $g(\mathbf{x}, 0) = 0\ \forall \mathbf{x} \in \Omega$, and determine a condition for the coefficients a_{ij} and the source term g under which the zero solution is stable in the $L_2(\Omega)$-norm.

Answer. $\dot{E} = -\int_{\Omega} u_t^2 d\Omega$. The conditions of uniform ellipticity, i.e.

$$\exists \alpha > 0 : \sum_{ij=1}^{n} a_{ij}(\mathbf{x}) y_i y_j \geq \alpha \sum_{i=1}^{n} y_i^2 \quad \forall \mathbf{x} \in \Omega,\ \forall y \in \mathbb{R}^n,$$

and

$$\xi g(\mathbf{x}, \xi) \geq 0 \qquad \forall (\mathbf{x}, \xi) \in \Omega \times \mathbb{R},$$

ensure the required stability of the zero solution.

Chapter three

ESTIMATES BASED ON FIRST ORDER INEQUALITIES 2: VOLUME INTEGRAL AND OTHER METHODS

3.1 THE VOLUME INTEGRAL METHOD

3.1.1 Introduction

This chapter deals with some simple problems for P.D.E. by deriving a first order differential inequality for an appropriate (non-negative) measure of the solution, and by deducing therefrom an inequality for the measure itself, in terms of the data for the problem. Our aim is to treat methodologies in a simple way. The first such issue considered is what may be called the "volume integral method" for a suitable Neumann (or other) problem.

What we call the "volume integral method" has its origin in the work of Knowles [99] and Toupin [100], whereby they obtained spatial decay estimates justifying the celebrated Saint-Venant principle of elasticity. [For a rough/verbal statement of this principle, see Appendix to this chapter.] For extensive surveys of the "volume integral method" – and of related matters in elasticity, together with many references – the reader is referred to the review article by Horgan and Knowles [101] and to an update thereto by Horgan [102].

We first give a general description of the method. We then show how it may be applied to a Neumann problem for a rectangular region. This problem is much simpler than that considered by Knowles and Toupin. Nevertheless, it may be viewed as a simple traction boundary value problem for an elastic region, which has some of the essential characteristics of a more general traction boundary value problem; this simple traction boundary value problem is discussed in the Appendix to this chapter in view of the origin in elastic theory of the volume integral method.

We also show how decay estimates of Phragmèn-Lindelöf type may be obtained for semi-infinite rectangular region. Roughly speaking, this means: a measure of the solution decays exponentially with distance as one moves towards infinity, if it is known that the measure does not increase asymptotically any faster than an exponential function of given exponent. Similar analyses for analogous regions defined by polar coordinates are also given. Estimates of the general type referred to in this paragraph were obtained in the context of linear elasticity by Flavin, Knops and Payne [103], and in the context of a quasi-linear equation (applicable to elasticity) by Horgan and Payne [104].

Finally we show the technique may be adapted to a time-dependent problem – a diffusion problem for a right cylinder.

3.1.2 Description of Method

Attention is confined to a three-dimensional region, and it is left to the reader to make the obvious adaptation to a two-dimensional context.

We consider a three-dimensional region (See Fig. 3.1) V whose boundary consists of three disjoint portions: C the lateral boundary, together with Γ_0, Γ_L which will presently be defined.

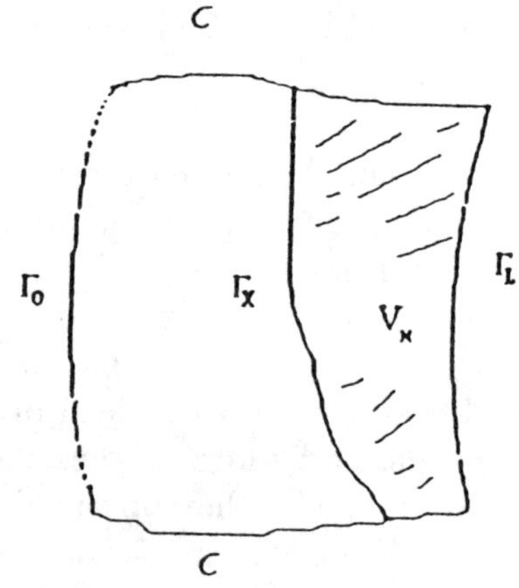

Figure 3.1

Suppose that a family of non self-intersecting surfaces Γ_x parametrized by the variable x $(0 \leq x \leq L)$ exists with the following characteristics:

(a) Γ_0, Γ_L (corresponding to $x = 0$, $x = L$ respectively) constitute part of the boundary of V as aforesaid. [In degenerate case, however, Γ_L may degenerate to a point.]

(b) Each surface Γ_x intersects the lateral surface in a simple closed curve or curves.

Suppose that u is a (scalar, vector etc.) function of position (only) which satisfies the P.D.E., or system of P.D.E.s,

$$L(u) = 0 \qquad \text{in } V.$$

Suppose that u satisfies a non-null boundary condition

$$B_0(u) = b_0 \qquad \text{on } \Gamma_0$$

where b_0 is a given function; and satisfies, in some sense, "null" or "zero" boundary conditions on the remainder of the boundary of V (i.e. on C and

Γ_L). Let V_x denote the domain (or "volume") contained between Γ_x, Γ_L and $\mathcal{C}$. Let $P(u)$ be a non-negative function of u (and/or its derivatives) and define a "volume integral measure" of the solution in V_x by

$$F(x) = \int_{V_x} P(u)dV .$$

The object of the volume integral method (as developed thus far) is to obtain a first-order differential inequality (cf. chapter two) for $F(x)$ whence an upper bound can be obtained therefor, in terms of x and the "data" term $F(0)$; it is understood that an upper bound, in terms of data, can be obtained for $F(0)$.

Typically one has a differential inequality of the type

$$F'(x) + h(x)F(x) \leq 0 \tag{3.1}$$

where $h(x)$ depends on the geometry of the cross-section Γ_x. This inequality easily integrates to give

$$F(x) \leq F(0)\, e^{-\int_0^x h(x)dx} . \tag{3.2}$$

Remark 3.1 - *Whereas the method as described has arisen in the context of "null" boundary conditions on the portion of the boundary complementing* Γ_0, *perhaps this is not a sine qua non.*

It is sometimes possible to adapt the broad general approach outlined, to (parabolic) problems which depend on the time t as well as on spatial variables. Suppose one has a P.D.E. or system of P.D.E. which includes a time derivative/time derivatives, together with boundary conditions of the type previously mentioned, and a "zero" initial condition. One might define a measure of the solution in V_x along the lines

$$F(x,t) = \int_0^t \int_{V_x} P(u)dV d\tau + \int_V Q(u)dV \tag{3.3}$$

where P, Q are appropriate non-negative functions (of u and/or its derivatives) with a view to obtaining a suitable differential inequality for $F(x,t)$. As before, the object is to obtain an upper bound for $F(x,t)$ in terms of x and $F(0,t)$.

3.1.3 A Neumann Problem

We now show how the method may be applied to a simple Neumann problem in a rectangle. Consider the rectangular region

$$R = 0 < x < L \quad , \quad 0 < y < h$$

where (x, y) are rectangular cartesian coordinates and L, h are constants. Suppose that $u(x, y)$ is a classical solution of

$$\nabla_1^2 u = 0 \qquad \text{in } R \tag{3.4$_1$}$$

subject to

$$u_y = 0 \qquad \text{on the edges } y = 0, h\,, \tag{3.4$_2$}$$

$$u_x = 0 \qquad \text{on the edge } x = L\,, \tag{3.4$_3$}$$

$$u_x = f(y) \qquad \text{on the edge } x = 0\,, \tag{3.4$_4$}$$

where (necessarily)

$$\int_0^h f(y)dy = 0 \tag{3.4$_5$}$$

– a consequence of (3.4_1) – (3.4_4) and the divergence theorem. It is known that u is determined to within an arbitrary additive constant. The problem has many physical interpretations e.g.

(i) an "antiplane" problem for an elastic block with rectangular cross-section where $u(x, y)$ denotes the component of displacement normal to the xy plane, induced by the following traction distribution: a suitable shear stress applied to the side $x = 0$ which is proportional to $f(y)$ and which has zero resultant, while the other three sides are free. [Further details are given in the Appendix.]

(ii) the problem of determining the steady state temperature u in a homogeneous isotropic rectangle, if the heat flux is specified on the edge and if the remaining edges are insulated.

Theorem 3.1 - *Let $R_z = R \cap x > z$, and defining the "volume measure" of the solution in R_z (see Figure 3.2) by*

$$F(z) = \int_{R_z} (\nabla_1 u)^2 dA \tag{3.5$_1$}$$

where u is the solution of the problem defined by (3.4), then

$$F(z) \le F(0)\exp(-2\pi z/h)\,. \tag{3.5$_2$}$$

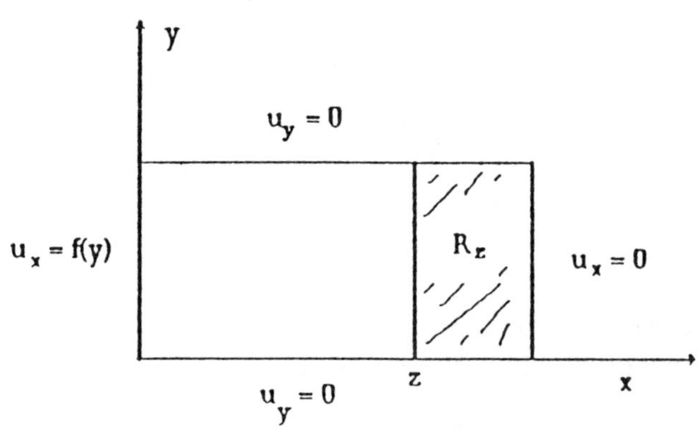

Figure 3.2

Proof: Let Γ_z denote the straight line cross-section of the rectangle with abscissa $x = z$. Using the divergence theorem applied to R_z together with (3.4) we find that

$$\int_{\Gamma_z} u_x dy = 0 \tag{3.6}$$

and

$$F(z) = -\int_{\Gamma_z} u u_x \, dy\,. \tag{3.7}$$

Furthermore it is evident that

$$F'(z) = -\int_{\Gamma_z} (\nabla_1 u)^2 dy \tag{3.8}$$

by elementary analysis. For any constant k

$$\begin{aligned} F'(z) + 2kF(z) &= -\int_{\Gamma_z} [(\nabla_1 u)^2 + 2kuu_x] dy \\ &= -\int_{\Gamma_z} [(u_x + ku)^2 + u_y^2 - k^2 u^2] dy \\ &= -\int_{\Gamma_z} (u_y^2 - k^2 u^2) dy\,. \end{aligned} \tag{3.9}$$

The solution u of the problem is determined to within an arbitrary additive constant only. The constant may be fixed by requiring that

$$\int_{\Gamma_0} u dy = 0\,. \tag{3.10}$$

In view of (3.6), (3.10), it follows that

$$\int_{\Gamma_z} u dy = 0\,. \tag{3.11}$$

Using (3.11) in the context of inequality (10) of appendix, case (b) we obtain

$$\int_{\Gamma_z} u_y^2 dy \geq \pi^2 h^{-2} \int_{\Gamma_z} u^2 dy\,. \tag{3.12}$$

This together with (3.9) gives

$$F'(z) + 2kF(z) \leq -(\pi^2 h^{-2} - k^2) \int_{\Gamma_z} u^2 dy\,. \tag{3.13}$$

Choosing $k = \pi h^{-1}$ we obtain the first order differential inequality

$$F'(z) + (2\pi/h) F(z) \leq 0\,. \tag{3.14}$$

Inequalities of this general type arose in chapter two. Integration of this inequality (for example by multiplying by $e^{2\pi z/h}$ in the first instance) gives the required result. This completes the proof.

There remains the issue of determining an upper bound for $F(0)$ in terms of data. This may be accomplished using a standard variational principle which we embody in the following theorem.

Theorem 3.2 - *Suppose that* $\mathbf{v}$ *is an arbitrary, continuously differentiable vector function defined in* $\overline{R}$ *such that*

$$\nabla_1 \cdot \mathbf{v} = 0 \qquad \text{on } R \tag{3.15$_1$}$$

together with

$$\begin{aligned} v_1 &= f \qquad \text{on } \Gamma_0, \qquad v_1 = 0 \qquad \text{on } \Gamma_L \\ v_2 &= 0 \qquad \text{on edges } y = 0, y = h. \end{aligned} \tag{3.15$_2$}$$

Then

$$\int_R \mathbf{v}^2 dA \geq \int_R (\nabla_1 u)^2 dA \equiv F(0)\,; \tag{3.16}$$

the equality sign obtains iff $\mathbf{v} = \nabla u$.

To prove this it is easily seen, using the divergence theorem, (3.4), (3.15), that

$$\int_R \mathbf{v} \cdot \nabla_1 u \, dA = \int_R (\nabla_1 u)^2 \, dA \tag{3.17}$$

and application of Schwarz's inequality thereto gives the required result (3.16). The circumstances in which the equality sign obtains are easily established.

A good choice of $\mathbf{v}$ in the foregoing, is as follows:

$$v_1 = f(y)\varsigma(x)\;,\; v_2 = -\{\int_0^y f(y)dy\}\varsigma'(x) \tag{3.18}$$

where ς is any continuously differentiable function such that

$$\varsigma(0) = 1 \quad , \quad \varsigma(L) = 0\,. \tag{3.19}$$

The optimum choice of ς may be obtained by variational methods. Thus we have shown how to make the estimate (3.5_2) fully explicit.

Remark 3.2 - *The conditions* (3.4_2)–(3.4_4) *– and* (3.4_4) *in particular – may be regarded as a perturbation of the boundary conditions of a more general problem (that of finding a harmonic function whose normal derivative has specified – generally non-zero – values on the boundary). The estimate embodied in Theorem* 3.1, *incorporating the upper bound for* $F(0)$ *in terms of "perturbations", may be regarded as a continuous dependence estimate: "small" perturbations in data on* $x = 0$, *give rise to "small" values of* $F(x)$, *for any given* x. *Moreover, it has echoes of exponential stability introduced in chapter two.*

Indeed, Saint-Venant – considering that he lived in the nineteenth century, well in advance of the development of the relevant mathematics – showed remarkable intuition in anticipating, in effect, the similarity between the two disparate phenomena, one an elastic problem and the other relating to a dynamical system; see [101], *for example.*

3.1.4 Phragmèn-Lindelöf (and other) Estimates

As a preliminary to discussing Phragmèn-Lindelöf estimates (and as a matter of independent interest), notice also the alternative, but equivalent, derivation of (3.14). Applying Schwarz's inequality to (3.7) followed by use of (3.12):

$$F(z) \le \left\{ \int_{\Gamma_z} u^2 dy \right\}^{1/2} \left\{ \int_{\Gamma_z} u_x^2 dy \right\}^{1/2}$$
$$\le \pi^{-1} h \left\{ \int_{\Gamma_z} u_y^2 dy \right\}^{1/2} \left\{ \int_{\Gamma_z} u_x^2 dy \right\}^{1/2} . \tag{3.20}$$

Using the arithmetic-geometric inequality and (3.8) leads once again to (3.14).

We now show how the volume integral method can be adapted to cater for a semi-infinite rectangular region $0 < x < \infty$, $0 < y < h$, all conditions being as previously stated except that no *a priori* condition is specified at infinity ($x \to \infty$). One may prove that, in a certain sense, the solution decays exponentially with respect to x if we merely know that it increases no faster than exponentially at infinity; this type of result is called a Phragmèn-Lindelöf result, by analogy with a theorem arising in function theory which bears these names.

In this context we *define* the measure of the solution by

$$\mathcal{F}(z) = -\int_{\Gamma_z} u u_x dy . \tag{3.21$_1$}$$

In the context we cannot assert, in general, that

$$\mathcal{F}(z) = \int_{R_z} (\nabla_1 u)^2 dA \tag{3.21$_2$}$$

as we do not know the behaviour of u at infinity *a priori*; indeed, we cannot even say that it is non-negative. However, if the context were as before ($u \to 0$ as $z \to \infty$) $\mathcal{F}(z)$ and $F(z)$ would be indistinguishable. Note that for $\delta > 0$

$$\mathcal{F}(z+\delta) - \mathcal{F}(z) = -\int_{\Gamma_{z+\delta}} u u_x dy + \int_{\Gamma_z} u u_x dy$$
$$= -\int_z^{z+\delta} \int_{\Gamma_\varsigma} (\nabla_1 u)^2 dy d\varsigma . \tag{3.22}$$

Letting $\delta \to 0$

$$\mathcal{F}'(z) = -\int_{\Gamma_z} (\nabla_1 u)^2 dy . \tag{3.23}$$

A repetition of the argument leading to (3.20), *mutatis mutandis*, gives

$$|\mathcal{F}(z)| \le \pi^{-1} h \left\{ \int_{\Gamma_z} u_y^2 dy \right\}^{1/2} \left\{ \int_{\Gamma_z} u_x^2 dy \right\}^{1/2} . \tag{3.24}$$

Use of the arithmetic – geometric inequality together with (3.23) yields (the analogue of (3.14)):

$$-\mathcal{F}'(z) \ge 2\pi h^{-1} |\mathcal{F}(z)| . \tag{3.25}$$

The latter implies the inequalities

$$\mathcal{F}'(z) \leq -2\pi h^{-1}\mathcal{F}(z) \tag{3.26$_1$}$$

and

$$-\mathcal{F}'(z) \geq 2\pi h^{-1}[-\mathcal{F}(z)]\,. \tag{3.26$_2$}$$

Suppose that there exists $z_1 : 0 \leq z_1 < \infty$ such that $-\mathcal{F}(z_1) > 0$. Then (3.26_2) implies $-\mathcal{F}(z) > 0$ for all $z > z_1$; indeed, (3.26_2) implies, for $z > z_1$, that

$$-\mathcal{F}(z)\exp\{-2\pi h^{-1}(z - z_1)\} \geq -\mathcal{F}(z_1) \tag{3.27}$$

(i.e. $\mathcal{F}(z)$ increases asymptotically at least exponentially fast). Let us assume the asymptotic condition

$$\lim_{z\to\infty} -\mathcal{F}(z)\exp\{-2\pi h^{-1}z\} = 0\,. \tag{3.28}$$

In view of (3.28), the assumption that there exists a point $z_1 : -\mathcal{F}(z_1) > 0$ is contradicted. In these circumstances, $\mathcal{F}(z) \geq 0$ and integration of (3.26_1) leads to

$$\mathcal{F}(z) \leq \mathcal{F}(0)\exp(-2\pi h^{-1}z)\,. \tag{3.29}$$

But as this implies that

$$\int_{\Gamma_z} uu_x\,dy \to 0$$

as $z \to \infty$, it follows that the representation (3.21_2) also holds in these circumstances.

The asymptotic condition (3.28) can be expressed in alternative form. By l'Hospital's rule it may be written (ratio of limit of two derivatives) as

$$\lim_{z\to\infty} -\mathcal{F}'(z)\exp(-2\pi h^{-1}z) = 0\,. \tag{3.30}$$

In view of (3.23) this is equivalent to

$$\lim_{z\to\infty}\int_{\Gamma_z}(\nabla_1 u)^2 dy\exp(-2\pi h^{-1}z) = 0\,. \tag{3.31}$$

If this asymptotic condition holds, the decay law (3.29) holds also. We may summarize our result thus:

Theorem 3.3 (of Phragmèn-Lindelöf type) - *Suppose that u is a classical solution of* (3.4_1)–(3.4_2), (3.4_2)–(3.4_5) *in the semi-infinite rectangle* $0 < x < \infty, 0 < y < h$, *that it satisfies the "normalization" condition* (3.10), *and the asymptotic condition*

$$\lim_{z\to\infty}\int_{\Gamma_z}(\nabla_1 u)^2 dy\cdot\exp(-2\pi h^{-1}z) = 0\,. \tag{3.32$_1$}$$

then the quantity $\mathcal{F}(z)$ *defined by* (3.21_1) *or* (3.22_2), *satisfies the decay law*

$$\mathcal{F}(z) \leq \mathcal{F}(0)\exp(-2\pi h^{-1}z)\,. \tag{3.32$_2$}$$

An upper bound for $\mathcal{F}(0)$, in terms of data, may be obtained in much the same way as before.

3.1.5 Regions with Curved Boundaries

The latter general approach to rectangular regions, finite or semi-infinite, is applicable to other geometries also. Consider the region defined by

$$r_0 \leq r \leq r_1 \quad , \quad 0 \leq \theta \leq \alpha$$

where (r, θ) are plane polar coordinates and r_1, r_0, and α are positive constants. Let (See Figure 3.3) Γ_r denote the cross-section of the region with radial coordinate r, and R_r the portion of the region with radial coordinate not less than r.

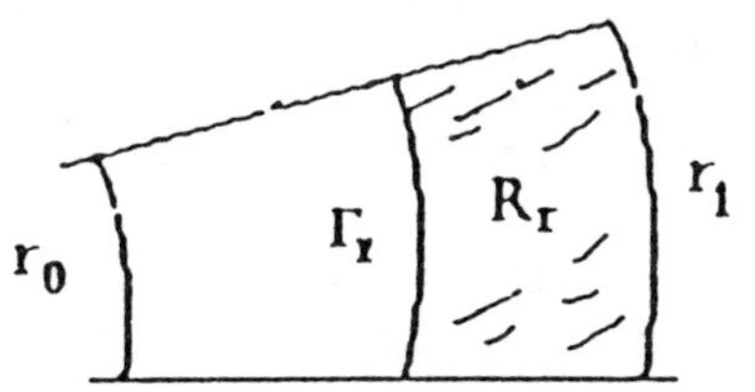

Figure 3.3

Suppose that $u(r, \theta)$ is a classical solution of

$$\begin{cases} u_{rr} + r^{-1}u_r + r^2 u_{\theta\theta} = 0\,, \\ u = 0 \quad \text{on edges} \quad r = r_1,\ \theta = 0,\ \theta = \alpha\,, \\ u \qquad \text{specified on} \quad r = r_0\,. \end{cases} \tag{3.33}$$

We define

$$\mathcal{F}(r) = -\int_{\Gamma_r} u u_r r d\theta \tag{3.34}$$

Use of the divergence theorem together with (3.33) gives the alternative expression

$$\mathcal{F} = \int\int_{R_r} (u_r^2 + r^{-2}u_\theta^2) r dr d\theta\,. \tag{3.35}$$

Furthermore, as before, it is evident that

$$\mathcal{F}'(r) = -\int_{\Gamma_r} (u_r^2 + r^{-2}u_\theta^2) r d\theta\,. \tag{3.36}$$

Applying to (3.34), Schwarz's inequality, the inequality ((10) of appendix, case (a))

$$\int_{\Gamma_r} u_\theta^2 d\theta \geq (\pi/\alpha)^2 \int_{\Gamma_r} u^2 d\theta\,, \tag{3.37}$$

followed by the arithmetic-geometric inequality together with (3.36), gives:

$$\begin{aligned}|\mathcal{F}(r)| &\leq \left\{\int_{\Gamma_r} u^2 r d\theta\right\}^{1/2}\left\{\int_{\Gamma_r} u_r^2 r d\theta\right\}^{1/2}\\ &\leq (\alpha/\pi) r\left\{\int_{\Gamma_r} r^{-2} u_\theta^2 r d\theta\right\}^{1/2}\left\{\int_{\Gamma_r} u_r^2 r d\theta\right\}^{1/2}\\ &\leq \frac{1}{2}(\alpha/\pi) r\int_{\Gamma_r}(u_r^2 + r^{-2}u_\theta^2) r d\theta\\ &= -\frac{1}{2}(\alpha/\pi) r\mathcal{F}'(r)\end{aligned}$$

or

$$\mathcal{F}'(r) + 2(\pi/\alpha) r^{-1}|\mathcal{F}(r)| \leq 0\,. \tag{3.38}$$

Since $\mathcal{F}(r) \geq 0$ this gives on integration

$$\{\mathcal{F}(r) r^{2\pi/\alpha}\}' \geq 0$$

leading to

$$\mathcal{F}(r) \leq \mathcal{F}(r_0)(r_0/r)^{2\pi/\alpha}\,. \tag{3.39}$$

Suppose now that the region extends to infinity ($r_1 \to \infty$) and that no ***a priori*** condition is prescribed thereat, but that all other conditions are as before. We now prove a Phragmèn-Lindelöf type result in this context. The quantity $\mathcal{F}(r)$ is once again defined by (3.34) but (3.35) cannot be concluded, in general, under the present conditions. However, as before, (3.36) and (3.38) continue to be valid. Suppose now that there exists $r_2 > 0 : 0 \leq \tau_2 < \infty$ such that $\mathcal{F}(\tau_2) < 0$. Then

$$\{-\mathcal{F}(r)\}' \geq 2(\pi/\alpha) r^{-1}\{-\mathcal{F}(r)\}\,, \tag{3.40}$$

implied by (3.38), implies that

$$\frac{-\mathcal{F}(r)}{(r/r_2)^{2\pi/\alpha}} \geq -\mathcal{F}(r_1)\,. \tag{3.41}$$

The aforementioned supposition is evidently invalid if

$$\lim_{r\to\infty} -\frac{\mathcal{F}(r)}{r^{2\pi/\alpha}} = 0\,. \tag{3.42}$$

By l'Hopital's rule, the latter is equivalent to

$$\lim_{r\to\infty} -\frac{\mathcal{F}'(r)}{r^{(2\pi/\alpha)-1}} = 0 \tag{3.43}$$

and, in view of (3.36), this is equivalent to

$$\lim_{r\to\infty} r^{-2(\pi/\alpha)+2}\int_{\Gamma_r}(u_r^2 + r^{-2}u_\theta^2)\,d\theta = 0\,. \tag{3.44}$$

Thus if the latter asymptotic condition holds, $\mathcal{F}(r) \geq 0$ and the same decay law (3.39) holds, as before. Moreover, the representation (3.35) is valid in these circumstances. We summarize the results obtained – both for finite and infinite regions – as follows:

Theorem 3.4 - *The problem* (3.33) – *with* r_1 *finite – is such that the (non-negative) measure of the solution, given by* (3.35) *satisfies the decay law*

$$\mathcal{F}(r) \leq \mathcal{F}(r_0)(r_0/r)^{2\pi/\alpha} . \tag{3.45$_1$}$$

The same decay law holds when $r_1 \to \infty$ *provided that (instead of* $u \to 0$ *as* $r \to \infty$ *) the asymptotic condition*

$$\lim_{r \to \infty} r^{-2(\pi/\alpha)+2} \int_{\Gamma_r} (u_r^2 + r^{-2} u_\theta^2) d\theta = 0 \tag{3.45$_2$}$$

holds.

3.1.6 An Initial Boundary - Value Problem

We now proceed to consider a problem which depends on the time variable as well as on spatial variables. Specifically, we consider an initial boundary value problem for diffusion in a right cylinder. Let R denote the interior of the cylinder, whose cross-section is bounded by a piecewise simple closed curve. Choose cartesian coordinates x_1, x_2, x_3 with the origin at the end of the cylinder, $x_3 = L$ at the other end, and the x_3 axis parallel to the generators. Let Γ_z denote the open cross-section of R for which $x_3 = z$, $z \geq 0$, with corresponding curvilinear boundary denoted by $\partial\Gamma_z$. Suppose that R is occupied by a homogeneous isotropic heat conductor of constant thermal diffusivity, and the lateral surface $\mathcal{L}$ of R is maintained at zero temperature. Let us assume, accordingly, that the temperature $u(\mathbf{x},t) = u(x_1, x_2, x_3, t)$ is a classical solution of

$$\nabla^2 u - u_t = 0 \qquad \text{in } R \times (0, \infty) \tag{3.46$_1$}$$

subject to the boundary conditions

$$u = f \quad \text{on } \Gamma_0 , \qquad u = 0 \quad \text{on } \Gamma_L , \tag{3.46$_2$}$$

(where f is a prescribed function)

$$u = 0 \quad \text{on } \mathcal{L} , \tag{3.46$_3$}$$

and to the initial condition

$$u(\mathbf{x}, 0) = 0 , \quad x \in R . \tag{3.46$_4$}$$

We now proceed to show, following Knowles [106], how the volume integral method can be extended to cater for this problem.

Theorem 3.5 - *Letting R_z denote the portion of the right cylinder R for which $x_3 > z$, the "volume integral measure" (in R_z) of the solution to the initial-boundary value problem* (3.46), *given by*

$$E(z,t) = \int_0^t \int_{R_z} (\nabla u)^2 dV d\tau + \frac{1}{2}\int_{R_z} u^2 dV\,, \qquad z \geq 0\,, t \geq 0\,, \tag{3.47}$$

satisfies the spatial decay estimate

$$E(z,t) \leq E(0,t)\exp\{-2(\lambda_1^{1/2})z\} \tag{3.48}$$

where λ_1 is the lowest (fixed membrane) eigenvalue of the cross-section i.e. of

$$\begin{cases} \nabla_1^2\phi + \lambda\phi = 0 & \text{on } \Gamma_z\,, \\ \phi = 0 & \text{on } \partial\Gamma_z\,. \end{cases} \tag{3.49}$$

We now proceed to prove this; however, we shall first derive a more general inequality which holds in the absence of the zero initial condition (3.46_4). Notice that the divergence theorem, the differential equation (3.46_1) and the boundary conditions (3.46_2)–(3.46_3) imply that

$$E(z,t) = -\int_0^t d\tau \int_{\Gamma_z} uu_{,3} dA + \frac{1}{2}\int_{R_z} u^2(\mathbf{x},0)dV\,. \tag{3.50}$$

Next, differentiation of (3.47) with respect to z yields

$$E_z(z,t) = -\int_0^t d\tau \int_{\Gamma_z} (\nabla u)^2 dA - \frac{1}{2}\int_{\Gamma_z} u^2 dA\,. \tag{3.51}$$

For any constant k, (3.50) and (3.51) imply that

$$E_z + 2kE = -\int_0^t d\tau \int_{\Gamma_z} \{(\nabla u)^2 + 2kuu_{,3}\}dA - \frac{1}{2}\int_{\Gamma_z} u^2 dA + k\int_{R_z} u^2(\mathbf{x},0)dV\,. \tag{3.52}$$

Now

$$\int_{\Gamma_z} \{(\nabla u)^2 + 2kuu_{,3}\}dA = \int_{\Gamma_z} (u_{,3} + ku)^2 dA + \int_{\Gamma_z} \{(\nabla_1 u)^2 - k^2u^2\}dA\,. \tag{3.53}$$

In view of inequality (15) of appendix, case (a) we have

$$\int_{\Gamma_z} (\nabla_1 u)^2 dA \geq \lambda_1 \int_{\Gamma_z} u^2 dA\,. \tag{3.54}$$

Writing $k = \sqrt{\lambda_1}$ in (3.53) (and henceforward) it is seen that the left hand side of (3.53) is non-negative. Thus (3.52) gives

$$E_z + 2kE \leq k\int_{R_z} u^2(\mathbf{x},0)dV\,. \tag{3.55}$$

This is easily integrated: it is left as an exercise to the reader to prove that

$$E(z,t) \le E(0,t)e^{-2kz} + k\int_0^z d\varsigma e^{2k(\varsigma - z)} \int_{R_\varsigma} u^2(x,0)dV \,. \tag{3.56}$$

Integration by parts gives

$$\begin{aligned} k\int_0^z d\varsigma e^{2k(\varsigma-z)} \int_{R_\varsigma} u^2(\mathbf{x},0)dV &= \\ = \frac{1}{2}\int_{R_z} u^2(\mathbf{x},0)dV - \frac{1}{2}e^{-2kz}\int_R u^2(\mathbf{x},0)dV + \frac{1}{2}\int_0^z d\varsigma e^{2k(\varsigma-z)}\int_{\Gamma_\varsigma} u^2(\mathbf{x},0)dA \\ \le \frac{1}{2}\int_{R_z} u^2(\mathbf{x},0)dV + \frac{1}{2}\int_0^z d\varsigma \int_{\Gamma_\varsigma} u^2(\mathbf{x},0)dA = \frac{1}{2}\int_R u^2(\mathbf{x},0)dV \,. \end{aligned} \tag{3.57}$$

Substitution of this in (3.56) yields

$$E(z,t) \le E(0,t)e^{-2kz} + \frac{1}{2}\int_R u^2(\mathbf{x},0)dV \,,$$

and the decay estimate (3.48) follows from this on noting (3.46_4).

3.2 MISCELLANEA

3.2.1 Introduction

First order differential inequalities arise naturally in many situations because of their simplicity. In addition to Liapunov techniques (of chapter two) and the volume integral method already considered, we consider here a miscellany of techniques. The choice of topics is dictated either by their novelty, their intrinsic interest, and/or their relevance to other topics covered in the book.

The topics dealt with can be divided into three categories:

(a) A lemma is proved which proves useful in deriving first order inequalities particularly for some diffusion type equations. It resembles another lemma – derived in a later chapter – which proves useful in deriving second order inequalities, and which is much used in the book. The lemma proved in this chapter is first used to prove a non-existence result for an initial boundary value problem for a nonlinear P.D.E.

(b) A number of aspects of one-dimensional linear diffusion are considered which use the lemma. Firstly, some simple monotonic (maximum type) principles are obtained for the spatial derivative/heat flux in one dimension. Secondly, an estimate is derived for the temperature in a one-dimensional deforming continuum whose velocity field is assumed known and which has moving boundaries.

(c) Some non-existence (and some other) results are obtained for some nonlinear diffusion equations, including a simple estimate for the time beyond which the solution to an initial boundary value problem for the backwards (in time) "porous" medium equation cannot exist.

3.2.2 A Simple Lemma

Consider a region R (See Figure 3.4) in the $t-y$ plane bounded

(i) by a straight line segment Γ_0 with abscissa $t=0$,

(ii) by two curves C_+, C_- defined respectively by

$$y = y_+(t)\,,\; y = y_-(t)$$

which are

(α) single-valued, continuously differentiable functions,

(β) such that $y_+(t) > y_-(t)$.

Let Γ_t denote the straight line segment contained in the region with abscissa t.

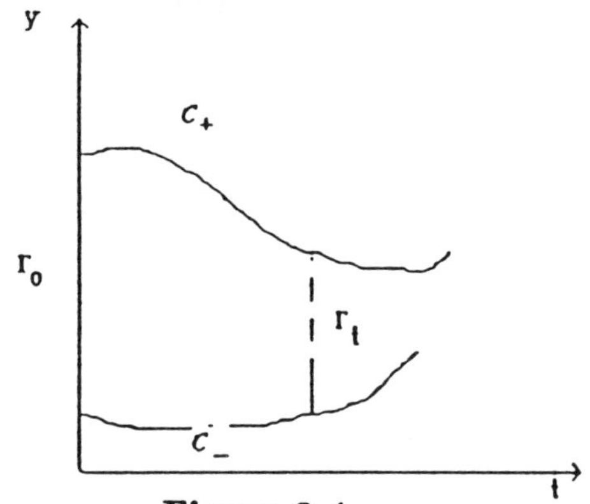

Figure 3.4

Let $\psi(t,y)$ be a twice continuously differentiable function defined in the region which take values $\mathcal{F}_+(t)$, $\mathcal{F}_-(t)$ on C_+, C_- respectively. Let $\phi(u)$ be a twice continuously differentiable function of its argument u. In what follows, a superposed dot or dots shall denote derivatives of ϕ with respect to its argument u, while a superposed prime shall denote differentiation with respect to t of functions thereof.

Lemma 3.1 - *In the context and notation outlined in the last two paragraphs, the quantity*

$$F(t) = \int_{\Gamma_t} \phi(\psi_y)dy \tag{3.58$_1$}$$

satisfies

$$F'(t) = -\int_{\Gamma_t} \ddot{\phi}\,\psi_t\psi_{yy}dy + \{\dot{\phi}\mathcal{F}' - (\dot{\phi}\psi_y - \phi)y'\}] \tag{3.58$_2$}$$

where the notation] means that the relevant quantity is evaluated at C_+ and from this is subtracted its value on C_-.

Proof. Leibnitz's Theorem states that, for continuously differentiable $f(t,y)$,

$$\frac{d}{dt}\int_{\Gamma_t} f(t,y)dy = \int_{\Gamma_t} f_t dy + fy'] \,.$$

Consequently

$$F'(t) = \int_{\Gamma_t} \dot{\phi}\psi_{yt} dy + \phi(\psi_y)y'] \,, \tag{3.59}$$

and on integration by parts we obtain

$$F'(t) = -\int_{\Gamma_t} \ddot{\phi}\psi_t\psi_{yy} dy + (\phi y' + \dot{\phi}\psi_t)] \,. \tag{3.60}$$

We now differentiate the boundary condition

$$\psi(t, y(t)) = \mathcal{F}(t)$$

(with $+,-$ appended as appropriate) with respect to t:

$$\psi_t + \psi_y y' = \mathcal{F}' \,. \tag{3.61}$$

Substitution of (3.61) in (3.60) completes the proof of the Lemma.

Interesting conclusions can be drawn from the Lemma if ϕ is chosen judiciously. In this connection, note that if $\ddot{\phi}(u) \geq 0$ (i.e. ϕ is a convex function of its argument) and if $\phi(0) = 0$, it is easily verified that

$$u\dot{\phi} - \phi \geq 0 \,; \tag{3.62}$$

the proof is left as an exercise to the reader [cf. (4.2), if necessary]. It will be noted that this combination occurs in (3.58_2).

The first application of the Lemma is to a simple non-existence theorem for an initial boundary value problem.

Theorem 3.6 - *Suppose $\phi(u)$ is a given non-negative, strictly convex function of its argument ($\ddot{\phi} > 0$ strictly) such that $\phi(0) = 0$; and suppose that $f(t,y)$ is a given function. Let $\psi(t,y)$ be a twice continuously differentiable solution of the initial boundary value problem consisting of*

$$\psi_t\psi_{yy} = \{\ddot{\phi}(\psi_y)\}^{-1} f(t,y) \tag{3.63$_1$}$$

in the region R, the initial condition

$$\psi (\text{or} \psi_y) \quad \text{specified on } \Gamma_0 \,, \tag{3.63$_2$}$$

together with the boundary conditions

$$\psi = \text{constant on } C_+ \,, C_- \,, \tag{3.63$_3$}$$

where these lateral boundaries are assumed to satisfy

$$y'_+(t) \geq 0 \quad , \quad y'_-(t) \leq 0 \,. \tag{3.63$_4$}$$

The solution fails to exist for values of t and for data such that

$$F(0) - \int_0^t \int_{\Gamma_t} f(t,y)dtdy < 0 \,. \tag{3.63$_5$}$$

The proposition follows immediately from $(3.58_2),(3.62)–(3.63)$ and the non-negativity of $F(t)$. It is possible to generalize this result to include non-constant boundary conditions on the lateral boundaries.

3.2.3 Linear diffusion equations

Some Monotonicity (Maximum) Principles. We now make use of Lemma 3.1 to derive some properties of the one-dimensional diffusion (heat-conduction) equation

$$\psi_t = \psi_{yy} \,.$$

Specifically, simple monotonicity principles (or maximum type principles) are derived for spatial measures (norms) of the temperature gradient for a spatial region which may vary with time; the principles do not appear to be well known. The aforementioned measures are reminiscent of Liapunov functions (cf. chapter two). The derivation of the results is somewhat analogous to the derivations by Bellman [107] of similar results for the temperature itself in a fixed region; see also [108] and relevant references therein. Apart from the intrinsic interest of the principles derived, they are similar in spirit to, and give a foretaste of, results for the harmonic equation and the Monge-Ampère equation which are derived in Part 2.

Theorem 3.7 - *Suppose (in the context of the preceding Lemma) that $\psi(t,y)$ is a classical solution of*

$$\psi_t = \psi_{yy} \tag{3.64$_1$}$$

in the region R, which satisfies the boundary conditions

$$\psi = \text{constant on} \quad y = y_+(t) \,, \, y = y_-(t) \tag{3.64$_2$}$$

(not necessarily the same constant). If the lateral boundaries are assumed to satisfy

$$y'_+(t) \geq 0 \quad , \quad y'_-(t) \leq 0 \,, \tag{3.64$_3$}$$

then both

$$F(t) = \int_{\Gamma_t} \psi_y^{2n}\, dy \qquad (n \text{ any positive integer})$$

and

$$\max_y |\psi_y(y,t)|$$

are monotonically decreasing functions of t.

Proof. We choose $\phi(u) = u^{2n}$ (n a positive integer) in connection with the Lemma. In view of $(3.64_1), (3.64_2)$, the Lemma gives

$$F'(t) = -2n(2n-1) \int_{\Gamma_t} \psi_y^{2n-2} \psi_{yy}^2 dy - (2n-1)\psi_y^{2n} y']\,. \tag{3.65}$$

In view of (3.64_3) the first part of the preceding proposition is thus established. The second part follows from this since (e.g. [108])

$$\lim_{n\to\infty} \{ \int_{\Gamma_t} f^{2n}(y)dy\}^{1/2n} = \max_y |f(y)|$$

where $f(y)$ is a continuous function, and where the integration is over a finite interval.

The preceding proposition resembles the well known one got by similar methods: $\max\limits_y |\psi(y,t)|$ is a monotonically decreasing function of t where ψ is a sufficiently smooth solution in $0 \le x \le 1$, $t \ge 0$ of the diffusion equation, which takes zero values on the boundaries $x = 0\,, x = 1$, e.g. [107,108]. We now follow with a somewhat more general proposition:

Theorem 3.8 - *Suppose (in the context of Lemma 3.1) that $\psi(t,y)$ is a classical solution of (3.64_1) subject to the (altered) boundary conditions*

$$\psi = \mathcal{F}_+(t) \quad \text{on } y = y_+(t) \quad , \quad \psi = \mathcal{F}_-(t) \quad \text{on } y = y_-(t)\,, \tag{3.66}$$

where the boundary restrictions (3.64_3) again obtain. Suppose that $\phi(u)$ is a non-negative (twice continuously differentiable) convex function of its argument such that $\phi(0) = 0$.
Then

(*i*)
$$F(t) = \int_{\Gamma_t} \phi(\psi_y)dy \tag{3.67_1}$$

satisfies

$$F'(t) \le \dot{\phi}\mathcal{F}']\,. \tag{3.67_2}$$

(ii) *Defining, for any (real) u*

$$_+u = \begin{cases} u & \text{if } u \geq 0\,, \\ 0 & \text{if } u < 0\,, \end{cases}$$

the quantities

$$\int_{\Gamma_t} ({}_+\psi_y)^n dy \quad (n \text{ any integer } > 2)$$

and

$$\max_y {}_+\psi_y(y,t)$$

are monotonically decreasing functions of t, provided

$$\mathcal{F}'_+(t) \leq 0 \quad , \quad \mathcal{F}'_-(t) \geq 0\,. \tag{3.68}$$

The inequality (3.67_2) follows from (3.58_2), (3.62) and the conditions satisfied by ψ. Part (ii) of the Theorem is seen to be a consequence of (3.67_2) as follows: Choose in connection therewith

$$\phi(u) = ({}_+u)^n$$

where n is any positive integer (>2). It is readily verified that this choice of ϕ has all the properties required, and that

$$\dot{\phi}(\psi_y) \geq 0\,.$$

The first part of (ii) thus follows from (3.67_2), (3.68), while the second part follows from this on using a limiting technique similar to that employed in connection with Theorem 3.7.

Remark 3.8 - *Whereas results such as those contained in Theorems* 3.7 *and* 3.8 – *in particular the last results contained therein – can be obtained via maximum principles [e.g.* (131)], *our objective in this section – and indeed in this book – is to explore properties of P.D.E.s by reference to the variation of integral measures.*

This completes the discussion of monotonicity principles.

Diffusion in a deforming medium. We now consider heat conduction (diffusion) in a deforming, one dimensional continuum with moving boundaries. An integral estimate is derived which uses Lemma 3.1.

We consider a deforming, one-dimensional heat conducting medium whose boundaries are defined by $y = y_-(t), y = y_+(t)$ for times $t \geq 0$; these functions are as specified in the Lemma. On the basis of a simple model which envisages, inter alia, that heat conduction dominates thermomechanical effects, the temperature distribution $\psi(y,t)$ is assumed to satisfy a P.D.E. of the type

$$\psi_t + v\psi_y = \psi_{yy} \qquad \text{in } y_- < y < y_+ \tag{3.69}$$

where $v(y,t)$ is the velocity field which we assume known. We shall suppose that (3.69) is subject to the boundary conditions

$$\psi = 0 \qquad \text{on } y = y_-(t)\,,\, y = y_+(t) \tag{3.70$_1$}$$

and to the initial condition

$$\psi = \text{specified}\,. \tag{3.70$_2$}$$

We now proceed to obtain integral estimates for ψ. What seems natural and analytically convenient informs our approach – rather than some notionally optimal approach which is analytically inconvenient. We seek an upper estimate for the measure

$$F(t) = \int \psi_y^2 dy\,. \tag{3.71}$$

The Lemma (with $\phi(u) = u^2$) together with (3.69) and integration by parts gives

$$F'(t) = -\int [2\psi_{yy}^2 + v_y \psi_y^2] dy\,. \tag{3.72}$$

In view of inequality (13) of appendix and the fact that $\int \psi_y dy = 0$, we have

$$\int \psi_{yy}^2 dy \geq \pi^2 l^{-2}(t) \int \psi_y^2 dy \tag{3.73}$$

where $l(t) = y_+(t) - y_-(t)$ is the width of the domain at time t. One derives from (3.72), (3.73) the inequality

$$F'(t) \leq -\left\{2\pi^2 l^{-2}(t) + \min_y v_y\right\} F(t)\,. \tag{3.74}$$

Integration leads to

Theorem 3.9 - *The integral measure* (3.71) *of the temperature distribution in a deforming, one-dimensional heat conducting medium, defined by* (3.69)–(3.70) *satisfies the estimate*

$$F(t) \leq F(0) \exp\left[-\int_0^t \{2\pi^2 l^{-2}(t) + \min_y v_y\} dt\right]\,. \tag{3.75}$$

Uniqueness follows from this estimate. We remark that pointwise bounds for $|\psi|$, which reflect position, follow by standard means (i.e. via Schwarz's inequality). Let us also remark that when $v_y \geq 0$, one may use variational methods to obtain an upper bound for the right hand side of (3.72) in terms of F.

3.2.4 Nonlinear Diffusion Equations: Nonexistence, etc.

Fairly simple arguments involving first order differential inequalities sometimes suffice in order to establish non-existence of solution.

The first example considered concerns the "porous medium" equation (one-dimensional) backwards in time. We consider solutions $u(x,t)$ – which may be considered non-negative – of the one dimensional "porous medium" equation, backwards in time:

$$u_t + (u^{2m} u_x)_x = 0 \quad , \qquad 0 < x < 1 \, , \tag{3.76$_1$}$$

wherein $m > 0$, subject to the boundary conditions

$$u(0,t) = u(1,t) = 0 \tag{3.76$_2$}$$

and to the initial condition

$$u(x,0) = \text{specified } (non - negative) \, . \tag{3.76$_3$}$$

We now consider the class of functions in which the solution (apart from being positive) must lie. Whereas the analysis given below is valid if one assumes the solution to be a classical one, it is unnecessary to so restrict it. Let us note in this connection that – for the corresponding, forward in time equation, at least – there are important solutions of the P.D.E. which have compact support. At the extremities of the support, which vary with time in general, all the relevant derivatives do not exist/are singular – a phenomenon associated with degeneracy in parabolicity ($u = 0$) of the P.D.E. In defining the concept of the solution, we shall ensure that it is sufficiently broad to accomodate singularities of this general type. The specific criteria adopted are dictated by two considerations: (a) the validity of the analysis carried out below, (b) the known, explicit solution for an instantaneous point source (see exercise 3.12).

Acceptable solutions are characterised thus: The solutions (in addition to satisfying $(3.76_2), (3.76_3)$) satisfy the P.D.E. everywhere in a classical sense, except possibly, at a finite number of points $x = x_i(t)$ $(1 \leq i \leq n,$ n being a positive integer) – which vary with time, in general: at each of these exceptional points, $u = 0$, u is a continuous function thereat, and the relevant (singular) derivatives are such that

$$\frac{\partial^\nu u}{\partial x^\nu} = o[|x - x_i|^{\{2(m+1)\}^{-1} - \nu}]$$

as $x \to x_i$, where $0 \leq \nu \leq 2$.

For comprehensive surveys of the porous medium equation, including existence of weak solutions and their smoothness – at least for the forward in time equation – and for surveys of related equations, the reader is referred to [109,110]. Moreover, the applicability of such equations to biology/population dynamics is discussed in [51], for example.

We now consider the ***non-existence*** question. Consider the measure of the solution (L^2 norm) of (3.76)

$$F(t) = \int_0^1 u^2 dx\,. \tag{3.77}$$

Differentiation and use of (3.76_1) yields

$$F'(t) = -2\int_0^1 u(u^{2m}u_x)_x dx\,. \tag{3.78}$$

Integration by parts using (3.76_2) etc. gives

$$F'(t) = 2\int_0^1 u^{2m}u_x^2 dx = \frac{2}{(m+1)^2}\int_0^1 \left(\frac{du^{m+1}}{dx}\right)^2 dx\,. \tag{3.79}$$

Application of inequality (10) of appendix, case (a) yields

$$F'(t) \geq \frac{2\pi^2}{(m+1)^2}\int_0^1 u^{2m+2} dx\,. \tag{3.80}$$

By Hölder's inequality we obtain therefrom

$$F'(t) \geq 2\pi^2(m+1)^{-2}F^{m+1}\,. \tag{3.81}$$

Integration of this inequality yields

$$F^{-m}(0) - 2\pi^2 m(m+1)^{-2}t \geq F^{-m}(t) \tag{3.82}$$

– which leads to

Theorem 3.10 - *The solution of the initial boundary value problem* (3.76) *does not exist for times* t *such that*

$$t \geq (2\pi^2 m)^{-1}(m+1)^2 F^{-m}(0) \tag{3.83}$$

where $F(0)$ *is determined by* (3.76_3) *and* (3.77).

Indeed the solution must "blow up" at some time less than or equal to the aforesaid critical time, or must cease to exist for some other reason. Let us note that a somewhat different estimate involving second order differential inequality techniques is available [Levine and Payne [111]]. Moreover, the reader is referred to the book by Straughan [35] for other analyses of the porous medium and of other equations; these include non-existence results for flows, backward in time, of fluids satisfying constitutive equations suggested by Ladyzhenskaya.

Finally let us note that, assuming existence up to the aforementioned critical time, (3.82) can be rewritten in the form

$$F^m(t) \geq \frac{F^m(0)}{1 - 2\pi^2 m(m+1)^{-2} F^m(0)t}. \tag{3.84}$$

To complete the study it is natural to ask what happens if a parallel analysis is applied to the forward in time equation

$$u_t - (u^{2m} u_x)_x = 0\,, \tag{3.85}$$

all other conditions being unaltered. One may readily establish, in almost identical fashion, that $F(t)$ – defined as in (3.77) – satisfies

$$F(t) \leq F(0)\left\{1 + 2\pi^2 m(m+1)^{-2} F^m(0)t\right\}^{-1/m}\;; \tag{3.86}$$

the proof is left as an exercise to the reader. On letting $m \to 0$ one obtains the decay law appropriate to linear diffusion

$$F(t) \leq F(0)\exp\left(-2\pi^2 t\right). \tag{3.87}$$

Nonexistence can be proved in a similar fashion for diffusion type equations backward in time other than (3.76_1). Consider

$$u_t + u_{xx} = u^{2m+1}\quad, \qquad 0 < x < 1\,, \tag{3.88$_1$}$$

(m being a positive integer) subject to the boundary conditions

$$u(0,t) = u(1,t) = 0 \tag{3.88$_2$}$$

and to the initial condition

$$u(x,0) = \text{specified} \tag{3.88$_3$}$$

i.e. a standard diffusion equation with a sink, backwards in time.

Again defining a measure of the solution by (3.77), one readily establishes that

$$F'(t) \geq 2\pi^2 F + 2F^{m+1}\,. \tag{3.89}$$

This is readily integrated (on ignoring the first term on the right) to give

$$F^{-m}(0) - 2mt \geq F^{-m}(t) \tag{3.90}$$

– which leads to

Theorem 3.11 - *The solution of the initial boundary value problem given by* (3.88), *cannot exist for times* t *such that*

$$t \geq (2m)^{-1} F^{-m}(0) \tag{3.91}$$

where $F(0)$ *is determined by* (3.77) *and* (3.88_3).

Note that if the first term on the right in (3.89) is retained, a better estimate for the critical time is obtainable on integration; there may be difficulty, however, in evaluating the relevant integral explicity – except in cases such as $m = 1$.

3.3 EXERCISES

Exercise 3.1 - Consider a region R bounded by a simple closed curve C, one portion of which consists of a straight line Γ_0 (See Figure 3.5). Suppose $u(x, y)$ is a classical solution of

$$\nabla_1^2 u = 0 \qquad \text{in } R$$

subject to the boundary conditions:

(normal derivative) $$\frac{\partial u}{\partial n} = 0 \qquad \text{on } C/\Gamma_0 \,,$$

$$u_x = f(y) \qquad \text{on } \Gamma_0$$

where (necessarily)

$$\int_{\Gamma_0} f(y)dy = 0 \,.$$

Defining $R_z = R \cap x > z$ and

$$F(z) = \int_{R_z} (\nabla_1 u)^2 dA \,,$$

prove that

$$F(z) \leq F(0)\exp\left[-2\pi \int_0^z h^{-1}(x)dx\right],$$

where $h(x)$ is the width of the region (in the y direction) at abscissa x. Discuss how a suitable upper bound is obtained for $F(0)$.

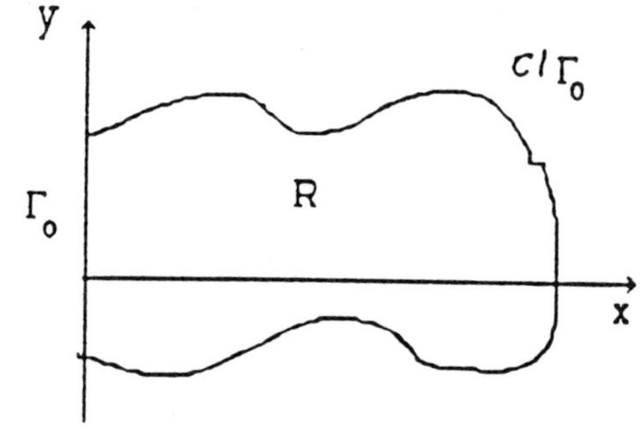

Figure 3.5

Exercise 3.2 - Prove a result analogous to the foregoing if the boundary conditions are

$$\begin{aligned} u &= 0 & \text{on } C/\Gamma_0 \,, \\ u &= f(y) & \text{on } \Gamma_0 \end{aligned}$$

with no integral restriction on $f(y)$. How is a suitable upper bound obtained for $F(0)$ in this case?

Exercise 3.3 - Prove analogous results in three dimensions: consider problems of the type dealt with in Question 1 and in Question 2 above; consider right cylinders in the first instance.

Hint. One requires the variational characterizations for ν_1 (free membrane eigenvalue) and λ_1 (fixed membrane eigenvalue) respectively for these problems.

Exercise 3.4 - Derive analogues of Theorems (3.1) and (3.2) where Laplace's equation (3.41) is replaced by the P.D.E.

$$k(y)u_{xx} + \{k(y)u_y\}_y = 0$$

where $k(y)$ is a given, sufficiently smooth, positive function, and the boundary conditions are

$$\begin{aligned} u_y &= 0 & \text{on the edges } y = 0, h \,, \\ u_x &= 0 & \text{on the edge } x = L \,, \\ k(y)u_x &= f(y) & \text{on the edge } x = 0 \,, \end{aligned}$$

where

$$\int_0^h f(y)dy = 0 \,.$$

Note. For the analogue of Theorem 3.1 one requires the variational characterization of the lowest (non-zero) eigenvalue of an eigenvalue problem of the type

$$\frac{d}{dy}\{k(y)\frac{d\phi}{dy}\} + \lambda k(y)\phi = 0 \qquad \text{in } y \in (0, h)$$

subject to

$$\frac{d\phi}{dy} = 0 \qquad \text{on } y = 0, h \,.$$

Exercise 3.5 - Prove in the notation associated with Theorem 3.1

$$\frac{d}{dz}\int_{R_z}\phi dA = -\int_{\Gamma_z}\phi dA$$

where ϕ is an arbitrary, sufficiently smooth function.

Exercise 3.6 - Consider the following modification of the problem defined by (3.33).
Consider the region

$$0 < r < r_1 \quad , \quad 0 < \theta < \alpha ,$$

in which (3.33_1) holds together with the boundary conditions

$$u = 0 \qquad \text{on the edges} \quad \theta = 0, \theta = \alpha$$

$$u \text{ specified on } \ r = r_1$$

but no *a priori* condition is prescribed as $r \to 0$. Derive an analogue of the second part of Theorem 3.3, which expresses decay of effects as r decreases (as one recedes from the edge $r = r_1$).

Exercise 3.7 - Show how the analysis of the initial boundary problem (3.46) considered above, may be adapted to altered boundary conditions: suppose the boundary conditions $(3.46_2), (3.46_3)$ are replaced by

$$\frac{\partial u}{\partial x_3} = g \quad \text{on } \Gamma_0 \quad , \quad \frac{\partial u}{\partial x_3} = 0 \quad \text{on } \Gamma_L$$

where

$$\int_{\Gamma_0} g dA = 0 ,$$

and

$$\frac{\partial u}{\partial n} = 0 \quad \text{on } \mathcal{L}$$

where $\partial/\partial n$ denotes the (outward) normal derivative.

Hint. Define the cross-sectional average value $\bar{u}$ of u:

$$\bar{u}(x_3, t) = \frac{1}{A}\int_{\Gamma_z} u dA$$

and prove that

$$\bar{u}_t - \bar{u}_{,33} = 0 \qquad (\alpha)$$

and

$$\bar{u}(x_3, 0) = 0. \qquad (\beta)$$

Integrate the P.D.E. over R_z to obtain

$$-\int_{\Gamma_z} u_{,3}\, dA = \int_{R_z} u_t\, dV$$

and deduce therefrom that

$$\bar{u}_{,3}(L, t) = 0 \qquad (\gamma)$$

and

$$\bar{u}_{,3}(0, t) = 0. \qquad (\delta)$$

Prove that $(\alpha), (\beta), (\gamma), (\delta)$ imply that

$$\bar{u}(x_3, t) = 0.$$

This result is needed in order to apply the variational characterization of ν_1 (the lowest, non-zero, 'free-membrane' eigenvalue for the cross-section).

Exercise 3.8 - Show how the treatment of Question 7 can be extended to a region whose cross-section varies with the x_3 coordinate.

Exercise 3.9 - The context of the following questions is that of the Lemma 3.1 where for convenience, however, t is replaced by x.

(i) Prove that for

$$F(x) = \int_{\Gamma_x} \psi_y^2\, dx$$

one has

$$F'(x) = \int_{\Gamma_x} (\psi_y \psi_{xy} - \psi_x \psi_{yy})\, dy$$

provided ψ=constant on C_+, C_-.

(ii) Defining (in the same context)

$$F_q(x) = \int_{\Gamma_x} \psi_y^{2q}\, dy \qquad (q \text{ any positive integer}),$$

prove that

$$F_q'(x) = (2q-1)\int_{\Gamma_x} \psi_y^{2q-2}(\psi_y \psi_{xy} - \psi_x \psi_{yy})\, dy.$$

[Notice that there are no 'boundary' terms and no boundary restrictions in either (i) or (ii).]

(iii) Consider two dimensional steady flow of an incompressible continuum which has streamlines C_+, C_-: the stream function ψ is such that

(α) (u, v), the x, y rectangular cartesian velocity components, satisfy $u = \psi_y$, $v = -\psi_x$,

(β) $\psi_y \psi_{xy} - \psi_x \psi_{yy} = a_x$, a_x being the (rectangular cartesian) x component of acceleration,

(γ) ψ takes constant values on C_+, C_-.

Establish that

$$F_q(x) = \int_{\Gamma_x} u^{2q} dy$$

satisfies

$$F_q'(x) = (2q-1) \int_{\Gamma_x} u^{2q-2} a_x dy$$

and use Hölder's inequality to prove that

$$F_q'(x) \le (2q-1)\{\int_{\Gamma_x} |a_x|^q dy\}^{1/q} F_q^{1-1/q} .$$

Integrate to obtain

$$\{F_q(x)\}^{1/q} \le \{F_q(0)\}^{1/q} + (2 - 1/q) \int_0^x \{\int_{\Gamma_x} |a_x|^q dy\}^{1/q} dx .$$

Discuss the limit as $q \to \infty$. Is there a simpler way of finding this limiting result?

Exercise 3.10 - Suppose, in the context of Lemma 3.1, that $\varphi(u) = u^2$; $y_+' > 0$, $y_-' < 0$ (strictly). Prove that

$$F'(t) \le -2 \int \psi_t \psi_{yy} dy + \frac{\mathcal{F}'^2}{y'}] .$$

Show that this leads to an upper bound, in terms of data, for $F(t)$ in the case of the simple, one-dimensional diffusion equation with $u(x,0)$ given; and that pointwise bounds for $|u(x,t)|$ may be deduced therefrom.

Hint. Prove that Inequality (13) of appendix may be transformed to give:
Any sufficiently smooth function $\varphi(y,t)$ satisfies an inequality of the form

$$\int_{y_-}^{y_+} \varphi_{yy}^2 dy \ge \pi^2 l^{-2} \{\int_{y_-}^{y_+} \varphi_y^2 dy - l^{-1}(\mathcal{F}])^2\}$$

where

$$l = y_+ - y_-$$

(the width of the region at any time t).

Exercise 3.11 - Prove the decay law (3.86) for the porous medium equation forward in time.
Deduce therefrom

(i) the decay law (3.87) for linear diffusion;

(ii) that, in a certain sense, decay for linear diffusion is more rapid than that for the porous medium equation ($m > 0$).

Exercise 3.12 - Let T, α be positive numbers and set $t^*=\alpha(T-t)^{[2(m+1)]^{-1}}$. Verify that the (non-negative) function $u^n(x,t)$ defined by

$$u^{2m}=\begin{cases} \dfrac{m}{2(m+1)}(T-t)^{-1}[\alpha^2(T-t)^{(m+1)^{-1}}-(x-\dfrac{1}{2})^2] & |x-\frac{1}{2}|\le t^* , \\ 0 & |x-\frac{1}{2}|>t^* , \end{cases}$$

for $t < T$, satisfies the P.D.E. (3.76_1) everywhere in R, in a classical sense, except at

$$|x-\frac{1}{2}|=t^* .$$

[This solution corresponds to an instantaneous point source occurring at time $t = T$, at the location $x = 1/2$].
Verify that the relevant criteria for an acceptable solution of (3.76_1) are satisfied by the above.
In what way (if any!) does the above solution illuminate the nature of the non-existence established in Theorem 3.10?

Exercise 3.13 - Give details of the non-existence arguments for (3.88) outlined in the text. In this connection obtain explicit improvements of (3.90) and (3.91) when m=1.

Exercise 3.14 - Show how the non-existence arguments for the backward diffusion equations (3.76_1) and (3.88_1) can be generalized to two dimensions e.g. consider

$$u_t+\nabla_1\cdot\{u^{2m}\nabla_1 u\}=0$$

and recall the variational characterization of λ_1, the lowest 'fixed membrane' eigenvalue. Note that the area of the region enters the calculations.

Exercise 3.15 - Show how the non-existence argument for (3.76) can be generalized to cater for the backwards in time diffusion equation

$$u_t+(u^{2m}u_x)_x=ku^{2n+1}$$

where m, n are assumed to be positive integers, and k is a positive constant.

3.4 APPENDIX

Historically the "volume integral method" was developed in the context of elasticity (Saint-Venant Principle). For this reason, the relevance of the Neumann Problem (4) to elasticity is discussed hereunder. It is generated by a simple traction problem (anti-plane problem):

Consider a homogeneous, isotropic (linear) elastic right cylinder whose cross-section (assumed simply connected) is R with boundary C; rectangular cartesian coordinates (x_1, x_2, x_3) are chosen such that the x_3 axis is parallel to the generators and the x_1, x_2 plane coincides with that of R. In order that the 'antiplane shear displacement field' with components u_1, u_2, u_3 such that

$$u_1 = u_2 = 0 \qquad u_3 = u(x_1, x_2) \tag{1_1}$$

obtain in the cylinder, one must have

$$\nabla_1^2 u = 0 \quad \text{in } R\,.$$

This follows on substituting (1_1) in the Navier displacement equations

$$(\lambda + \mu) u_{k,ki} + \mu u_{i,kk} = 0 \tag{1_2}$$

λ, μ being the Lamé constants, $\mu(> 0)$ being the shear modulus. The only stress components, which do not vanish identically, accompanying the aforesaid displacement field are

$$\tau_{31} = \mu u_{,1} \quad , \quad \tau_{32} = \mu u_{,2}\,. \tag{1_3}$$

This follows from (1_1) and the stress-strain relations

$$\tau_{ij} = \lambda u_{k,k}\delta_{ij} + \mu(u_{i,j} + u_{j,i})\,.$$

An 'antiplane displacement field' is induced in the cylinder if suitable tractions are applied to its surfaces. In addition to suitable tractions to its plane ends which will be discussed later, the traction components which need to be applied to the lateral boundary C of the cylinder are

$$(T_1, T_2, T_3) = (0, 0, T(x_1, x_2)) \tag{1_4}$$

where

$$T(x_1, x_2) = \tau_{31} n_1 + \tau_{32} n_2 = \mu(u_{,1} n_1 + u_{,2} n_2) = \mu \frac{\partial u}{\partial n}\,. \tag{1_5}$$

This follows from (1_3) and $T_i = \tau_{ij} n_j$.

Supposing that $T(x_1, x_2)$ is assigned, we have a Neumann boundary value problem for $u(x_1, x_2)$ which satisfies (1_1) in R and the boundary condition (1_5) on C. We note that

$$(i) \qquad \int_C T\,ds = 0$$

in view of $(1_2)(1_5)$ and the divergence theorem. This is sometimes expressed as follows: these tractions are self-equilibrated;

(ii) $\quad u$ is determined to within an arbitrary additive constant only.

This Neumann problem (specified by (1_1) and (1_5) and (i)) has the essential characteristic of a general traction boundary value problem in elasticity; such a problem has properties which echo (i) and (ii). Finally we note that the tractions which need to maintain the deformation are of two types: the tractions on the curved surface $T(x_1, x_2)$, and tractions on the plane ends determined by (1_3) from the displacement u.

We conclude with a rough/verbal statement of *Saint-Venant Principle* (for linear elastic bodies): When such a body is deformed by a self-equilibrated load applied to portion of its surface, when the remainder of its surface is free and when there are no body forces, the stress field decays rapidly with distance away from the loaded portion of the surface.

Chapter four

ESTIMATES BASED ON SECOND ORDER INEQUALITIES

4.1 INTRODUCTION

In the case of many important problems for partial differential equations which arise in mathematical physics, it is sometimes possible to define an integral measure (norm) of the solution, which is found to satisfy a second order differential inequality, together with appropriate side conditions. The principal purposes of this chapter are as follows: to discuss upper and/or lower bounds to which such second order differential inequalities give rise; to give representative, simple and interesting examples of partial differential equations which give rise to these; and briefly to survey the essential conclusions that can be drawn therefrom.

The chapter is arranged in sections which generally – though not exclusively – correspond to the principal classes of differential inequality arising; for the most part, these are ordinary differential inequalities although partial differential inequalities also occur.

Section 4.2 begins with a brief survey of simple convexity and concavity of a function – corresponding to the simplest second order differential inequality. Upper and lower bounds for such a function are given, and these form the basis for a number of the analyses occurring in this chapter. By way of direct application, two initial boundary value problems arising in physics are considered. An integral measure (norm) of the solution is defined for each, and these measures are shown to be (essentially) concave functions of time. In each of the two cases, the concavity property is used to prove that the solution fails to exist beyond a certain critical time, provided that the data are suitably restricted.

Section 4.3 discusses logarithmically convex functions (i.e. ones with convex logarithms) and functions which satisfy differential inequalities similar to those satisfied by such functions. As an application, a Cauchy (ill-posed) problem for a rectangular strip is considered, and a cross-sectional measure of the solution is shown to be a logarithmically convex function, or one of a similar nature.

The principal conclusion drawn from this is: assuming that the solution of the problem satisfies an ***a priori*** upper bound, it is established that the solution depends continuously on data, in a certain sense.

The "method of cross-sections" forms the principal subject of Section 4.4; very often, it ultimately depends upon a differential inequality which defines "generalized convexity" – another generalization of simple convexity. Following a general description of the method of cross-sections, and of generalized convexity, we discuss an application of the method to a Dirichlet problem for a rectangular region occupied by inhomogeneous material. We then consider a particularly useful cross-sectional measure for two-dimensional regions; it arises in a number of contexts and is particularly appropriate to regions with convex (or concave) boundaries. With a view to facilitating the analysis of such a measure a number of lemmas are proved; a number of applications of these to Dirichlet problems for harmonic functions is then considered, and the influence of the boundary convexity (or otherwise) on the spatial variation of the measure is discussed. The section concludes with the analysis of a simple B.V.P. which involves the use of curvilinear cross-sections, and to an initial boundary value problem for diffusion in a right cylinder.

4.2 CONVEXITY/CONCAVITY; TWO EXAMPLES OF CONCAVITY

Convexity and concavity are complementary, mirror-image concepts, which are defined by the simplest second order differential inquality. Many of the analyses in this chapter are ultimately based on these concepts.

Definition 4.1 - *A function $f(x)$ defined in the interval $[a,b]$ which satisfies therein*

$$f''(x) \geq 0\,, \tag{4.1}$$

is said to be a convex function (in the interval). If the inequality is strict, the function is said to be strictly convex.

Definition 4.2 - *If $-f(x)$ is a convex function, $f(x)$ is said to be a concave function of x. If $-f(x)$ is a strictly convex function, $f(x)$ is said to be strictly concave.*

Geometrically, it is evident that the curve representing a convex function must

(i) not be above the straight line segment joining any two points on the curve;

and

(ii) not be below the tangent drawn at any point to the curve. [See Figures 4.1 and 4.2].

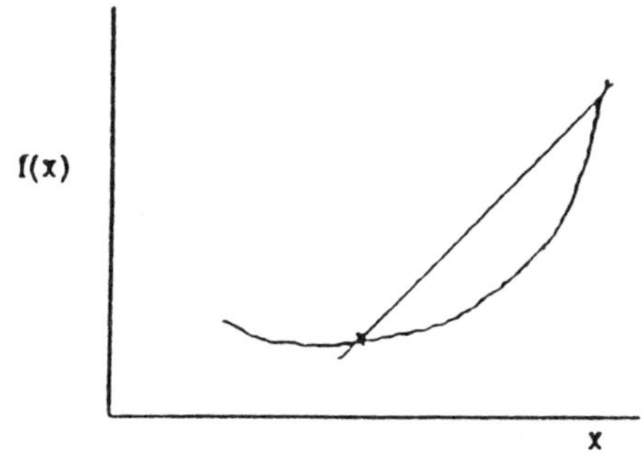

Figure 4.1

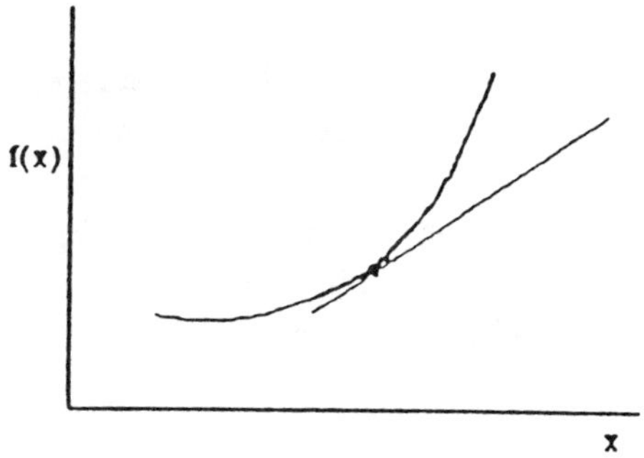

Figure 4.2

The analytic counterparts of these are essentially contained in the following continued inequality:

$$f(a) + \frac{f(b) - f(a)}{b - a}(x - a) \geq f(x) \geq f(a) + f'(a)(x - a) \tag{4.2}$$

The proofs of these are elementary and are left as exercises to the reader (see exercises). Moreover, more general versions of these inequalities – corresponding to a more general concept called "generalized convexity" – are given in Section 4.4 together with proofs (or reference to proofs). Let us note, also, that the upper bound in (4.2) continues to hold if (4.1) merely holds on the *open* interval $a < x < b$, while f is continuous in the *closed* interval $a \leq x \leq b$; and that the lower bound in (4.2) continues to hold in any neighbourhood of $x = a$ provided merely that $f'' \geq 0$ holds in the neighbourhood, *open* at $x = a$, while $f'(x)$ is continuous in the neighbourhood, *closed* at $x = a$.

Functions $F(x)$ arise in subsequent sections of this chapter such that $log F(x)$ is convex, $F^{-\gamma}(x)\,(\gamma > 0)$ is concave, etc. It is simple matter to write down counterparts of (4.2) for such functions.

Two examples are now given which depend on concavity. The first has to do with the Schrödinger equation containing a nonlinear forcing function, on the (entire) real line. An integral property (norm) of the solution of the initial value problem, is shown to be a concave function of time, and this is used to characterize data which give rise to non-existence of solution. This phenomenon is connected with Langmuir wave collapse in a plasma [112]. Perhaps it should be pointed out that while the differential inequality obtained in this example is among the simplest that can arise, the analysis by which it is derived is relatively long. The second example has to do with the longitudinal motion of a nonlinear elastic rod – an example of a type of nonlinear wave motion. A certain integral property of the assumed motion is shown to be a concave function of time. This is used to show that for rods of a certain constitution, subject to certain initial and boundary conditions, a norm of the assumed solution is bounded below by a function which becomes unbounded in finite time; that is to say, the assumed solution fails to exist ("blows up") after a finite time. Though not strictly necessary, attention is confined to one spatial dimension in the examples considered as it is our intention to provide simple illustrations of important principles.

4.2.1 Nonlinear Schrödinger Equation

We consider the initial problem for the nonlinear Schrödinger equation: $u(x,t)$ is a classical solution of

$$\begin{cases} iu_t = u_{xx} + f(|u|^2)u\,, & x \in \mathbb{R}\,, t > 0\,, \\ u(x,0) = \phi(x)\,, \end{cases} \tag{4.3}$$

where ϕ and f are given functions of their arguments, which are specified more completely later. It is assumed that u together with its derivatives decay sufficiently rapidly as $|x| \to \infty$; this is required for the convergence of, and the validity of operations on, all integrals arising in the analysis, and, of course, has implications for the behaviour of ϕ. Here and subsequently, subscripts denote partial differentiation with respect to the relevant independent variables, while the usual notation for complex functions is adopted: $u = u_1 + iu_2$ where u_1, u_2, are real, $\overline{u} = u_1 - iu_2$ and $|u|^2 = u_1^2 + u_2^2$. As usual, we shall denote the norm on $L^2(\mathbb{R})$ by $\|\cdot\|$ e.g. $\|u\|^2 = \int |u|^2 dx$, where integration in this, and in all other integrals arising, is over $\mathbb{R}$.

The programme which we have described above is essentially due to Glassey [113] who, however, gives a much more extensive analysis of the issues. To implement the programme, we define a measure of the solution by

$$F(t) = \int x^2 |u|^2 dx\,. \tag{4.4}$$

We proceed to prove that it is, under certain restrictions on the data, a concave function. The proof depends upon the following lemma which embraces four separate equations:

Lemma 4.1

(i) $$\|u\|^2 = \|\phi\|^2\,; \tag{4.5$_1$}$$

(ii)
$$\frac{d}{dt}\int x^2|u|^2dx = -4\,Im\int x\bar{u}u_x dx\,; \tag{4.5$_2$}$$

(iii)
$$\|u_x\|^2 - \int g(|u|^2)dx = E_0 \quad \text{(a constant)} \tag{4.5$_3$}$$

where g, the potential for f, is defined by

$$g(s) = \int_0^s f(a)da\,,$$

and where E_0 may be interpreted as the left-hand side of (4.5_3) *with u replaced by ϕ;*

(iv)
$$\frac{d}{dt}\{Im\int x\bar{u}u_x dx\} = -2\|u_x\|^2 + \int\{f(|u|^2)|u|^2 - g(|u|^2)\}dx\,. \tag{4.5$_4$}$$

The proofs of the foregoing propositions are derived by multiplying the P.D.E. (4.3_1) by suitable functions, by integration by parts using the posited decay at infinity (together with the initial condition). These proofs are left as formal exercises to the reader: see accompanying exercises where further details are given.

Using $(4.4), (4.5_2), (4.5_4)$ yields

$$F'(t) = -4\,Im\int x\bar{u}u_x dx\,, \tag{4.6}$$

$$F''(t) = 8\|u_x\|^2 - 4\int\{f(|u|^2)|u|^2 - g(|u|^2)\}dx\,. \tag{4.7}$$

In order to establish concavity of $F(t)$ with a view to obtaining non-existence results, we place the following restrictions on data:

(i) the forcing function f is restricted by

$$sf(s) - g(s) \geq \alpha g(s) \tag{4.8}$$

for all $s \geq 0$ and a constant $\alpha \geq 2$;

(ii) the initial conditions are restricted by

$$E_0 \leq 0 \tag{4.9$_1$}$$

and

$$Im\int x\bar{\phi}\phi_x dx > 0\,. \tag{4.9$_2$}$$

Use of $(4.5_3), (4.7), (4.8)$ gives

$$F''(t) \leq 4\alpha E_0 - 4(\alpha - 2)\|u_x\|^2\,. \tag{4.10}$$

Since $\alpha \geq 2$ and $E_0 \leq 0$ are assumed ((4.8) and (4.9)) it follows that

$$F''(t) \leq 0 \tag{4.11}$$

i.e. $F(t)$ is a concave function. Use of the lower bound in (4.2) gives

$$F(t) \leq F(0) + F'(0)t\,. \tag{4.12}$$

Since $F'(0) < 0$ (in view of (4.6) and (4.9_2)), (4.12) implies that the solution fails to exist if $t > -F(0)/F'(0)$ as the (intrinsically non-negative) $F(t)$ would then be negative; Figure 4.3 provides a geometrical interpretation.

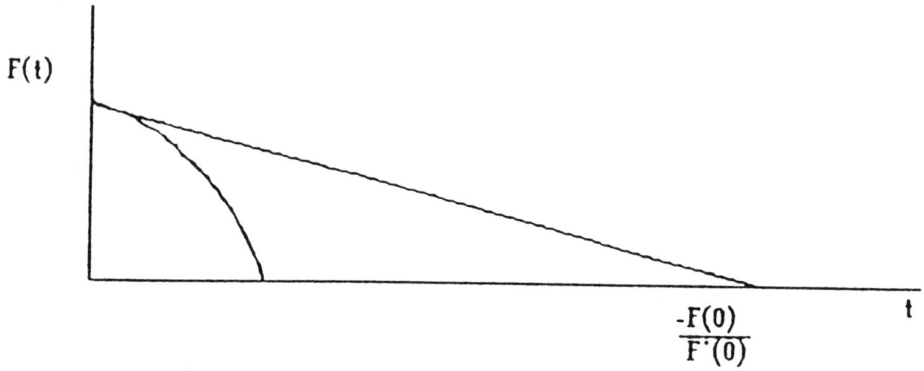

Figure 4.3

Moreover, (4.12) implies that, provided the solution exists long enough, then, for some time t_c, $F(t) \to 0$ as $t \to t_c \leq -F(0)/F'(0)$. We may embody the *principal* conclusion in

Theorem 4.1 - *The solution of the initial problem* (4.3) *for the nonlinear Schrödinger equation, where the data are restricted by* (4.8), (4.9), *fails to exist for times t such that*

$$t > -F'(0)/F(0) \quad (> 0)$$

where F, F' *are calculable from* (4.3_2), (4.4), (4.6).

It is not hard to modify the foregoing analysis to cater for different restrictions on the data. For example, if the restrictions on the initial conditions (4.9_1), (4.9_2) are replaced by $E_0 < 0$ (strictly), it is easy to deduce from (4.10) that

$$F(t) \leq F(0) + F'(0)t + 2\alpha E_0 t^2\,. \tag{4.13}$$

It is evident from this that the upper bound becomes zero at a certain, sufficiently large, time irrespective of the sign of $F'(0)$; the detailed argument is left as an exercise to the reader.

In conclusion, we proceed to the interesting conclusion that if $F(t) \to 0$ as $t \to t_c - 0$, then $\|u_x\|^2 \to \infty$ thereat (i.e. the x derivative "blows up" in L^2 norm). Again assuming appropriate decay as $|x| \to \infty$, we obtain

$$\int u^2 dx = -2\int x u u_x dx \tag{4.14}$$

on integrating by parts. Applying Schwarz's inequality thereto and using (4.4), (4.5_1) we obtain

$$\|\phi\|^4 \leq 4F(t)\|u_x\|^2 \,. \tag{4.15}$$

The stated conclusion follows from this.

There follows an example where a "blow up" phenomenon is established by a different technique - but one which also depends on concavity.

4.2.2 Longitudinal Motion of a Nonlinear Elastic Rod

Consider, for simplicity, a one-dimensional elastic continuum (e.g. a rod). Let $u(x,t)$ be the (finite) displacement at time t of a point whose position in the reference configuration B is x. The underformed rod is supposed to occupy $0 \leq x \leq 1$. The Piola-Kirchoff stress σ satisfies the equation of motion

$$\frac{\partial \sigma}{\partial x} = \rho_0 u_{tt} \tag{4.16_1}$$

where ρ_0, assumed positive, is the mass density in the reference configuration. The material is supposed to satisfy the constitutive relation

$$\sigma = \frac{\partial W}{\partial u_x} \quad , \quad W = W(u_x) \tag{4.16_2}$$

where W is the strain energy function per unit length of the reference configuration. It is supposed that homogeneous data are assigned on the ends:

$$\begin{cases} \text{either} \quad u = 0 \qquad \textit{or} \qquad \sigma = 0 \\ \text{on each end} \quad x = 0 \quad x = 1 \,. \end{cases} \tag{4.17}$$

We assume that the rod is deformed at time $t=0$ and that the classical solutions obtain subsequently ($0 < t < T$, T being a constant).

On multiplying (4.16_1) by u_t, integrating over B, using integration by parts and (4.17), the energy conservation law is derived:

$$E(t) = \frac{1}{2}\int_0^1 \rho_0\, u_t^2\, dx + \int_0^1 W\, dx = E(0) \,. \tag{4.18}$$

Let us postulate in relation to the strain energy function W, that there exists a constant $\alpha > 2$ such that

$$\int_0^1 (\alpha W - u_x \frac{\partial W}{\partial u_x}) dx \geq 0 \,. \tag{4.19}$$

The non-existence issue for the problem defined above is intended to exemplify a phenomenon characteristic of a certain type of nonlinear wave motion. With a view to establishing the required non-existence, we define a global measure of deformation by

$$F(t) = \int_0^1 \rho_0 u^2 dx + \beta(t + t_0)^2 \quad , \qquad 0 < t < T \tag{4.20}$$

where β and t_0 are positive constants yet to be chosen, and proceed to prove that it satisfies $(F^{-\gamma})'' \leq 0$ where γ is a positive number. Elementary manipulation using the hypotheses yields

$$F'(t) = 2\int_0^1 \rho_0 u u_t dx + 2\beta(t+t_0)\,, \tag{4.21}$$

$$F''(t) = (2+\alpha)\int_0^1 \rho_0 u_t^2 dx - 2\alpha E(0) + 2\int_0^1 \{\alpha W - u_x \frac{\partial W}{\partial u_x}\} dx + 2\beta\,. \tag{4.22}$$

Schwarz's inequality together with (4.19), (4.20), leads to

$$FF'' - \left(\frac{2+\alpha}{4}\right) F'^2 \geq -\alpha(\beta + 2E(0))F\,. \tag{4.23}$$

Since $\alpha > 2$, this differential inequality may be put in the equivalent form

$$(F^{-\gamma})'' \leq 2\gamma(1+2\gamma)F^{-(\gamma+1)}(\beta + 2E(0)) \tag{4.24}$$

where

$$\gamma = (\alpha - 2)/4\,.$$

We now proceed to establish circumstances giving rise to non-existence of the posited solution. Suppose that the initial conditions are such that $E(0) \leq 0$, and choose $\beta + 2E(0) = 0$. As $-F^{-\gamma}$ is convex, the lower bound occurring in (4.2) may be written

$$F^{\gamma}(t) \geq F^{\gamma}(0)\left(1 - \gamma t\frac{F'(0)}{F(0)}\right)^{-1}\,. \tag{4.25}$$

Irrespective of the initial data, the quantity t_0 occurring in (4.20) can be chosen so that $F'(0) > 0$. In all these circumstances, the right hand side of (4.25) becomes unbounded in finite time. We embody this conclusion in

Theorem 4.2 - *The solution of the initial value problem* $(4.16_1), (4.16_2)$, (4.17) *ceases to exist in finite time provided that*
(i) *the strain energy function W is such that a constant $\alpha > 2$ exists such that* (4.19) *is satisfied*
and
(ii) *the intial conditions are such that the total energy $E(0)$ – defined by* (4.18) *– is negative.*

Thus we have exhibited another example of non-existence of solution. This approach is often called the method of 'blow up'. We remark that rather than viewing (4.19) as a constitutive assumption which gives rise to the catastrophic behaviour described, one might take the view that such material behaviour is inadmissible in order to avoid such behaviour.

An upper bound, complementing (4.25), is also obtainable for times prior to the critical one just described. Again assuming initial conditions such that

$E(0) \le 0$ and choosing $\beta + 2E(0) = 0$, the upper bound in (4.2) appropriate to $(-F^{-\gamma})'' \le 0$ takes the form

$$F^{\gamma}(t) \le \frac{F^{\gamma}(0)F^{\gamma}(T)}{(1-t/T)F^{\gamma}(T) + (t/T)F^{\gamma}(0)}, \tag{4.26}$$

where it is understood that the solution exists in $0 \le t \le T$. It follows from this that the null solution to the initial boundary value problem described is both unique and depends "Hölder continuously" on the data provided $F(T)$ is bounded.

In conclusion, we remark that our intention in the above example is to illustrate techniques. It would have been more appropriate to consider weak rather than classical solutions, and the restrictions to a one dimensional context is not necessary. These deficiencies may be rectified without too much difficulty.

The methodology employed in the above analysis has been used by Knops, Levine and Payne [114], Levine [115], Payne [34], Knops and Straughan [116] *inter alia.* The monograph of Straughan [35] – which contains a comprehensive list of references – treats this and other second order differential inequality techniques (e.g. convergent integral methods) by which non-existence may be established. See, also, Levine [117] which contains a comprehensive list of relevant references.

4.3 LOGARITHMIC CONVEXITY AND RELATED CONCEPTS

4.3.1 Introduction and Basics

This section treats functions with convex logarithms – called logarithmically convex functions – and, also, functions which satisfy differential inequalities similar to that satisfied by logarithmically convex functions. It is shown how to obtain upper and lower bounds for such functions using the corresponding properties of simple convex functions. To illustrate the application of these ideas, we consider a Cauchy (ill-posed) problem for Laplace's equation in a rectangular strip: physically speaking, that of determining the steady state temperature distribution therein if the temperature and its normal derivative (heat flux) are given on one edge, if the two adjacent edges are held at zero temperature, and if a universal bound is given for the absolute value of the temperature. An integral measure of the temperature over the cross-section (L^2 cross-sectional measure) is proved to be a logarithmically convex function of the distance from the end upon which data are assigned, or a function with similar properties. Upper and lower bounds are obtainable for the measure, and are used to establish that the solution is unique, and that it is, in a certain sense, continuous with respect to data. Extensions of the problem are also dealt with.

Examples of the earliest work on logarithmic convexity and its applications may be found in Agmon [118], Agmon and Nirenberg [119], Levine [120], Knops and Payne [121-123]. One should, also, consult [34] which contains a comprehensive list of references. Examples of more recent work are Levine [124], Galdi and Rionero [125], Payne and Straughan [126-129], Straughan [130] together with references contained therein.

Definition 4.3 - *By a logarithmically convex function is meant a positive function whose logarithm is convex.*

Consider, accordingly, a function $F(x) > 0$ defined in $x_1 \le x \le x_2$ such that

$$f(x) = lnF(x) \tag{4.27}$$

is convex. Elementary computation shows that

$$FF'' - F'^2 \ge 0 \quad , \qquad x_1 \le x \le x_2 ; \tag{4.28}$$

and that the inequalities satisfied by F implied by (4.2), are

$$\{F(x_1)\}^{(x_2-x)/(x_2-x_1)}\{F(x_2)\}^{(x-x_1)/(x_2-x_1)} \\ \ge F(x) \ge F(x_1)\exp\left\{\frac{F'(x_1)}{F(x_1)}(x-x_1)\right\} . \tag{4.29}$$

A number of generalizations of this concept arise in practice but only one of these dealt with: suppose a function $F(x) > 0$ is defined in $x_1 \le x \le x_2$ and satisfies therein the differential inequality

$$FF'' - F'^2 \ge -a_1 FF' - a_2 F^2 \tag{4.30}$$

where a_1, a_2 are constants. It is readly verified that (4.30) is expressible in the form

$$(ln\, F)'' + a_1(ln\, F)' + a_2 \ge 0 . \tag{4.31}$$

Assume first that $a_1 \ne 0$; making the change of variable

$$\sigma = e^{-a_1 x} \tag{4.32}$$

it is found that (4.30) transforms into

$$\frac{d^2}{d\sigma^2} ln\left[F(\sigma)\sigma^{-a_2/a_1^2}\right] \ge 0 \tag{4.33$_1$}$$

in $\sigma_1 \le \sigma \le \sigma_2$, where

$$\sigma_\alpha = e^{-a_1 x_\alpha} \qquad (\alpha = 1, 2) . \tag{4.33$_2$}$$

The upper bound occurring in (4.2), implies in the current context:

$$F(x) \le e^{-(a_2/a_1)x}\{F(x_1)e^{(a_2/a_1)x_1}\}^\delta\{F(x_2)e^{(a_2/a_1)x_2})\}^{1-\delta} \tag{4.34$_1$}$$

in $x_1 \le x \le x_2$, where

$$\delta = \frac{e^{-a_1 x} - e^{-a_1 x_2}}{e^{-a_1 x_1} - e^{-a_1 x_2}} . \tag{4.34$_2$}$$

The lower bound which corresponds to that occurring in (4.2) turns out to be

$$F(x) \geq F(x_1) \cdot \\ \cdot \exp\left[\frac{F'(x_1) + (a_2/a_1)F(x_1)}{a_1 F(x_1)}\{1 - e^{-a_1(x-x_1)}\} - (a_2/a_1)(x - x_1)\right] . \tag{4.35}$$

The special case where $a_1 = 0$ is left as an exercise to the reader. In these circumstances, the differential inequality (4.30) may be expressed as

$$[ln\,\{F\exp(a_2 x^2/2 + \alpha x + \beta\}]'' \geq 0 \tag{4.36}$$

where α, β are arbitrary constants. Upper and lower bounds for $F(x)$ are easily derived therefrom.

4.3.2 Cauchy (ill-posed) Problem

We now give an example where some of the foregoing inequalities arise. We consider the Cauchy problem for the Laplace equation in the rectangular strip $0 < x < X$ $(< \infty), 0 < y < 1$ where X is fixed:

$$u(x, y) \in C^2(0 \leq x \leq X \times 0 \leq y \leq 1)$$

satisfies

$$u_{xx} + u_{yy} = 0, \qquad 0 < x < X \times 0 < y < 1 \tag{4.37$_1$}$$

subject to the boundary conditions

$$u(x, 0) = u(x, 1) = 0 \tag{4.37$_2$}$$

and the initial conditions

$$u(0, y) = f(y) \quad , \quad u_x(0, y) = g(y) , \tag{4.37$_3$}$$

where f, g are specified functions. The quantity u may be interpreted as the steady state temperature in a (homogeneous, isotropic) strip.

The problem is ill-posed because the solution does not depend continuously upon the data. It is possible, however, to obtain continuous dependence – in a suitable sense – if a constant M is known such that

$$|u| \leq M . \tag{4.38}$$

This *a priori* bound does not need to be a sharp one; it might, for example, be the melting temperature of the material in question.

We pause for a moment to recall [cf. chapter one] the essence of the ill-posedness of the problem specified by (4.37) in the preceding paragraph. An example due to Hadamard deals with the issue: the function

$$u_n = \frac{\cosh n\pi x \sin n\pi y}{n} \tag{4.39}$$

where n is a (positive, say) integer, satisfies (4.37_1) and (4.37_2) and the initial conditions

$$u_n(0,y) = \frac{\sin n\pi y}{n} \quad , \quad \frac{\partial u_n}{\partial x}(0,y) = 0 . \tag{4.40}$$

The initial data $(u_n(0,y))$ can be made arbitrarily small by taking n sufficiently large. However, for given x, u_n can be made arbitrarily large by taking n sufficiently large. This means that the solution to the problem defined by (4.37) is ill-posed: even assuming that the solution exists, it does not depend continuously on data. In effect, this means that some other condition such as (4.38) needs to be imposed if there is to be any chance of obtaining continuous dependence in some sense. Moreover, we remark that it is known that the data in (4.37_3) needs to be analytic if a solution is to exist. In order to establish continuous dependence it is sufficient to prove that (a suitable measure of) the solution of (4.37_1)–(4.37_3) can be made arbitrarily small provided that (a suitable measure of) the data is made sufficiently small. With a view to establishing this, assuming a constraint (4.38) – and, also, with a view to establishing uniqueness in the absence of (4.38) – we define the following measure of the solution:

$$F(x) = \int u^2 dy + \beta(x + x_0)^2 \tag{4.41}$$

where β is a non-negative constant and x_0 is a constant, both of which are yet to chosen, and where the limits of integration are understood throughout. Differentiation yields

$$F'(x) = 2 \int u u_x dy + 2\beta(x + x_0) \tag{4.42}$$

and

$$F''(x) = 2 \int (u_x^2 + u_y^2) dy + 2\beta , \tag{4.43}$$

where integration by parts and the boundary conditions (4.37_2) have been used. The following conservation equation – an analogue of the conservation of energy principle for the wave equation – is easily proved using $(4.37_1), (4.37_2)$:

$$\int u_x^2 dy - \int u_y^2 dy = E(\text{constant}) . \tag{4.44}$$

Combining this with (4.43) yields

$$F''(x) = 4 \int u_x^2 dy + 2(\beta - E) . \tag{4.45}$$

In view of (4.41), (4.42), (4.45) we obtain

$$\begin{aligned} FF'' - F'^2 = [\{ \int u^2 dy + \beta(x + x_0)^2 \} \{ 4 \int u_x^2 dy + 4\beta \} \\ - 4\{ \int u u_x dy + \beta(x + x_0) \}^2 - 2(\beta + E)F] . \end{aligned} \tag{4.46}$$

In view of Schwarz's inequality (as applied to the first set of terms) we obtain

$$FF'' - F'^2 \geq -2(\beta + E)F. \tag{4.47}$$

We now discuss uniqueness and continuous dependence. The uniqueness issue is that of proving $u \equiv 0$ if the initial data $f = g = 0$; assumption (4.38) is not required for this purpose. Assuming this, we have $E \equiv 0$, and let us choose $\beta = 0$. Suppose that $u \neq 0$ in the interval $(0 <) x_1 < x < x_2$. Then (4.29) and (4.47) imply that

$$F(x) \leq \{F(x_1)\}^{(x_2-x)/(x_2-x_1)}\{F(x_2)\}^{(x-x_1)/(x_2-x_1)}. \tag{4.48}$$

Let x_1 be the lowest value of x beyond which u ceases to be identically zero. Then $F(x) \equiv 0$ in $x_1 < x < x_2$ – giving a contradiction. Uniqueness is thus proved.

We now discuss continuous dependence. Two cases arise: when initial conditions are such that $E \leq 0$ and when they are such that $E > 0$. Suppose, first, that $E \leq 0$ and choose $\beta = 0$. One obtains from (4.29), (4.38) and (4.47):

$$F(x) \leq M^{2x/X}\{F(0)\}^{1-x/X}. \tag{4.49}$$

Thus (Hölder) continuous dependence on data is established.

Consider now the case where the initial conditions are such that $E > 0$. Suppose β is positive; and it is useful, though not necessary, to envisage β/E as "large". The following is implied by (4.41) and (4.47):

$$FF'' - F'^2 \geq -\frac{(2+\varepsilon)}{(x+x_0)^2}F^2, \tag{4.50$_1$}$$

or equivalently,

$$[ln\, F(x+x_0)^{-(2+\varepsilon)}]'' \geq 0 \tag{4.50$_2$}$$

where

$$\varepsilon = 2E/\beta. \tag{4.50$_3$}$$

The following continuous dependence result follows from (4.2), (4.38), (4.50)

$$\begin{aligned} F(x)(x+x_0)^{-(2+\varepsilon)} \\ \leq \{F(0)x_0^{-(2+\varepsilon)}\}^{1-x/X}[\{M^2+\beta(X+x_0)^2\}(X+x_0)^{-(2+\varepsilon)}]^{x/X}. \end{aligned} \tag{4.51}$$

The uniqueness and continuous dependence issues for the Cauchy problem discussed, are embodied in

Theorem 4.3 - *The (ill-posed) Cauchy problem* (4.37) *has a unique solution. Provided that the a priori bound* (4.38) *obtains, the problem exhibits continuous dependence on the initial data, in a manner conveyed by* (4.49) *and* (4.51), *when the initial data* (4.37_3) *are such that* $E \leq 0$ *and* $E > 0$ *respectively, where* E *is* *defined by* (4.44) *and* (4.37$_3$).

It should be pointed out that similar results can be obtained for an initial value problem for more general operator equations of the type

$$M\frac{d^2u}{dx^2} + Nu = 0 \quad , \qquad x \in [0, X]$$

$$u(0) = u_0 \quad , \quad \frac{du(0)}{dx} = v_0$$

where M and N are suitable linear, symmetric operators (see Payne [34]).

We now make a series of observations with a view to amplifying and extending the foregoing results for Laplace's equation:

(1) In order that $E \leq 0$ (in the penultimate context) it is sufficient that the initial conditions (4.37$_3$) satisfy

$$\int_0^1 g^2\,dy \leq \pi^2 \int_0^1 f^2\,dy\,. \tag{4.52}$$

This follows from the inequality (10) of appendix, case (a).

(2) Lower bounds for $F(x)$, in the two contexts, complementing (4.49) and (4.51) are readily obtainable, and these are left as exercises to the reader. In some circumstances, they imply upper limits for X (the length of the rectangular strip) in terms of M.

(3) The previous analysis (context with $E \leq 0$) for a region of constant width is equally applicable ***mutatis mutandis*** to a region wich narrows monotonically with respect to x. Suppose the lateral boundaries are given by

$$y = y_+(x) \quad , \quad y = y_-(x)\,, \qquad 0 \leq x \leq X\,, \tag{4.53$_1$}$$

where these are single-valued, non-intersecting, continuously differentiable curves such that

$$\begin{cases} y_+(x) \geq y_-(x)\,, \\ y_+'(x) \leq 0\,, \quad y_-'(x) \geq 0\,, \\ y_+(0) = 1\,, \quad y_-(0) = 0\,. \end{cases} \tag{4.53$_2$}$$

We suppose that the conditions $u = 0$ again obtain on the (generally non-parallel) lateral edges, and that the remaining conditions are as before. Central to the modified analysis is the following: denoting by Γ_x the cross-section with abscissa x, one may prove that

$$E(x) \overset{\text{defn}}{=} \int_{\Gamma_x} u_x^2\,dy - \int_{\Gamma_x} u_y^2\,dy \tag{4.54$_1$}$$

satisfies

$$E'(x) \leq 0\,. \tag{4.54$_2$}$$

This may be proved using Leibnitz's theorem, integration by parts, and using zero condition on the lateral boundaries in differentiated form $u_x + u_y y' = 0$ with suffices $+, -$ appended as appropriate. One obtains

$$\frac{d}{dx}\int_{\Gamma_x}(u_x^2 - u_y^2)dy = (u_x^2 + u_y^2)y'] \tag{4.55$_1$}$$

where

$$f(x,y) = f(x,y^+(x)) - f(x,y^-(x)) \tag{4.55$_2$}$$

– from which (4.54_2) follows readily.

If one redefines

$$F(x) = \int_{\Gamma_x} u^2 dy \tag{4.56}$$

in the altered context, if the initial conditions are such that $E(0) \leq 0$, one finds that $FF'' - F'^2 \geq 0$ and uniqueness, and continuous dependence (as in (4.49)), are recovered once more.

(4) Reverting to the original problem for the rectangular strip – for the solution of which, however, we now posit an enhanced regularity – let us redefine

$$F(x) = \int_0^1 u_y^2 dy\,. \tag{4.57}$$

It is possible to show that $F(x)$ is logarithmically convex provided that the initial conditions are such that

$$\int u_{xy}^2(0,y)dy - \int u_{yy}^2(0,y)dy \leq 0\,. \tag{4.58}$$

(This is left as an exercise to the reader.) Suppose, in addition, that the universal bound (4.38) is replaced by the stronger one

$$|u_y| \leq M_1 \tag{4.59}$$

(e.g. an *a priori* upper bound for the heat flux) one has the analogue of (4.49) in the current context

$$F(x) \leq M_1^{2x/X}\{F(0)\}^{1-x/X}\,. \tag{4.60}$$

It is possible, by standard means, to deduce therefrom upper bounds for $u(x,y)$ which depend on M_1, $F(0)$ and which reflect position; thus we may say that there is "pointwise continuous dependence" in the circumstances envisaged. Once more the details are left as an exercise to the reader.

4.4 THE METHOD OF CROSS-SECTIONS AND GENERALIZED CONVEXITY

4.4.1 General Principles

What we term "generalized convexity" is a natural extension of the simple convexity considered in Section 4.2. For the most part, the "method of cross-sections" – which is the principal subject of this section – ultimately depends upon "generalized convexity"; the method may, however, give rise to differential inequalities other than that of "generalized convexity". In this subsection, we first discuss an outline of the method and follow with a discussion of "generalized convexity".

The example considered in Section 4.3 (as well as those considered in Section 3.2) have the following characteristics:

(i) it considers a (transparently) non-negative measure of the solution in the form of an integral over "cross- sections" which depends on one spatial variable;

(ii) a differential inequality is derived for this measure;

(iii) bounds in terms of the data of the problem, are derived from this inequality.

We use the term "the method of cross-sections" – or "the method of cross-sectional integrals" – to describe a method of analyzing a problem for a P.D.E., or system of P.D.E.s, which have the aforementioned general characteristics. We now proceed to describe it in more detail. We shall work in three dimensional space but the adaptation to two-dimensions is obvious.

We consider a three-dimensional region V whose boundary consists of three disjoint portions (Figure 4.4): the "lateral boundary" C together with Γ_0, Γ_L whose definitions follow immediately.

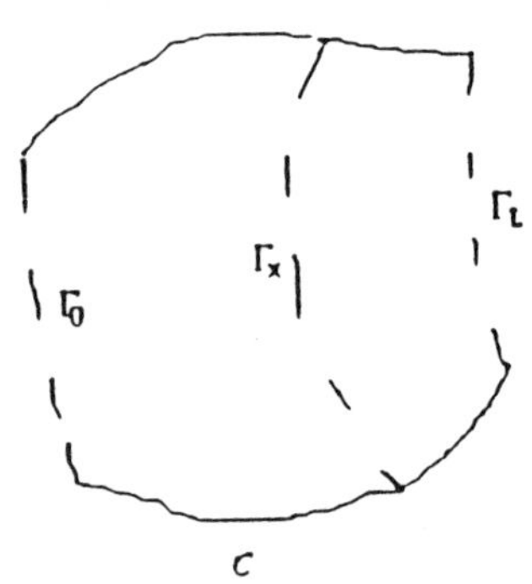

Figure 4.4

Suppose that a family of non self-intersecting surfaces Γ_x ("cross-sections") parametrized by the variable x $(0 \le x \le L; L$ a constant) exists with the following characteristics:

(a) Γ_0, Γ_L – corresponding to $x = 0, x = L$ – constitute portions of the

boundary of V as aforesaid (in limiting cases, however, Γ_0 and/or Γ_L may degenerate/vanish),
(b) each surface Γ_x intersects the 'lateral surface' C in a simple closed curve or curves.
[In a typical case, the 'cross-sections' Γ_x are planes and x is a rectangular cartesian coordinate.]

Suppose first that u is a (scalar, vector etc.) function of position only, which satisfies the P.D.E. (or system of P.D.E.s)

$$L(u) = f \qquad \text{in } V\,, \tag{4.61$_1$}$$

together with

$$B(u) = g \qquad \text{on } C \tag{4.61$_2$}$$

where f, g are assigned data, and L, B are (generally) differential operators. Suppose, also, either
(a) that some suitable property of u and/or its derivatives are specified on Γ_0, Γ_L respectively:

$$B_0(u) = b_0 \quad \text{and} \quad B_L(u) = b_L \tag{4.61$_3$}$$

where b_0 and b_L are assigned data,

or (more exceptionally)

(b) that some such properties are specified on Γ_0 alone. It is, of course, assumed that the problem for u is properly defined, in some sense.

We define a cross-sectional measure of the solution by

$$F(x) = \int_{\Gamma_x} P(u)dx \tag{4.62}$$

where $P(u)$ is a non-negative function of u (and/or its derivatives) and where the integral is evaluated over the cross-section Γ_x; the integral is, of course, a function of x. The object of the cross-sectional method is to choose a function $P(u)$ which will give rise to a differential inequality for $F(x)$, and to supplementary conditions on $F(x)$ at $x = 0$ and/or $x = L$, whence bounds (usually upper bounds) for $F(x)$, may be obtained in terms of the data of the problem.

It is sometimes possible to modify the method to cater for problems where u depends on t, the time variable, in addition to space variables. In this case, conditions of the type (4.61$_1$)–(4.61$_3$) are supplemented by initial conditions, at $t = 0$, of the type

$$I(u) = i\,, \tag{4.63}$$

while the cross-sectional measure of the solution

$$F(x,t) = \int_{\Gamma_x} P(u,t)dx \tag{4.64}$$

now depends both upon x and t. One again seeks a (partial) differential inequality for $F(x,t)$ – for suitable P – together with suitable boundary and initial conditions therefor, whence bounds, in terms of data, are obtainable for $F(x,t)$.

In the purely spatial case, a differential inequality typically arises which corresponds to "generalized convexity" – a natural extension of (simple) convexity considered at the beginning of the chapter.

Definition 4.4 - *Let $F(x)$ be a function satisfying*

$$F'' \geq h(x, F, F') \tag{4.65$_1$}$$

in the interval $0 \leq x \leq L$, where $h(x,y,z)$ is a given function which has the properties:

(i) $h(x,y,z), \dfrac{\partial h}{\partial x}(x,y,z), \dfrac{\partial h}{\partial y}(x,y,z)$ *are continuous functions throughout their domains of definition,*

(ii)
$$\frac{\partial h}{\partial y} \geq 0\,, \tag{4.65$_2$}$$

then $F(x)$ is said to exhibit generalized convexity in $0 \leq x \leq L$.

A function so defined has a property which extends the "curve under chord" property associated with simple convex functions. It is expressed in the following Theorem.

Theorem 4.4 - *A function $F(x)$ which exhibits generalized convexity satisfies*

$$F(x) \leq G(x) \tag{4.66$_1$}$$

provided $G(x)$ satisfies

$$G'' = h(x, G, G')\,, \tag{4.66$_2$}$$

subject to

$$G(0) = F(0) \quad , \quad G(L) = F(L)\,. \tag{4.66$_3$}$$

The proof may be based on the following lemma – a particular case of Theorem 4.4 (see [62], for example).

Lemma 4.2 - *A function $z(x)$ which satisfies*

$$z'' \geq a(x)z' + b(x)z \quad , \quad 0 \leq x \leq L\,, \tag{4.67$_1$}$$

where $a(x), b(x)$ are given continuous functions such that

$$b(x) \geq 0\,, \tag{4.67$_2$}$$

together with

$$z(0) = z(L) = 0, \tag{4.67$_3$}$$

satisfies

$$z \le 0. \tag{4.67$_4$}$$

We first prove the Lemma and then how it leads to the Theorem. The proof of the lemma proceeds thus. One may write (4.67_1) in the following form, on multiplying across by an integrating factor $\exp\{-\int a(x)dx\}$:

$$\{c(x)z'\}' - d(x)z \ge 0 \tag{4.67$_5$}$$

where $c(x) > 0, d(x) \ge 0$. Suppose, contrary to what is to be proved, that $z > 0$ at some point. By continuity, there exists an interval $x_1 \le x \le x_2$ in the interior of which $z > 0$, and such that $z(x_1) = z(x_2) = 0$. Multiply (4.67_5) by z and integrate by parts using the latter conditions to obtain

$$-\int_{x_1}^{x_2} c(x)z'^2 dx \ge \int_{x_1}^{x_2} d(x)z^2 dx .$$

Since $c > 0, d \ge 0$ there is an evident contradiction, and the lemma is proved.

We now show how this leads to the proof of Theorem 4.4. Writing $z = F - G$, we find using (4.65_1) etc. and (4.66_2) etc. together with the mean-value theorem, that

$$z'' \ge \frac{\partial h}{\partial G'}(x, G+\theta z, G'+\theta z')z' + \frac{\partial h}{\partial G}(x, G+\theta z, G'+\theta z')z, \quad 0 \le x \le L \tag{4.68$_1$}$$

where $0 \le \theta \le 1$; furthermore

$$z(0) = z(L) = 0. \tag{4.68$_2$}$$

In view of the assumption (4.65_2), it is evident that the results of the lemma are applicable to $z = F - G$, and the theorem is proved. Readers interested in other theorems similar to Theorem 4.5 are referred to books such as [131]. We mention one such result.

Theorem 4.5 - *A function $F(x)$ which exhibits generalized convexity satisfies*

$$F(x) \le G(x) \tag{4.69$_1$}$$

provided $G(x)$ is any function satisfying

$$G'' \le h(x, G, G') \quad , \quad 0 \le x \le L \tag{4.69$_2$}$$

subject to

$$G(0) \ge F(0) \quad , \quad G(L) \ge F(L). \tag{4.69$_3$}$$

It is possible to construct a proof analogous to that of Theorem 4.4; the details are left to the reader.

Remark 4.1 - *Theorems* 4.4 *and* 4.5 *can be established under less restrictive assumptions using the maximum principle* [e.g. [131]]. *Instead of requiring that the relevant differential inequalities and differential equations hold in the closed interval* $0 \leq x \leq L$, *it is sufficient that they be satisfied in the open interval* $0<x<L$ *provided that the relevant functions* F, G *are continuous in the closed interval.*

Comparison theorems for (4.65_1) when $F(0), F'(0)$ are known – generalizing the lower bound in (4.2) – are also available; see [131].

Perhaps the first use of the method cross-sections, envisaged above, arose in connection with a (time-dependent) unsteady heat flow problem [132]; this is essentially outlined in subsection 4.4.5. The use of the method of cross-sections, principally for elliptic problems, arises in the work of Flavin and Knops [133-134], Flavin [135], Flavin and Rionero [136] *inter alia*.

It is also appropriate to mention that a differential inequality of the type (4.89) – typical of generalized convexity – was derived for Laplace's equation by Carleman [137], who used it to elucidate some properties of analytic functions. Indeed, the study of integral "means" over concentric circles is a well developed theme in the repertoire of complex analysts.

Whereas what is meant by the method of cross-sections for the purposes of this book has been described above, it would not be an abuse of language to, also, so style a number of the analyses occurring in chapter three, e.g. Phragmèn-Lindelöf analyses, Lemma 3.2, etc.

4.4.2 Rectangular Region: Dirichlet Problem

We illustrate the "method of cross-sections" by a simple example: consider a rectangular region $V = 0 < x < L \times 0 < y < l$. Suppose

$$u \in C(\overline{V}) \cap C^2(V)$$

satisfies

$$\frac{\partial}{\partial x}\{h(y)\frac{\partial u}{\partial x}\} + \frac{\partial}{\partial y}\{h(y)\frac{\partial u}{\partial y}\} = 0\,, \tag{4.70$_1$}$$

subject to

$$u = 0 \text{ on the lateral boundaries } y = 0, l \tag{4.70$_2$}$$

and

$$u = u_0(y) \text{ on the edge } x = 0 \quad ; \quad u = 0 \text{ on the edge } x = L\,. \tag{4.70$_3$}$$

In the above $h(y)$ and $u_0(y)$ are specified, sufficiently smooth functions, the former being positive. This boundary value problem has many applications; it corresponds, for example, to a steady state diffusion problem where u denotes the temperature, and h the variable diffusivity, in a medium which exhibits inhomogeneity with respect to the y coordinate (the medium may be thought of as one consisting of a large number of very thin homogeneous horizontal layers whose properties (diffusivities) vary in a regular manner). Note that the zero boundary condition on $x = L$ is adopted merely for simplicity.

In this example the "cross-sections" are the straight lines parametrized by the rectangular coordinate x. We define a cross-sectional measure of the solution by

$$F(x) = \int_0^l h(y)u^2\,dy\,. \tag{4.71}$$

Successive differentiations (bearing in mind the constant width and/or zero boundary conditions in the lateral boundaries) yield

$$F'(x) = 2\int h\,u\,u_x\,dy\,, \tag{4.72}$$

$$F''(x) = 2\int hu_x^2\,dy + 2\int h\,u\,u_{xx}\,dy\,. \tag{4.73}$$

The limits of integration are to be understood here and subsequently.

Considering the last term in (4.73) yields

$$\int huu_{xx}\,dy = -\int u(hu_y)_y\,dy = \int hu_y^2\,dy \tag{4.74}$$

where $(4.70_1), (4.70_2)$ and integration by parts have been used. In view of (4.73), (4.74) we obtain

$$F''(x) = 2\int h(u_x^2 + u_y^2)\,dy\,. \tag{4.75}$$

As h is positive, we have the first (trivial) result

$$F''(x) \geq 0\,, \tag{4.76}$$

i.e. $F(x)$ is a convex function of x. A simple upper bound (in terms of data) follows from this and (4.2):

$$F(x) \leq F(0)(1 - x/L)\,, \tag{4.77}$$

where $F(0)$ is expressible in terms of $u_0(y) : F(0) = \int u_0^2\,dy$.

The inequality (4.76) can be sharpened by further analysis, and improved upper bounds for $F(x)$ obtained; one may, for example, use Schwarz's inequality in conjunction with the above analysis to prove the stronger result $(F^{1/2})'' \geq 0$. *Two* different approaches, both of which are instructive, are described, and the results embodied in Theorems.

First approach. The conservation principle (analogue of conservation of energy for the wave equation) is easily derived from $(4.70_1), (4.70_2)$:

$$\int hu_x^2\,dy - \int hu_y^2\,dy = E(\text{constant})\,. \tag{4.78}$$

The proof is left as an exercise. Since $u = 0$ on the edge $x = L$, it is apparent that $E \geq 0$, whence

$$\int hu_x^2 dy \geq \int hu_y^2 dy\,. \tag{4.79}$$

Use of this in (4.75) yields

$$F''(x) \geq 4 \int hu_y^2 dy\,. \tag{4.80}$$

In view of inequality (14) of appendix,

$$\int hu_y^2 dy \geq \lambda_1 \int hu^2 dy \tag{4.81}$$

where λ_1 is the lowest eigenvalue of

$$\begin{cases} \dfrac{d}{dy}\{h(y)\dfrac{d\phi}{dy}\} + \lambda h(y)\phi = 0 \qquad \text{in } 0 < y < l\,, \\ \\ \phi(0) = \phi(l) = 0\,. \end{cases} \tag{4.82}$$

We note that when $h(y) =$ constant, $\lambda_1 = \pi^2 l^{-2}$.

It follows from (4.80)–(4.82) that

$$F''(x) - 4\lambda_1 F(x) \geq 0\,. \tag{4.83}$$

In view of Theorem 4.3 with $h(x, F, F') = 4\lambda_1 F$, we obtain

Theorem 4.6 - *The cross-sectional measure $F(x)$, defined by (4.71), of the solution of the boundary value problem (4.70) satisfies*

$$F(x) \leq G(x) \tag{4.84$_1$}$$

where $G(x)$ satisfies

$$\begin{cases} G''(x) - 4\lambda_1 G(x) = 0\,, \\ \\ G(0) = F(0) = \displaystyle\int u_0^2 dy\,, \\ G(L) = 0\,. \end{cases} \tag{4.84$_2$}$$

It is left to the reader to obtain $G(x)$ explicitly, and to prove that as $L \to \infty$, one has

$$F(x) \leq F(0)\exp(-2\sqrt{\lambda_1}x)\,. \tag{4.85}$$

This completes the first approach.

Second approach: Use of (4.75), (4.81), (4.82) yields

$$F''(x) \geq 2\lambda_1 F + 2 \int hu_x^2 dy\,. \tag{4.86}$$

We introduce $F^{1/2}(x)(=\sqrt{F(x)})$ and readily verify that

$$(F^{1/2})'' = \frac{1}{2}F^{-3/2}[FF'' - \frac{1}{2}F'^2]. \tag{4.87}$$

This together with (4.72) and (4.86), yields

$$(F^{1/2})'' - \lambda_1 F^{1/2} \geq F^{-3/2}[\int hu^2 dy \int hu_x^2 dy - \{\int huu_x dy\}^2]. \tag{4.88}$$

The right hand side is non-negative (by Schwarz's inequality), and we obtain

$$(F^{1/2})'' - \lambda_1 F^{1/2} \geq 0. \tag{4.89}$$

Analogous to (4.84), we obtain (using Theorem 4.5)

Theorem 4.7 - *The cross-sectional measure $F(x)$, defined by (4.71), of the solution of the boundary value problem (4.70) satisfies*

$$F^{1/2}(x) \leq G(x) \tag{4.90}$$

where $G(x)$ satisfies

$$G'' - \lambda G = 0,$$

$$G(0) = \{F(0)\}^{1/2} = \{\int u_0^2 dy\}^{1/2} \quad , \quad G(L) = 0. \tag{4.91}$$

This completes the second approach.

It is left to the reader to obtain (this) $G(x)$ explicitly, and to prove that (4.85) is obtained, once again, in the limiting case $L \to \infty$. Other points left as exercises to the reader are:

(i) to prove that the bound represented by (4.90), (4.91) is not inferior to that represented by $(4.84_1), (4.84_2)$;

(ii) to prove that the equality sign occurs in (4.89), (4.90) when $u_0(y)$ is (proportional to) the eigenfunction corresponding to the lowest eigenvalue of (4.82);

(iii) and that equality occurs in (4.85) in the latter circumstances (when $L \to \infty$).

It is not difficult to prove that (4.89)–(4.90) remain valid when curved lateral boundaries replace straight (parallel to the x axis) boundaries provided that λ_1 is modified slightly: it is the analogous eigenvalue but will in the new circumstances depend on x.

Confining attention to the case $h(y)$ = constant for the moment, one can prove, without difficulty, for the modified cross-sectional measure

$$F(x) = \int u_y^2 dy,$$

that (4.90), (4.91) remain valid *mutatis mutandis*. Pointwise upper bounds for $|u|$ which reflect position are deducible from these upper bounds; see Appendix. These are left as exercises.

It is possible to use methods of a similar type to those already discussed, to gain some insight into the solutions of (4.70_1) when the rectangular region extends to infinity in the $\pm x$ directions. The analysis which follows is suggested by the remarks of Ericksen [138] as to how the Saint-Venant solutions for an elastic cylinder might be characterized; also see [139] in this connection. Suppose $-\infty < x < \infty$, and suppose

$$u_x = 0$$

or

$$u = \text{constant} \tag{4.92}$$

on the lateral boudaries $y = 0, l$. Defining the cross-sectional measure

$$F(x) = \int h(y) u_x^2 dy \ , \tag{4.93}$$

one readily proves that

$$F''(x) = 2 \int h(u_{xx}^2 + u_{yx}^2) dy \,. \tag{4.94}$$

Once again, $F(x)$ is seen to be a convex function of x since $h > 0$. We come to the essential point: Suppose we posit – as seems reasonable – that $F(x)$ is everywhere finite – at $x = \pm\infty$ in particular. It is clear from the convexity of F that $F \equiv$ constant. Thus $F'' = 0$, and, in view of (4.94), we obtain $u_{xx} = u_{xy} = 0$. It readily follows that $u = u(y)$ and satisfies

$$\frac{d}{dy}\{h(y)\frac{du}{dy}\} = 0 \,. \tag{4.95}$$

This is, in principle, easily solved for u under the conditions (4.92).

4.4.3 Lemmas relevant to Regions with Convex (Concave) Boundaries

We have mentioned some examples (e.g. paragraphs preceding (4.92)) where upper bounds in terms of data may be obtained for cross-sectional integral measures of the type $\int \psi_y^2 dy$. An important aspect of these is that pointwise estimates for ψ, which reflect position, are deducible therefrom. We proceed to prove a lemma which proves very useful in the analysis of cross-sectional measures of this and similar types in various contexts. It bears a considerable resemblance to Lemma 3.1 – a resemblance heralded in subsection 3.2.2. We also give auxiliary lemmas which are interesting special cases of the foregoing.

Let us first describe the general *context*. A simply connected region R (see Figures 4.5 and 4.6) in the xy (rectangular cartesian) plane is

(i) bounded by straight line segments Γ_0, Γ_L with abscissae $x = 0, x = L$ respectively, either or both of which may, in limiting case, degenerate to a point/vanish,

and

(ii) bounded by two curves C_+ (upper) and C_- (lower); these curves $y = y_+(x), y = y_-(x)$ respectively (say) have the properties:

(α) they are single-valued functions defined in $0 \leq x \leq L$,

(β) they are twice continuously differentiable with respect to x in $0 < x < L$,

(γ) $y_+(x) \geq y_-(x)$ and the equality sign can only occur therein – in degenerate cases – at $x = 0$ and/or $x = L$.

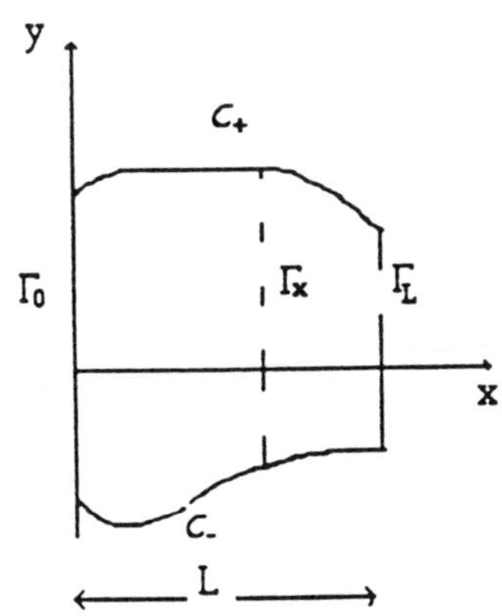

Figure 4.5

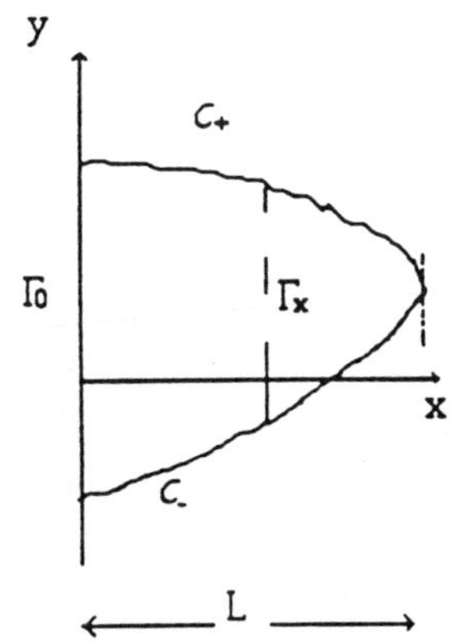

Figure 4.6

Let Γ_x be the straight line segment contained in R whose points have abscissa x.

A function ψ is defined in the foregoing region R which is

(a) three times continuously differentiable in $\overline{R}/(\Gamma_0 \cup \Gamma_L)$,

(b) such that

$$\psi = \mathcal{F}_+(x) \text{ on } C_+ \quad , \quad \psi = \mathcal{F}_-(x) \text{ on } C_- , \tag{4.96}$$

where $\mathcal{F}_+, \mathcal{F}_-$, are assigned functions.

Let $\phi(u)$ be a twice differentiable function of the single variable u and let a superposed dot(s) denote differentiation(s) with respect to u.

Lemma 4.3 - *In the context and notation outlined in the last three paragraphs, the quantity*

$$F(x) = \int_{\Gamma_x} \phi(\psi_y) dy \tag{4.97}$$

satisfies, for $0 < x < L$,

$$F'(x) = \int_{\Gamma_x} \dot{\phi}\psi_{yx} dy + \phi y'] \,, \tag{4.98}$$

$$F''(x) = \int_{\Gamma_x} \ddot{\phi}(\psi_{xy}^2 - \psi_{xx}\psi_{yy})dy + \{-(\psi_y \dot{\phi}(\psi_y) - \phi(\psi_y))y'' + \dot{\phi}\mathcal{F}''\}] \tag{4.99}$$

wherein the symbol $-]$ *is defined thus:*

$$g(x,y)] = g(x, y^+(x)) - g(x, y^-(x)) \,,$$

it being understood that subscripts $+,-$ *are appended to* $\mathcal{F}$, *as appropriate.* (i.e. $\dot{\phi}\mathcal{F}'']$ *is the difference between values of* $\dot{\phi}\mathcal{F}''$ *at the extremities of* Γ_x).

Proof. The proof depends largely on the readily verified relations (chain rule and Leibnitz rule)

$$\frac{d}{dx} g(x, y(x)) = g_x + g_y y' \,, \tag{4.100$_1$}$$

$$\frac{d}{dx} \int \phi(x,y)dy = \int_{\Gamma_x} \phi_x dy + \phi y'] \,. \tag{4.100$_2$}$$

Straightforward use of these yields (4.98) together with

$$\begin{aligned} F'' &= \int (\ddot{\phi}\psi_{yx}^2 + \dot{\phi}\psi_{yxx})dy + \{\dot{\phi}(2\psi_{xy}y'^2 + \psi_{yy}y'^2) + \phi y''\}] \\ &= \int \ddot{\phi}(\psi_{yx}^2 - \psi_{xx}\psi_{yy})dy + \{\dot{\phi}(\psi_{xx} + 2\psi_{xy}y' + \psi_{yy}y'^2) + \phi y''\}] \end{aligned} \tag{4.101}$$

where integration by parts has been used in the last line. Here and subsequently the domain of integration will be understood.

Differentiating relations (4.96) twice – and omitting the subscripts – we obtain, on using (4.100$_1$):

$$\psi_x + \psi_y y' = \mathcal{F}'(x) \,, \tag{4.102$_1$}$$

$$\dot{\phi}(\psi_{xx} + 2\psi_{xy}y' + \psi_{yy}y'^2 + \psi_y y'') = \dot{\phi}\mathcal{F}''(x) \,. \tag{4.102$_2$}$$

Substitution of (4.102$_2$) into (4.101) gives (4.99). This completes the proof. Let us note that this Lemma and subsequent ones remain valid if the posited smoothness for ψ is somewhat relaxed.

In view of the importance of the special case $\phi(u) = u^2$ we record the appropriate result separately; two cases are considered, and the results embodied in the following two lemmas.

Lemma 4.3a - *In the context of the preceding Lemma except that*

$$\psi = 0 \qquad \text{on } C_+, C_- \,, \tag{4.103}$$

and defining

$$F(x) = \int_{\Gamma_x} \psi_y^2 dy$$

one has

$$F'(x) = 2\int \psi_y \psi_{xy} dy + \psi_y^2 y'] \,, \tag{4.104}$$

$$F''(x) = 2\int (\psi_{xy}^2 - \psi_{xx}\psi_{yy})dy - \psi_y^2 y''] \,. \tag{4.105}$$

The results follow automatically from Lemma 4.3.

Lemma 4.3b - *In the context of Lemma* 4.3, *suppose that*

(*i*)
$$y_+''(x) < 0 \quad , \quad y_-''(x) > 0 \,, \tag{4.106}$$

(i.e. *the curves* C_+, C_- *are* <u>*strictly*</u> *convex*),

or

(*ii*)
$$y_+''(x) > 0 \quad , \quad y_-''(x) < 0 \,, \tag{4.107}$$

(i.e. *the curves* C_+, C_- *are* <u>*strictly*</u> *concave*).

Defining

$$F(x) = \int \psi_y^2 dy \,, \tag{4.108}$$

one has

$$F''(x) \geq 2\int (\psi_{xy}^2 - \psi_{xx}\psi_{yy})dy + \left(\frac{\mathcal{F}''^2}{y''}\right)] \tag{4.109$_1$}$$

(case (i) above),

$$F''(x) \leq 2\int (\psi_{xy}^2 - \psi_{xx}\psi_{yy})dy + \left(\frac{\mathcal{F}''^2}{y''}\right)] \tag{4.109$_2$}$$

(case (ii) above).

The equality sign occurs in $(4.109_1), (4.109_2)$ *when* ψ *is linear in* x *and* y ($\psi = c_1 x + c_2 y + c_3, c.$ *being constants*).

Proof. It follows from (4.99) that

$$F'' = 2\int (\psi_{xy}^2 - \psi_{xx}\psi_{yy})dy + \{(-y'')\psi_y^2 + 2\psi_y \mathcal{F}''\}] \,. \tag{4.110}$$

As the boundary terms are quadratic in ψ_y the results required follow on completing the square. It is left as an exercise to the reader to investigate the circumstances in which the equality signs occur in $(4.109_1), (4.109_2)$.

A lower bound for $F(x)$, defined above, in the general context of Lemma 4.3, is easily derived from Schwarz's inequality:

Lemma 4.4 - *In the general context of Lemma* 4.3

$$\int \psi_y^2 dy \geq \frac{\{\mathcal{F}]\}^2}{l(x)} \tag{4.111}$$

where $l(x)$ denotes the length of Γ_x and where $\mathcal{F}] = \mathcal{F}_+ - \mathcal{F}_-$. The equality sign holds iff

$$\psi = a(x) + b(x)y \tag{4.112}$$

where a, b are arbitrary functions.

This completes the discussion of the Lemmas *per se*. For illustrative purposes we apply some of the foregoing considerations to a harmonic function. Suppose, in the context of Lemma 4.3, that ψ is harmonic:

$$\nabla_1^2 \psi = 0\,; \tag{4.113}$$

that ψ is specified on Γ_0, Γ_L where these are non-degenerate; and that C_+, C_- are *strictly* convex. Lemma 4.3b yields

$$F''(x) \geq \left.\frac{\mathcal{F}''^2}{y''}\right] . \tag{4.114}$$

The equality sign occurs if ψ is linear in x. Thus for the Dirichlet problem for the harmonic equation in the context outlined, we have given in the following:

Theorem 4.8 - *In the general context of Lemma* 4.3, *suppose that ψ is a harmonic function* (i.e. *satisfies* (4.113)) *in R, that the lateral boundaries C_+, C_- are strictly convex, and that ψ is specified on the segments Γ_0, Γ_L (assumed non-degenerate), then the cross-sectional measure*

$$F(x) = \int \psi_y^2 dy$$

satisfies the continued inequality

$$G(x) \geq F(x) \geq \frac{\{\mathcal{F}]\}^2}{l(x)}\,, \tag{4.115}$$

where $G(x)$ satisfies

$$G''(x) = \left.\frac{\mathcal{F}''^2}{y''}\right] \tag{4.116$_1$}$$

subject to

$$G(0) = F(0) \quad , \quad G(L) = F(L)\,, \tag{4.116$_2$}$$

and where $l(x)$ *denotes the length of* Γ_x.

Both upper and lower bounds for $F(x)$ *are attained when* ψ *is any linear function of* x, y.

Proof. The upper bound follows from (4.114) together with Theorem 4.4 and the lower bound follows from (4.111). That the equality signs hold in (4.115) when ψ is any linear function of x, y, is obvious from previous remarks.

As usual, upper bounds for $|\psi|$, which reflect position, are deducible from $G(x)$.

We now discuss a similar result for harmonic functions, which has interesting ramifications. Lemma 4.3a gives rise to the following simple result for harmonic functions.

Theorem 4.9 - *In the general context of Lemma* 4.3, *suppose that* ψ *is harmonic in* R (i.e. *satisfies* (4.113) *in* R); *that the lateral boundaries* C_+, C_- *are convex* (*outwards*)

$$y''_+(x) \leq 0 \quad , \quad y''_-(x) \geq 0\,, \tag{4.117$_1$}$$

and that

$$\psi = 0 \qquad \text{on } C_+, C_-\,. \tag{4.117$_2$}$$

Then

$$F(x) = \int \psi_y^2 \, dy$$

satisfies

$$F''(x) \geq 0 \tag{4.117$_3$}$$

(i.e. *is a convex function of* x).

We note the following consequence of the foregoing proposition: If ψ is a specified (continuously differentiable) function of y on the vertical edges Γ_0, Γ_L – assumed non-degenerate – $F(0), F(L)$ are computable, and the explicit upper bound for $F(x)$ follows:

$$F(x) \leq F(0) + \{F(L) - F(0)\}(x/L)\,. \tag{4.118}$$

[Obviously $F(0) = 0$ or $F(L) = 0$ if Γ_0 or Γ_L (respectively) is degenerate.] It is evident that (4.118) follows from (4.117$_3$) and (4.2). Naturally, upper bounds for $|\psi(x, y)|$ in terms of data, reflecting position, are readly deducible from (4.118).

An interesting question arises in relation to the boundary convexity requirement in the foregoing proposition: does the requirement that the lateral boundaries be convex [(4.117$_1$)] reflect a reality? The answer is in the affirmative in that it prevents "re-entrant angles" – which are known to give rise to singularities in the derivative of ψ. To fix ideas, imagine the following extreme situation:

the boundary C_+ is extended by slitting it vertically (parallel to the y axis), and suppose that condition $\psi = 0$ continues to apply along the slit. It is known that ψ_y becomes singular at the slit tip; indeed, if r denotes the distance from the slit tip, one has $|\psi_y| = O(r^{-1/2})$. [See exercise 4.17 in this connection.] Plainly $\int \psi_y^2 dy$ evaluated along the line Γ_x of the slit is divergent (having a logarithmic singularity due to the presence of a term like r^{-1} in the integrand).

The manner in which the convexity of $F(x)$ fails at the slit tip is illustrated in Figure 4.7.

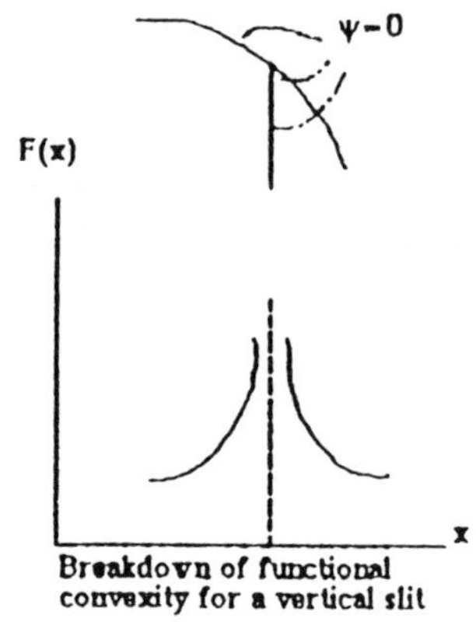

Figure 4.7

The condition that the lateral boundaries be convex prevents a breakdown in the convexity of $F(x)$ arising in this manner. Whereas it is true that the slit as envisaged is inconsistent with the boundary smoothness posited, nevertheless one can have a boundary of the required smoothness which is, in a certain sense, arbitrarily close to the slit. Whereas $F(x)$ is not singular (infinite) at the apex of the slit-like portion of the boundary in these circumstances, convexity of $F(x)$ may be expected to break down thereat i.e. the graph exhibits a 'bleep' (see Figure 4.8).

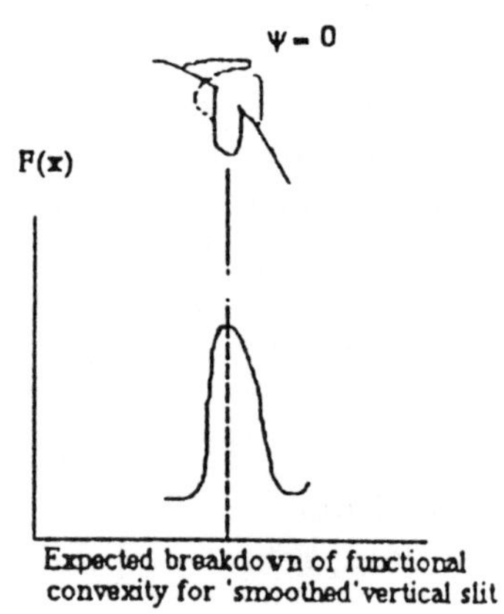

Figure 4.8

These considerations do not imply, of course, that convexity of the lateral boundaries is necessary for the convexity of $F(x)$. Indeed, convexity of $F(x)$ may be recovered if the lateral boundaries are concave, but there is a limit to the measure of concavity allowed (as the previous discussion shows). To gain insight into this, and to give an estimate of the allowable measure of concavity in the context, consider the following: suppose C_+ or a portion thereof, is (stictly) concave i.e. $y''_+ > 0$, and, for definiteness, that $y_- \equiv 0$ (the boundary C_- concides with the x axis). Lemma 4.3a implies that

$$F''(x) \geq 0$$

provided that

$$2\int \psi_{xy}^2 \, dy \geq y''_+ \psi_y^2(x, y_+)$$

– which is seen to be equivalent to

$$2\int \psi_{xy}^2 \, dy \geq \frac{y''_+}{y'^2_+} \psi_x^2(x, y_+) \qquad (y'_+ \neq 0)\,, \tag{4.119}$$

on using $\psi = 0$ on C_+ in differentiated form (cf.(4.102$_1$)). Schwarz's inequality implies that

$$[\int \psi_{xy}^2 \, dy] y_+(x) \geq \psi_x^2(x, y_+)\,. \tag{4.120}$$

The inequalities (4.119) and (4.120) imply that $F(x)$ is convex provided that the boundary concavity is limited by

$$2y_+^{-1} \geq \frac{y''_+}{y'^{\,2}_+} \qquad (y'_+ \neq 0) \tag{4.121}$$

under the stated conditions. An interesting issue is: what is the maximum boundary concavity – in some sense – which is consistent with the convexity of $F(x)$?

4.4.4 Non-Rectilinear Cross-sections: An Example

We now mention an example in which the "sections" are not rectilinear. Consider the region R: $r_0 \leq r \leq r_1,\, 0 < \theta < \alpha$ where (r, θ) are plane polar coordinates and where r_0, r_1, α are given constants. Suppose ψ is a sufficiently smooth function defined therein which satisfies

$$\begin{cases} \psi_{rr} + r^{-1}\psi_r + r^{-2}\psi_{\theta\theta} - n^2\psi = 0 & \text{in } R\,, \\ \psi = 0 \quad \text{on } \theta = 0\,, \qquad \psi_\theta = 0 \quad \text{on } \theta = \alpha\,, \\ \psi = \phi_0(\theta) \quad \text{on } r = r_0\,, \qquad \psi = \phi_1(\theta) \quad \text{on } r = r_1\,, \end{cases} \tag{4.122}$$

where n is a given constant.

The quantity ψ may be interpreted as the small transverse deflection of an elastic transverse force (whose strength is proportional to n^2); the deflection is specified on three edges and the fourth $\theta = \alpha$ is free.

Defining

$$F(r) = \int_0^{\alpha} \psi^2 \, d\theta \tag{4.123}$$

(equivalent to $\int_0^{\alpha}\psi^2 r d\theta$) one may prove – analogous to the analysis leading to (4.89) – that

$$(F^{1/2})'' + r^{-1}(F^{1/2})' - (\lambda r^{-2} + n^2)F^{1/2} \geq 0 \tag{4.124}$$

where

$$\lambda = \pi^2/(4\alpha^2)$$

and where primes denotes differentiation with respect to r. By making the substitution $\sigma = \log r$ in (4.124), one sees that Theorem 4.4 implies

Theorem 4.10 - *The cross-sectional measure* (4.123) *of the boundary value problem* (4.122) *satisfies the inequality*

$$F^{1/2} \leq G \tag{4.125$_1$}$$

where G satisfies

$$G'' + r^{-1}G' - (\lambda r^{-2} + n^2)G = 0 \tag{4.125$_2$}$$

where

$$\lambda = \pi^2/(4\alpha^2)\,,$$

subject to

$$G(r_0) = \{\int_0^{\alpha} \phi_0^2(\theta)d\theta\}^{1/2} \quad , \quad G(r_1) = \{\int_0^{\alpha} \phi_1^2(\theta)d\theta\}^{1/2} . \tag{4.125$_3$}$$

It is left as an exercise to the reader to prove the various steps, and to solve explicitly for G in terms of Hankel functions $H_{\pm n}(\lambda^{1/2}r)$ etc.

4.4.5 Time-dependent Problem

We now apply the method of cross-sections to a problem whose solution depends on time as well as on spatial variables; see [132] for a related analysis. We consider an initial boundary value problem for the heat conduction equation in a right cylinder R.

Suppose that x_i denotes rectangular cartesian coordinates such that the x_3 axis is parallel to the generators, and that the plane ends correspond to $x_3 = 0, x_3 = L$. Let Γ_{x_3} denote the open cross-section of R with abscissa x_3 and $\partial\Gamma_{x_3}$ its boundary, and let the lateral surface of the cylinder be denoted by C. Let us suppose that R is occupied by a homogeneous, isotropic heat conductor, that its temperature is specified at a given time $t=0$, and that its entire surface is subject to zero temperature from $t = 0$ onwards: specifically, suppose that the (time dependent) temperature field $u(x_1, x_2, x_3, t)$ is a classical solution of the following initial boundary value problem:

$$\nabla^2 u - u_t = 0 \qquad \text{in } R \times (0,\infty) \tag{4.126$_1$}$$

subject to

$$u = 0 \qquad \text{on } C \cup \Gamma_{x_0} \cup \Gamma_{x_L} \times [0, \infty) \tag{4.126$_2$}$$

and

$$u(x_1, x_2, x_3, 0) = \phi(x_1, x_2)(\sin n\pi x_3 / L), \tag{4.126$_3$}$$

n being a positive integer and ϕ a given function.

Introduce the cross-sectional measure of the solution

$$F(x_3, t) = \int_{\Gamma_{x_3}} u^2 dA \tag{4.127}$$

and obtain by straightforward differentiation:

$$F_t = \int 2uu_t dA, \tag{4.128$_1$}$$

$$F_{,3} = \int 2uu_{,3} dA, \tag{4.128$_2$}$$

$$F_{,33} = \int (2u_{,3}^2 + 2uu_{,33}) dA. \tag{4.128$_3$}$$

Use of $(4.126_1), (4.126_2), (4.127), (4.128_1), (4.128_3)$ together with integration by parts, followed by use of A4(i)(a), gives

$$F_{,33} \geq 2 \int u_{,3}^2 dA + 2\lambda_1 F + F_t \tag{4.129}$$

where λ_1 is the lowest "fixed-membrane" eigenvalue for the cross-section – that is to say, the lowest eigenvalue of

$$\nabla_1^2 \varphi + \lambda \varphi = 0 \qquad \text{in } \Gamma_{x_3}, \tag{4.129a}$$

$$\varphi = 0 \qquad \text{on } \partial\Gamma_{x_3}.$$

Now

$$\begin{aligned}(F^{1/2})_{,33} &= \frac{1}{2} F^{-3/2} [F F_{,33} - \frac{1}{2} F_{,3}^2] \\ &\geq \frac{1}{2} F^{-3/2} [2\{ \int u^2 dA \int u_{,3}^2 dA - (\int uu_{,3} dA)^2 \}] + \lambda_1 F^{1/2} + (F^{1/2})_t.\end{aligned} \tag{4.130}$$

In view of Schwarz's inequality we obtain

$$(F^{1/2})_t - (F^{1/2})_{,33} + \lambda_1 F^{1/2} \leq 0. \tag{4.131}$$

Defining

$$\mathcal{F} = e^{\lambda_1 t} F^{1/2}, \tag{4.132}$$

we obtain

$$\mathcal{F}_t - \mathcal{F}_{,33} \leq 0. \tag{4.133}$$

One may prove – using a comparison theorem similar to Theorem 4.4 (see A5) – that

$$\mathcal{F} \leq G \tag{4.134$_1$}$$

where $G(x_3, t)$ satisfies

$$G_t - G_{,33} = 0 \quad , \qquad x_3 \in (0, L) \times t \geq 0 , \tag{4.134$_2$}$$

$$G(0,t) = G(L,t) = 0 , \tag{4.134$_3$}$$

$$G(x_3, 0) = \{ \int \phi^2 dA \}^{1/2} \sin \frac{n\pi x_3}{L} . \tag{4.134$_4$}$$

Combining (4.132), (4.134$_1$) together with the explicit solution of the problem defined by (4.134$_2$)–(4.134$_4$) we obtain

Theorem 4.11 - *The cross-sectional measure* (4.127) *of the initial boundary value problem defined by* (4.126) *etc. satisfies the inequality*

$$F(x_3, t) \leq \{ \int \phi^2 dA \} e^{-2(\lambda_1 + n^2\pi^2 L^{-2})t} \sin^2(n\pi x_3 / L). \tag{4.135}$$

4.4.6 Closure

Let us conclude with the following observation: the cross- sectional method – as developed here – is relatively new, and its potential is not yet fully realized. It is a continuing challenge, therefore, to find further, more revealing analyses – not least the finding of the cross-sectional measures themselves – in order more effectively to elucidate natural phenomena modelled by P.D.E.s.

4.5 EXERCISES

Exercise 4.1

(a) Prove the lower bound (4.2) by successive integrations of $f''(x) = 0$.

(b) Show that the upper bound (4.2) may be proved by establishing that for $H(x) \in C^2[a,b]$ which satisfy

$$H''(x) \geq 0\,,$$

$$H(a) = H(b) = 0\,,$$

one has

$$H(x) \leq 0\,.$$

Prove that this may be established thus:

(i) Suppose, *per contra*, that $\exists x_0\,(a < x_0 < b)$ such that $H(x_0) > 0$;

(ii) prove that $\exists$ an interval $x_0^{(1)} < x_0 < x_0^{(2)}$ within which $H(x) > 0$, and such that $H(x_0^{(1)}) = H(x_0^{(2)}) = 0$;

(iii) consider

$$\int_{x_0^{(1)}}^{x_0^{(2)}} HH''\,dx$$

with a view to obtaining a contradiction [see "Generalized Convexity", Section 4.4, if necessary].

Exercise 4.2

(i) Prove (4.5_1) as follows: multiply (4.3_1) by $2\overline{u}$, take the imaginary part of the resulting expression, to obtain*

$$\frac{\partial}{\partial t}|u|^2 = 2\{(u_1 u_{2x} - u_2 u_{1x})\}_x\,;$$

assuming $u_\alpha, u_{\alpha x}$ decay sufficiently rapidly as $|x| \to \infty$, prove that integration over $\mathbb{R}$ leads to the required result.
Give explicit criteria for the rapidity with which the aforementioned quantities need to decay.

(ii) Prove (4.5_2) in a similar manner: multiply* in (i) preceding by x^2 and integrate.
Give explicit criteria for the rapidity with which u_α and relevant derivatives need to decay.

(iii) Prove (4.5_3) as follows: multiply (4.3_1) by $2\overline{u}_t$; integrate the real part of the corresponding identity over $\mathbb{R}$ to obtain

$$0 = -\frac{d}{dt}\|u_x\|^2 + \int f(|u|^2)\frac{\partial}{\partial t}|u|^2 dx$$

on assuming sufficiently rapid decays as $|x| \to \infty$.
Give explicit criteria for the required rapidity of decay.

(iv) Prove (4.5_4) as follows: use (4.3_1) and integration to establish that

$$\frac{d}{dt}\int \operatorname{Im} x\bar{u}_x u dx + \int (u_x\bar{u}_x - f(|u|^2)|u|^2)dx = \operatorname{Re}\int 2ix\bar{u}_x u_t dx$$
$$= \operatorname{Re}\int 2x\bar{u}_x u_{xx} dx + \operatorname{Re}\int 2xf\,\bar{u}_x u dx\,.$$

Apply a further integration to each of the latter two terms.

Exercise 4.3 - Give details of the derivation of (4.13) and justify the conclusion that can be drawn therefrom (mentioned in text).

Exercise 4.4 - Derive a lower bound for $F(t)$, implied by **(4.11)**, in terms of $F(0), t, t_c$. Derive a lower bound, in similar terms, which complements (4.13).

Exercise 4.5 - If $f(|u|^2) = |u|^{2m}$ where m is a positive integer, what conditions must be satisfied by m and α in order that (4.8) be satisfied?

Exercise 4.6 - Prove that the analysis leading to (4.25) continues to be valid if, instead of (4.18), the following is posited:

$$\frac{1}{2}\int_0^1 \rho_0 \dot{u}^2 dx + \int_0^1 W dx \le E(0)$$

where $E(0)$ is a given constant.
[The latter hypothesis is employed in connection with weak solutions.]

Exercise 4.7 - If $W(u_x) = \sum_{n=1}^{N} c_n u_x^{2n}$ where c_n are given constants, determine conditions on c_n sufficient for the validity of (4.19).

Exercise 4.8 - In the case where $F(x)$ satisfies (4.30) with $a_1 = 0$, derive upper and lower bounds therefor, analogous to those of (4.29). [Inequality (4.36) should prove useful.]

Exercise 4.9

(i) Prove the 'conservation law' (4.44).

(ii) Prove the more general result (4.54) for the context outlined.

(iii) What modification must be made in (4.54) if the defining context is altered in one respect: the sense of the inequality restrictions on the slopes of the lateral sides is reversed – that is to say, they become

$$y'_+(x) \ge 0 \quad , \quad y'_-(x) \le 0\,.$$

Exercise 4.10 - Referring to the modified Cauchy problem (essentially) defined by (4.37)–(4.38), formulate the corresponding problem for a right cylinder of arbitrary cross-section and give the analogous analysis.

Exercise 4.11 - Derive a lower bound which complements (4.49) in the context $E \leq 0$ with $\beta = 0$. Assuming $\int_0^1 fgdy > 0$, prove that the maximum allowable value of x (i.e. X) cannot exceed

$$\frac{1}{2}\left\{\frac{\displaystyle\int_0^1 f^2 dy}{\displaystyle\int_0^1 fgdy}\right\} \log\left\{\frac{M^2}{\displaystyle\int_0^1 f^2 dy}\right\} .$$

Exercise 4.12 - In connection with the modified Cauchy problem outlined in Observation 3, ((4.53) etc.) carry out an analysis analogous to that given in the text. In particular, prove uniqueness and continuous dependence.

Exercise 4.13 - The partial differential equation

$$u_t + u_{xx} = 0 \qquad 0 < x < 1 \quad , \quad 0 < t < T ,$$

can be envisaged as that governing the temperature u at position x at time t prior to an initial instant $t = 0$.
Consider the initial boundary-value problem consisting of this P.D.E. subject to the boundary conditions

$$u(0,t) = 0 \quad , \quad u(1,t) = 0 ,$$

and the initial condition

$$u(x,0) = f(x)$$

[a classical solution is assumed].

(i) Prove that the latter problem is ill-posed by first proving that

$$u = e^{n^2\pi^2 t} \sin n\pi x/n ,$$

(n being a positive integer) satisfies the P.D.E. (and boundary conditions).

(ii) Define a measure of the solution by

$$F(t) = \int_0^1 u^2 dx$$

and prove that is satisfies

$$F'(t) = \int_0^1 2uu_t dx$$

$$F''(t) = 4\int_0^1 u_t^2 dx .$$

Use the foregoing together with Schwarz's inequality to prove that $F(t)$ is a logarithmically convex function. Hence prove that the solution of the initial boundary value problem is unique (assuming it exists).

(iii) Suppose that the conditions defining the initial boundary value problem are supplemented by the "stabilizing condition"

$$|u| \leq M \qquad \text{for} \qquad 0 \leq t \leq T$$

where T, M are positive constants (e.g. the melting temperature material is M). Use logarithmic convexity to prove the estimate

$$F(t) \leq M^{2t/T}\,[F(0)]^{1-t/T} \qquad 0 \leq t \leq T\,;$$

that is to say, that Hölder continuous dependence on the initial condition is recovered – as a consequence of the stabilizing assumption.

(iv) Derive the lower bound for $F(t)$:

$$F(t) \geq \int_0^1 f^2(x)dx \, \exp\left[\frac{2\displaystyle\int_0^1 f'^2(x)dx}{\displaystyle\int_0^1 f^2(x)dx}t\right]$$

assuming $f(0) = f(1) = 0$. Hence prove that the maximum time T for which the analysis of the "stabilized" initial boundary value problem is valid is such that

$$\frac{1}{2\pi^2}\log\left\{\frac{M^2}{\displaystyle\int_0^1 f^2(x)dx}\right\} \geq T\,.$$

[In connection with this example, see the remarks of Clerk Maxwell mentioned in chapter five.]

Exercise 4.14 - Prove Theorem 4.5.

Exercise 4.15 - Referring to the boundary value problem specified by (4.70):

(i) Obtain the explicit solution of (4.84_2), and verify, in the limiting case of a semi-infinite strip ($L \to \infty$), that (4.85) is obtained.

(ii) Obtain the explicit solution of (4.91), and prove, in the limiting case of $L \to \infty$, that (4.85) is again obtained.

(iii) Prove that the upper bound (4.84) is not better than (4.90) – irrespective of the value of L.

(iv) Prove that the equality sign is realized in (4.90), and in (4.84) when $L \to \infty$, when $u_0(y)$ is the eigenfunction corresponding to the lowest eigenvalue of (4.82).

(v) Confining attention to the case $h(y)$=constant, prove that inequalities of the general type $(4.84_1), (4.90)$ are valid when $F(x)$ is redefined by

$$F(x) = \int u_y^2 dy\,.$$

Deduce an upper bound for $|u(x,y)|$ which reflects position.

Exercise 4.16

(a) Suppose a function $F(x)$ satisfies a modified "generalized convexity" condition

$$F''(x) - mF(x) \geq f(x) \quad , \quad 0 \leq x \leq L,$$

where m is a constant such that

$$m > -\pi^2/L^2 .$$

Use inequality A1(i)(a) in connection with a method previously used to prove that

$$F(x) \leq G(x)$$

where $G(x)$ satisfies

$$G''(x) - mG(x) = f(x) \quad , \quad 0 \leq x \leq L,$$

subject to

$$G(0) = F(0) \quad , \quad G(L) = F(L) .$$

[It may be proved that the above principle continues to hold under less restrictive conditions: it is sufficient that the differential inequality and equation for F, G respectively, hold in the *open* interval provided that these functions are continuous in the *closed* interval e.g. see [131]].

(b) Suppose that the problem defined by (4.70) is altered in one respect only: the P.D.E. is changed to

$$h(y)u_{xx} + \{h(y)u_y\}_y + w^2 u = 0 ,$$

w being a constant.
Show how (a) *supra* may be used to modify the various analyses in the text when the constant w satisfies

$$\lambda_1 < w^2 < \lambda_1 + \pi^2 L^{-2} .$$

[This problem is relevant to an inhomogeneous rectangular region one end of which is subject to time-harmonic excitation; e.g. [133].]

Exercise 4.17 - The function

$$\psi = r^{N\pi/\alpha} \sin\left(\frac{N\pi\theta}{\alpha}\right) ,$$

where N is any positive integer, is a harmonic function in the wedge-shaped region $0 < r$, $0 < \theta < \alpha$ ((r, θ) being plane polars). How does this fact motivate the statement, in the paragraph following (4.118), that $\psi = O(r^{-1/2})$ at the slit tip?

Exercise 4.18 - In the context of *Lemma* 4.3a, suppose that C_- is the reflection of C_+ in the x axis, and that conditions are such the ψ is a (harmonic) function even in y. Prove the following analogue of (4.121): if the lateral boundary is (strictly) concave ($y_+'' > 0$) and if its concavity is restricted by

$$2y_+^{-1} \geq y_+'' \qquad (y_+ \neq 0),$$

then $F(x)$ is a convex function of x.

Exercise 4.19 - Defining (or redefining)

$$F(x) = \int h(y)\psi_y^2 dy$$

in the context of *Lemma* 4.3a, $h(y)$ being a positive, continuously differentiable function, derive an expression for $F''(x)$ analogous to (4.105). Suggest uses to which this may be put if ψ satisfies the P.D.E.

$$h(y)\psi_{xx} + \{h(y)\psi_y\}_y = 0 .$$

Exercise 4.20 - Verify (4.124), and obtain the solution of $(4.125_2), (4.125_3)$.

Exercise 4.21

(i) Solve (4.134_2)–(4.134_4) explicitly.

(ii) Under what conditions is the equality sign realized in (4.135)?

Exercise 4.22 - Consider the initial boundary value problem for a two dimensional rectangular strip

$$0 < x_1 < 1 \quad , \quad 0 < x_3 < L$$

analogous to that defined by (4.126). Define

$$F(x_3, t) = \int_0^1 u_{,1}^2 dx_1$$

and show how to obtain an analogous upper bound for it, in terms of data. Show how to deduce 'pointwise' bounds for $|u(x_1, x_3, t)|$.

Chapter five

IDENTITIES HOLDING ALONG SOLUTIONS TO PARTIAL DIFFERENTIAL EQUATIONS: LAGRANGE IDENTITY METHOD

5.1 INTRODUCTION

The Lagrange identity method gives interesting identities governing the solutions of certain classes of partial differential equations (linear ones as hitherto developed) under suitable boundary conditions; the method is applicable to some other equations also, including integral equations. In its simplest manifestation, it depends on two sets of integration by parts. It hinges crucially on the notion of a symmetric operator whose defining symmetry property is often styled a Lagrange identity.

The method is introduced by applying it to two simple partial differential equations, one a diffusion equation (forwards or backwards in time), the other a wave equation or Laplace type equation, under homogeneous boundary conditions. Both treatments are given side by side as we believe that this contributes to the understanding of the method. Having derived the identities we show how they may be applied to prove uniqueness for suitable initial boundary value problems for the two equations, under minimal restrictions.

We proceed to generalize these identities to include some general operator equations. More general identities for pairs of solutions of these equations are then derived, and certain reciprocal relations are deduced therefrom. The forms which they take for the P.D.E. mentioned in the last paragraph are discussed. Two applications of the identities are then given. Firstly, the initial boundary value problem of the diffusion (heat) equation backwards in time is considered; if the temperature is supposed to be bounded above by a given positive constant, the relevant identity is shown to furnish continuous dependence of the solution on data – a property which the problems lacks in the absence of such an *a priori* bound. Secondly, in connection with an initial boundary value problem for the one-dimensional wave equation, the relevant identity is used to prove the principle of equipartition of energy: that is to say, that the time-averaged values of the associated kinetic and potential energies tend to coincidence for large times.

Finally, a second identity for a general "wave like" operator equation is proved. The identity which is in the same spirit as the one already derived in this context, proceeds from what has also been called a Lagrange identity [140]. In the context of the simple one-dimensional wave equation, this second identity may be used as a pivot upon which to mount alternative proofs of the uniqueness issue and the issue of equipartition of energy.

The Lagrange identity method is generally attributed to Brun [141,142] who was the first to use it in the context of elastodynamics. In the context of uniqueness for linear elastodynamics, the method enabled Brun to dispense with major restrictions on the elastic coefficients which were needed up to then – typically, (though not exclusively) that the strain energy density be sign-definite.

In addition to [140-142], one should also consult [34], [143-145] for treatments of Lagrange identity methods and for applications thereof of the general type discussed in this chapter.

5.2 LAGRANGE IDENTITY METHOD FOR TWO P.D.E.

The essence of this method is perhaps best explained, in the first instance, by reference to two simple partial differential equations, each governing a (real) function u of two independent variables x, t. For convenience, and to facilitate reference, we refer to x as a 'space' variable and to t as a 'time' variable; there is no need, of course, to interpret these appellations literally. We consider classical solutions of

$$Ku_t + u_{xx} = 0 \qquad \text{(case (i))}, \tag{5.1}$$

$$Ku_{tt} + u_{xx} = 0 \qquad \text{(case (ii))}, \tag{5.2}$$

where K is a constant (positive or negative). We suppose that

$$0 \leq x \leq 1 \quad , \quad 0 \leq t \leq 2\tau < T$$

where τ and T are constants; and that boundary conditions of the type

$$\alpha u_x + \beta u = 0 \tag{5.3}$$

obtain on $x = 0, x = 1$, where α, β are (real) constants (generally differing on the two ends).

Let us first note that integration by parts, using (5.3), yields

$$\int_0^1 \{u(x, 2\tau - \eta)u_{xx}(x, \eta) - u(x, \eta)u_{xx}(x, 2\tau - \eta)\}dx = 0; \tag{5.4}$$

it is supposed that $0 \leq \eta \leq 2\tau$. This is often termed a Lagrange identity: it is a particular example of

$$\int \{u(\mathbf{x}, 2\tau - \eta) Lu(\mathbf{x}, \eta) - u(\mathbf{x}, \eta) Lu(\mathbf{x}, 2\tau - \eta)\} d\mathbf{x} = 0 \,,$$

where L is a symmetric, linear operator, and where there may be more than one 'space' variable.

The method considers the (similar) identities which are obvious consequences of (5.1) and (5.2)

$$\int_0^\tau \int_0^1 u(\cdot, 2\tau - \eta)\{K\dot{u}(\cdot, \eta) + u_{xx}(\cdot, \eta)\} dx d\eta$$
$$- \int_0^\tau \int_0^1 u(\cdot, \eta)\{K\dot{u}(\cdot, 2\tau - \eta) + u_{xx}(\cdot, 2\tau - \eta)\} dx d\eta = 0 \,, \qquad (5.5)$$

(case(i))

and

$$\int_0^\tau \int_0^1 u(\cdot, 2\tau - \eta)\{K\ddot{u}(\cdot, \eta) + u_{xx}(\cdot, \eta)\} dx d\eta$$
$$- \int_0^\tau \int_0^1 u(\cdot, \eta)\{K\ddot{u}(\cdot, 2\tau - \eta) + u_{xx}(\cdot, 2\tau - \eta)\} dx d\eta = 0 \,, \qquad (5.6)$$

(case(ii)) .

Here and subsequently

(a) the 'spatial' variable within brackets is omitted, and replaced by a centrally placed dot;

(b) a superposed dot or dots denote partial differentiations with respect to the relevant 'time' argument (second argument)

$$\text{e.g.} \quad \dot{u}(\cdot, 2\tau - \eta) = \partial u(\cdot, 2\tau - \eta)/\partial(2\tau - \eta) \qquad [= -\partial u(\cdot, 2\tau - \eta)/\partial\eta] \,.$$

In view of (5.4), the identities (5.5) and (5.6) reduce to

$$\int_0^\tau \int_0^1 \{u(\cdot, 2\tau - \eta)\dot{u}(\cdot, \eta) - u(\cdot, \eta)\dot{u}(\cdot, 2\tau - \eta)\} dx d\eta = 0 \qquad (5.7)$$

(case(i)) ,

and

$$\int_0^\tau \int_0^1 \{u(\cdot, 2\tau - \eta)\ddot{u}(\cdot, \eta) - u(\cdot, \eta)\ddot{u}(\cdot, 2\tau - \eta)\} dx d\eta = 0 \qquad (5.8)$$

(case(ii)) .

We note that both of these identities are independent of K. This is an important observation as the solutions of the P.D.E. change character quite radically as K changes its sign.

We now make the crucial observation that the integrands in (5.7), (5.8) may be expressed as partial derivatives with respect to η: (5.7), (5.8) reduce to

$$\int_0^\tau \int_0^1 \frac{\partial}{\partial \eta}\{u(\cdot, 2\tau - \eta)u(\cdot, \eta)\}dx d\eta = 0 \qquad (5.9)$$

(case(i)),

and

$$\int_0^\tau \int_0^1 \frac{\partial}{\partial \eta}\{u(\cdot, 2\tau - \eta)\dot{u}(\cdot, \eta) + u(\cdot, \eta)\dot{u}(\cdot, 2\tau - \eta)\}dx d\eta = 0 \qquad (5.10)$$

(case(ii)).

As

$$\int_0^\tau \int_0^1 \frac{\partial}{\partial \eta}\{...\}d\eta = \int_0^\tau \frac{d}{d\eta}\left[\int_0^1 \{...\}dx\right] d\eta \qquad (5.11)$$

– an obvious, but important observation – integration with respect to η of (5.9), (5.10) yields

Theorem 5.1 - *Solutions* $u(x,t)$ *of* (5.1) (*case* (i)) *and of* (5.2), (*case* (ii)), *each subject to the boundary conditions* (5.3), *satisfy the following identities*

$$\int_0^1 u^2(x,\tau)dx = \int_0^1 u(x, 2\tau)u(x,0)dx \qquad (5.12)$$

(case(i)),

and

$$\int_0^1 2u(x,\tau)\dot{u}(x,\tau)dx = \int_0^1 \{(u(x,2\tau)\dot{u}(x,0) + u(x,0)\dot{u}(x,2\tau)\}dx \qquad (5.13)$$

(case(ii)).

Let us note that the left hand side of (5.13) may be expressed as $d/d\tau\{\int_0^1 u^2 dx\}$.

Perhaps the simplest application of the foregoing is to uniqueness:

Theorem 5.2 - *The initial boundary value problem consisting of case* (i): (5.1) *subject to the boundary conditions* (5.3) *and to the initial condition*

$$u(x,0) = \textit{specified},$$

and
case (ii): (5.2) *subject to the boundary condition* (5.3) *and to the initial conditions*

$$u(x,0) = \textit{specified} \quad , \quad u_t(x,0) = \textit{specified},$$

have each, at most, one solution.

Proof. Because of the linearity of the problems, it suffices to prove that initial conditions which vanish identically imply solutions which vanish identically for all time. In case (i), 'null' conditions together with (5.12) imply that

$$\int_0^1 u^2(x,\tau)dx = 0\,.$$

Since u is assumed continuous, this implies that $u \equiv 0$ as required. In case (ii), 'null' conditions together with (5.13) imply that

$$\int_0^1 2u(x,\tau)\dot{u}(x,\tau)dx = 0\,.$$

This implies in turn that

$$\frac{d}{d\tau}\int_0^1 u^2(x,\tau)dx = 0$$

(since the spatio-temporal region is rectangular/or $u=0$ on the boundaries $x=$constant). This implies that the integral is constant; the constant is plainly zero and uniqueness follows as in case (i).

The most interesting aspect of these proofs – as compared to elementary uniqueness proofs that are often given – is that they are independent of the signs of these quantities.

5.3 OPERATOR EQUATIONS

The identities (5.12), (5.13) are easily generalized to more general operator equations. Consider two operator equations each governing a (scalar, vector etc.) function u of (up to three) 'space' variables denoted by $\mathbf{x}$, and of a 'time' variable t

$$Ku_t + Lu = 0 \qquad \text{(case (i))}, \tag{5.14}$$

$$Ku_{tt} + Lu = 0 \qquad \text{(case (ii))}, \tag{5.15}$$

where K is a constant (positive or negative) and L is an operator whose properties are described below; it is supposed again that

$$0 \leq t \leq 2\tau < T$$

where T, τ are constants. It is supposed that the domain of definition of the space variables does not vary with time. We envisage a class of functions of position $\mathbf{x}$ and time t – defined therein – which, for each time, constitutes a real vector space H, and that a scalar product $<,>$ is defined therefor [see Appendix of chapter]. It is supposed, for all functions $f(\mathbf{x},t), g(\mathbf{x},t)$ arising, that the *differentiation rule*

$$\frac{d}{dt} < f(\mathbf{x},t), g(\mathbf{x},t) >=< f(\mathbf{x},t), g_t(\mathbf{x},t) > + < f_t(\mathbf{x},t) g(\mathbf{x},t) > \qquad (5.16_1)$$

holds. By L is meant *a real operator* which is *symmetric* i.e.

$$< Lu_1, u_2 >=< u_1, Lu_2 > \qquad (5.16_2)$$

for any pair of appropriate functions.

In the matter of notation, we shall typically write $u(\mathbf{x},\eta) = u(\cdot,\eta)$ i.e. we replace 'spatial' variables by a centralized dot, and, once again, superposed dots denote spatial differentiations with respect to the relevant time (last placed) variable.

Analogous to (5.5), (5.6) one considers the obvious identities, corresponding to (5.14), (5.15), respectively:

$$\begin{aligned} &<u(\cdot, 2\tau - \eta), \{K\dot{u}(\cdot,\eta) + Lu(\cdot,\eta)\} > \\ &\quad - < u(\cdot,\eta), \{K\dot{u}(\cdot, 2\tau - \eta) + Lu(\cdot, 2\tau - \eta)\} >= 0 \end{aligned} \qquad (5.17)$$

(case(i)),

and

$$\begin{aligned} &<u(\cdot, 2\tau - \eta), \{K\ddot{u}(\cdot,\eta) + Lu(\cdot,\eta)\} > \\ &\quad - < u(\cdot,\eta), \{K\ddot{u}(\cdot, 2\tau - \eta) + Lu(\cdot, 2\tau - \eta)\} >= 0 \end{aligned} \qquad (5.18)$$

(case(ii)).

In view of (5.16), (5.17) (case (i)) reduces to

$$< u(\cdot, 2\tau - \eta), \frac{\partial}{\partial \eta} u(\cdot,\eta) > + < u(\cdot,\eta) \frac{\partial}{\partial \eta} u(\cdot, 2\tau - \eta) >= 0.$$

or

$$\frac{d}{d\eta} < u(\cdot, 2\tau - \eta), u(\cdot, \eta) >= 0 \, ; \tag{5.19}$$

on using properties (5.16_1). Integration between $\eta = 0, \eta = \tau$ gives

Theorem 5.3 (i) - *A solution of the operator equation* (5.14), *under the stated assumptions, satisfies the identity*

$$< u(\cdot, \tau), u(\cdot, \tau) >=< u(\cdot, 2\tau), u(\cdot, 0) > \; . \tag{5.20}$$

A parallel result may be obtained from (5.18) as follows: In view of (5.16), (5.18) (case (ii)) reduces to

$$< u(\cdot, 2\tau - \eta), \ddot{u}(\cdot, \eta) > - < u(\cdot, \eta), \ddot{u}(\cdot, 2\tau - \eta) >= 0 \tag{5.21$_1$}$$

or, more transparently,

$$\frac{d}{d\eta} < u(\cdot, 2\tau - \eta), \dot{u}(\cdot, \eta) > + \frac{d}{d\eta} < u(\cdot, \eta), \dot{u}(\cdot, 2\tau - \eta) >= 0 \, . \tag{5.21$_2$}$$

Integration between $\eta = 0$ and $\eta = \tau$ gives

Theorem 5.3 (ii) - *A solution of the operator equation* (5.15), *under the stated assumptions, satisfies the identity*

$$2 < u(\cdot, \tau), \dot{u}(\cdot, \tau) >=< u(\cdot, 2\tau), \dot{u}(\cdot, 0) > + < u(\cdot, 0), \dot{u}(\cdot, 2\tau) > \; . \tag{5.22}$$

The reader should verify that, with suitable identification of operators, scalar product etc., (5.20) and (5.22) reduce to (5.12), (5.13) respectively in the special cases considered earlier.

We now proceed to generalized (5.20), (5.22): For each of the two equations (5.14), (5.15) we obtain an identity connecting pairs of solutions v, w; and, in conclusion, we deduce therefrom reciprocal relations connecting v, w.

Let us suppose that $v(\mathbf{x}, t), w(\mathbf{x}, t)$ denote (generally different) solutions of the operator equation (5.14) (case (i)), or of the operator equation (5.15) (case (ii)), as appropriate. We start with the obvious identities (generalizing (5.17), (5.18) respectively):

$$\begin{aligned} &< w(\cdot, 2\tau - \eta), \{K\dot{v}(\cdot, \eta) + Lv(\cdot, \eta)\} > \\ &\quad - v(\cdot, \eta), \{K\dot{w}(\cdot, 2\tau - \eta) + Lw(\cdot, 2\tau - \eta)\} >= 0 \end{aligned} \tag{5.23}$$

(case(i)),

$$\begin{aligned} &< w(\cdot, 2\tau - \eta), \{K\ddot{v}(\cdot, \eta) + Lv(\cdot, \eta)\} > \\ &\quad - < v(\cdot, \eta), \{K\ddot{w}(\cdot, 2\tau - \eta) + Lw(\cdot, 2\tau - \eta)\} >= 0 \end{aligned} \tag{5.24}$$

(case(ii)).

In view of (5.16), these reduce respectively to

$$< w(\cdot, 2\tau - \eta), \dot{v}(\cdot, \eta) > - < v(\cdot, \eta), \dot{w}(\cdot, 2\tau - \eta) > = 0 \tag{5.25$_1$}$$

or

$$\frac{d}{d\eta} < v(\cdot, \eta), w(\cdot, 2\tau - \eta) > = 0 \tag{5.25$_2$}$$

(case(i)) ,

and

$$< w(\cdot, 2\tau - \eta), \ddot{v}(\cdot, \eta) > - < v(\cdot, \eta), \ddot{w}(\cdot, 2\tau - \eta) > = 0 \tag{5.26$_1$}$$

or

$$\frac{d}{d\eta} < w(\cdot, 2\tau - \eta), \dot{v}(\cdot, \eta) > + \frac{d}{d\eta} < v(\cdot, \eta), \dot{w}(\cdot, 2\tau - \eta) > = 0 \tag{5.26$_2$}$$

(case(ii)) .

Integrating these between $\eta = 0$ and $\eta = \tau$ gives generalizations of Theorem 5.3(i) and Theorem 5.3(ii):

Theorem 5.4 (i) - *Solutions* $v(\mathbf{x}, t)$ *and* $w(\mathbf{x}, t)$ *of the operator equation* (5.14), *under the stated assumptions, satisfy*

$$< v(\cdot, \tau), w(\cdot, \tau) > = < v(\cdot, 0), w(\cdot, 2\tau) > \tag{5.27}$$

and

Theorem 5.4 (ii) - *Solutions* $v(\mathbf{x}, t)$ *and* $w(\mathbf{x}, t)$ *of the operator equation* (5.15), *under the stated assumptions, satisfy*

$$\begin{aligned} &< v(\cdot, \tau), \dot{w}(\cdot, \tau) > + < w(\cdot, \tau), \dot{v}(\cdot, \tau) > = \\ &\qquad < v(\cdot, 0), \dot{w}(\cdot, 2\tau) > + < w(\cdot, 2\tau), \dot{v}(\cdot, 0) > . \end{aligned} \tag{5.28}$$

Notice that the left hand side of (5.28) can be expressed as a derivative with respect to τ. Moreover, it is clear that (5.27), (5.28) reduce to (5.20), (5.22) on putting $v = w = u$. Certain 'reciprocal relations' follow immediately from (5.27), (5.28) on interchanging v and w and on noting $< f, g > = < g, f >$. We have

Theorem 5.5 - *If* $v(\mathbf{x}, t)$ *and* $w(\mathbf{x}, t)$ *are solutions of the operator equation* (5.14), *then the following reciprocal relation holds*

$$< v(\mathbf{x}, 0), w(\mathbf{x}, 2\tau) > = < v(\mathbf{x}, 2\tau), w(\mathbf{x}, 0) > , \tag{5.29}$$

while if
solutions of (5.15), *then the following reciprocal relation holds*

$$
\begin{aligned}
&< v(\mathbf{x},0), \overset{\bullet}{w}(\mathbf{x},2\tau) > + < \overset{\bullet}{v}(\mathbf{x},0), w(\mathbf{x},2\tau) >= \\
&< v(\mathbf{x},2\tau), \overset{\bullet}{w}(\mathbf{x},0) > + < \overset{\bullet}{v}(\mathbf{x},2\tau), w(\mathbf{x},0) > .
\end{aligned} \tag{5.30}
$$

It is instructive to see the forms which the above identities, and the above reciprocal relations, take when the context is that of equations (5.1), (5.2) under the conditions accompanying these, and when the scalar product is identified as in the Appendix to this chapter. The identities are

$$
\int_0^1 v(\cdot,\tau)w(\cdot,\tau)dx = \int_0^1 v(\cdot,0)w(\cdot,2\tau)dx \tag{5.31}
$$

(case(i)),

and

$$
\begin{aligned}
&\int_0^1 v(\cdot,\tau)\overset{\bullet}{w}(\cdot,\tau)dx + \int_0^1 w(\cdot,\tau)\overset{\bullet}{v}(\cdot,\tau)dx = \\
&\int_0^1 v(\cdot,0)\overset{\bullet}{w}(\cdot,2\tau)dx + \int_0^1 w(\cdot,2\tau)\overset{\bullet}{v}(\cdot,0)dx
\end{aligned} \tag{5.32}
$$

(case(ii)).

The reciprocal relations are

$$
\int_0^1 v(\cdot,0)w(\cdot,2\tau)dx = \int_0^1 v(\cdot,2\tau)w(\cdot,0)dx \tag{5.33}
$$

(case(i)),

and

$$
\begin{aligned}
&\int_0^1 v(\cdot,0)\overset{\bullet}{w}(\cdot,2\tau)dx + \int_0^1 \overset{\bullet}{v}(\cdot,0)w(\cdot,2\tau)dx = \\
&\int_0^1 v(\cdot,2\tau)\overset{\bullet}{w}(\cdot,0)dx + \int_0^1 \overset{\bullet}{v}(\cdot,2\tau)w(\cdot,0)dx
\end{aligned} \tag{5.34}
$$

(case(ii)).

Verification of these is left as an exercise to the reader: one may deduce them from (5.29), (5.30), or one may prove them directly starting from (5.1)–(5.3).

5.4 APPLICATION TO BACKWARDS HEAT EQUATION: CONTINUOUS DEPENDENCE

We consider the one-dimensional heat equation backwards in time: the temperature $u(x,t)$ is supposed to be a classical solution of

$$u_t + u_{xx} = 0 \qquad \text{in } 0 < x < 1\,, 0 < t < T\,, \tag{5.35$_1$}$$

subject to the boundary conditions

$$u(0,t) = u(1,t) = 0\,, \tag{5.35$_2$}$$

and to the initial condition

$$u(x,0) = f(x)\,. \tag{5.35$_3$}$$

Uniqueness of solution has already been established for this problem. It is, however, known to be ill-posed in that the solution does not depend continuously on the data. It is possible to recover continuous dependence – in a certain sense – if it is known that

$$|u| < \frac{1}{2}M \qquad (t \leq T) \tag{5.35$_4$}$$

where M and T are known, positive constants. In the case of heat flow in a solid, the *a priori* bound $\frac{1}{2}M$ might, for example, denote the melting temperature; in any case, it does not have to be a sharp bound.

Before discussing the recovery of continuous dependence, it is of interest to quote the beautifully expressed reflections of Clerk Maxwell [146] on temperature distributions backward in time:

Sir William Thompson has shown, in a paper published in the 'Cambridge and Dublin Mathematical Journal' in 1844, how to deduce, in certain cases, the thermal state of a body in past time from its observed condition at present.

For this purpose, the present distribution of temperature must be expressed (as it always may be) as the sum of a series of harmonic distributions. Each of these harmonic distributions is such that the difference of the temperature of any point from the final temperature diminishes in a geometrical progression as the time increases in arithmetical progression, the ratio of the geometrical progression being the greater the higher the degree of the harmonic.

If we now make t negative, and trace the history of the distribution of temperature up the stream of time, we shall find each harmonic increasing as we go backwards, and the higher harmonics increasing faster then the lower ones.

If the present distribution of temperature is such that it may be expressed in a finite series of harmonics, the distribution of temperature at any previous time may be calculated; but if (as is generally the case) the series of harmonics is infinite, then the temperature can be calculated only when this series is convergent. For present and future time it is always convergent, but for past time it becomes ultimately divergent when the time is taken at a sufficiently remote epoch. The negative value of t for which the series becomes ultimately divergent, indicates a certain date in past time such that the present state of things cannot be deduced from any distribution of temperature occurring previously to that date, and becoming diffused by ordinary conduction. Some other event besides ordinary conduction must have occurred since that date in order to produce the present state of things.

This is only one of the cases in which a consideration of the dissipation of energy leads to the determination of a superior limit to the antiquity of the observed order of things.

We proceed to show how to recover continuous dependence. Suppose that the temperature distribution changes from $u_1(x,t)$ to $u_2(x,t)$ when the initial data changes from $f_1(x)$ to $f_2(x)$ respectively. In order to establish continuous dependence we need to show – in a suitable sense (norm) – that the difference between u_1 and u_2 ($u = u_1 - u_2$) may be made arbitrarily small if the difference between f_1 and f_2 ($f = f_1 - f_2$) is made sufficiently small. In view of the linearity of the problem this is equivalent to proving – in a suitable sense (norm) – that the solution u of (5.35_1)–(5.35_3) may be made arbitrarily small provided $f(x)$ is made sufficiently small, assuming that $|u| < M$ (in view of the fact that $|u_1 - u_2| \leq |u_1| + |u_2|$).

We may rewrite (5.12) as

$$\int u^2(x,\tau)dx = \int u(x,2\tau)f(x)dx \qquad (\tau \leq \frac{T}{2}). \tag{5.36}$$

Two general types of continuous dependence results ((a),(b) hereunder) follow from this. The first is obtainable from Hölder's inequality while the second follows from Schwarz's inequality used recursively.

Applying Hölder's inequality to (5.36) we obtain

$$\int u^2(x,\tau)dx \leq \{\int |f(x)|^p dx\}^{1/p}\{\int |u(x,2\tau)|^q dx\}^{1/q} \tag{5.37}$$

where p, q are arbitrary positive numbers which satisfy

$$p^{-1} + q^{-1} = 1.$$

This leads to (a) of the following theorem while (b) is left as an exercise:

Theorem 5.6 - *The solution to the problem* (5.35_1)–(5.35_3) *where* u *satisfies the a priori bound*

$$|u| < M,$$

M being a positive constant, satisfies the continuous dependence estimates

$$\text{(a)} \qquad \int u^2(x,\tau)dx \le M\{\int |f(x)|^p dx\}^{1/p} \qquad (\tau \le \frac{T}{2}); \tag{5.38}$$

$$\text{(b)} \qquad \|u(T/2^n)\| \le M^{2^{-n}} \|f\|^{1-2^{-n}},$$

$$\|u(T(1-2^{-n}))\| \le M^{1-2^{-n}} \|f\|^{2^{-n}}, \tag{5.39}$$

where n is any positive integer and where the usual norm notation is used [e.g. $\|u(t)\| = \{\int_0^1 u^2(x,t)dx\}^{1/2}$ etc.]

Let us note that the following corollary is obtainable from (5.38) on letting $p \to \infty$:

$$\int u^2(x,\tau)dx \le M \max_{x\in(0,1)} |f(x)| \tag{5.40}$$

– as $f(x)$ is assumed to be a continuous function.

5.5 EQUIPARTITION OF ENERGY

We now turn to our second example: the Lagrange identity method can be made to furnish a nice proof of the principle of equipartition of energy. We choose the simplest context and leave it to the reader to generalize it to cater for other contexts. The argument given below is due to Levine [147] who, however, considered a much more general context.

Let us consider classical solutions of the initial boundary value problem for the wave equation: $u(x,t)$ satisfies

$$\begin{cases} u_{tt} - u_{xx} = 0, & 0 < x < 1 \quad , \quad t > 0 \\ u(0,t) = u(1,t) = 0, & \\ u(x,0) = u_0(x), \quad u_t(x,0) = v_0(x). & \end{cases} \tag{5.41}$$

For example, $u(x,t)$ could denote the small transverse deflection of a string streched to unit tension, which occupies the portion $0 \le x \le 1$ of the x axis when in the undeflected state; suppose that the string is of unit density (per unit length) and that its ends are pinned at the original level for all subsequent time.

The principle of equipartition of energy appropriate to the context is embodied in

Theorem 5.7 - *The mean kinetic and potential energies, over the time interval $0 \le t \le \tau$, defined by*

$$\begin{cases} \overline{K}(\tau) = \frac{1}{\tau}\int_0^\tau\{\int_0^1 \frac{1}{2}\dot{u}^2(\cdot,\eta)dx\}d\eta \\ \overline{V}(\tau) = \frac{1}{\tau}\int_0^\tau\{\int_0^1 \frac{1}{2}u_x^2(\cdot,\eta)dx\}d\eta \end{cases} \tag{5.42}$$

respectively, satisfy

$$\lim_{\tau\to\infty} \overline{K}(\tau) = \lim_{\tau\to\infty} \overline{V}(\tau)\,. \tag{5.43}$$

Proof. We first consider

$$I = \int_0^1 \int_0^\tau \{u(\cdot,\eta)\dot{u}(\cdot,\eta)\}^\bullet \, dx d\eta$$

and obtain two different expressions for it.

(a) On the one hand, integrating first w.r.t. η:

$$I = \int_0^1 u(\cdot,\tau)\dot{u}(\cdot,\tau)dx - \int_0^1 u_0 v_0 \, dx\,; \tag{5.44$_1$}$$

(b) on the other, integrating first w.r.t. x:

$$\begin{aligned} I &= \int_0^1 \int_0^\tau [u(\cdot,\eta)\ddot{u}(\cdot,\eta) + \dot{u}^2(\cdot,\eta)] dx d\eta \\ &= \int_0^1 \int_0^\tau [u(\cdot,\eta)u_{xx}(\cdot,\eta) + \dot{u}^2(\cdot,\eta)] dx d\eta \\ &= \int_0^1 \int_0^\tau [\dot{u}^2(\cdot,\eta) - u_x^2(\cdot,\eta)] dx d\eta \end{aligned} \tag{5.44$_2$}$$

where the P.D.E., integration by parts and boundary conditions have been used. Equating the foregoing two expressions, we obtain

$$\int_0^\tau \int_0^1 [\dot{u}^2(\cdot,\eta) - u_x^2(\cdot,\eta)] dx d\eta = \int_0^1 u(\cdot,\tau)\dot{u}(\cdot,\tau)dx - \int_0^1 u_0 v_0 \, dx\,. \tag{5.45}$$

The Lagrange identity method [(5.13)] gives

$$2\int_0^1 u(\cdot,\tau)\dot{u}(\cdot,\tau)dx = \int_0^1 [u(\cdot,2\tau)v_0 + \dot{u}(\cdot,2\tau)u_0] dx\,, \tag{5.46}$$

on using initial conditions. On combining (5.45), (5.46) we obtain

$$\begin{aligned} 2\int_0^\tau \int_0^1 &[\dot{u}^2(\cdot,\eta) - u_x^2(\cdot,\eta)] dx d\eta = \\ &= \int_0^1 \{u(\cdot,2\tau) - u_0\} v_0 \, dx + \int_0^1 \{\dot{u}(\cdot,2\tau) - v_0\} u_0 \, dx \end{aligned} \tag{5.47}$$

or, in terms of $\overline{K}, \overline{V}$ (defined by (5.42)),

$$\overline{K}(\tau) - \overline{V}(\tau) = \frac{1}{4\tau}\int_0^1 \{u(\cdot,2\tau) - u_0\} v_0 \, dx + \frac{1}{4\tau}\int_0^1 \{\dot{u}(\cdot,2\tau) - v_0\} u_0 \, dx\,. \tag{5.48}$$

The proof of equipartition of energy hinges upon this identity and upon the conservation of energy which follows. Energy conservation follows on multiplying the P.D.E. by u_t and integrating by parts using the boundary conditions (the details are left to the reader):

$$\int_0^1 \{\frac{1}{2}\dot{u}^2(x,\eta) + \frac{1}{2}u_x^2(x,\eta)\}dx \equiv E(0) \stackrel{\text{defn}}{=} \int_0^1 [\frac{1}{2}v_0^2 + \frac{1}{2}\{u_0'(x)\}^2]dx\,. \tag{5.49}$$

The essence of the proof is as follows: the two integrals occurring in (5.48) are bounded above in terms of (finite!) initial data by using (5.49), and letting $\tau \to \infty$. We consider the two terms on the right hand side of (5.48) successively and prove that each tends to zero as $\tau \to \infty$. Let us consider the (absolute value of) the second term first: Schwarz's inequality and (5.49) give

$$\begin{aligned} |I_2(\tau)| \leq & \frac{1}{4\tau}\int_0^1 |\{\dot{u}(\cdot,2\tau) - v_0\}u_0|dx \leq \frac{1}{4\tau}\{\int_0^1 |\dot{u}(\cdot,2\tau)u_0|dx + \int_0^1 |u_0 v_0|dx\} \\ \leq & \frac{1}{4\tau}[\{\int_0^1 \dot{u}(\cdot,2\tau)dx\}^{1/2}\{\int_0^1 u_0^2 dx\}^{1/2} + \int_0^1 |u_0 v_0|dx] \\ \leq & \frac{1}{4\tau}[\{2E(0)\}^{1/2}\{\int_0^1 u_0^2 dx\}^{1/2} + \int_0^1 |u_0 v_0|dx]\,. \end{aligned} \tag{5.50}$$

It follows that

$$\lim_{\tau\to\infty} |I_2(\tau)| = 0\,. \tag{5.51}$$

The consideration of the first term in (5.48) proceeds along similar lines. In this case, however, we need an inequality of the type

$$\int_0^1 u^2(x,2\tau)dx \leq c\int_0^1 u_x^2(x,2\tau)dx \tag{5.52}$$

where c is a constant which may be taken to be $1/\pi^2$ (its optimum value); see (10) of appendix, case (a). It is left to the reader to prove that

$$\begin{aligned} I_1(\tau) = & \frac{1}{4\tau}\int_0^1 \{u(\cdot,2\tau) - u_0\}v_0\,dx \\ \leq & \frac{1}{4\tau}[\{2cE(0)\}^{1/2}\{\int_0^1 v_0^2 dx\}^{1/2} + \int_0^1 |u_0 v_0|dx]\,. \end{aligned} \tag{5.53}$$

Plainly

$$\lim_{\tau\to\infty} |I_1(\tau)| = 0\,. \tag{5.54}$$

The required proposition is proved in view of (5.48), (5.50), (5.51), (5.53)–(5.54).

5.6 A SECOND IDENTITY FOR WAVE-LIKE EQUATIONS

We now derive a second identity – additional to (5.28) – appropriate to the wave-like equation (5.15). It follows from (5.15) – in the context and notation previously discussed – that

$$\int_0^\tau [< v_\eta(\cdot,\eta), Kw_{\eta\eta}(\cdot,\eta) + Lw(\cdot,\eta) > + \\ < w_\eta(\cdot,\eta), Kv_{\eta\eta}(\cdot,\eta) + Lv(\cdot,\eta) >] d\eta = 0 . \tag{5.55}$$

Payne [140] speaks of this as a (slightly different) Lagrange identity; however, Payne is concerned with a more general equation in which K is replaced by a general, symmetric operator. It may be expressed in alternative form in view of the symmetry of the operator L, the constancy of K, and the commutativity of the scalar product:

$$\int_0^\tau [K < v_\eta(\cdot,\eta), w_{\eta\eta}(\cdot,\eta) > + K < v_{\eta\eta}(\cdot,\eta), w_\eta(\cdot,\eta) > \\ + < v_\eta(\cdot,\eta), Lw(\cdot,\eta) > + < v(\cdot,\eta), Lw_\eta(\cdot,\eta) >] d\eta = 0$$

or

$$\int_0^\tau [K \frac{d}{d\eta} < v_\eta(\cdot,\eta), w_\eta(\cdot,\eta) > + \frac{d}{d\eta} < v(\cdot,\eta), Lw(\cdot,\eta) >] d\eta = 0$$

in view of (5.16_1). Straightforward integration leads to

$$\{K < v_\eta(\cdot,\eta), w_\eta(\cdot,\eta) > + < v(\cdot,\eta), Lw(\cdot,\eta) >\}]_{\eta=0}^{\eta=\tau} = 0$$

or, using the "dot" notation, this leads to

Theorem 5.8 - *Any solutions $v(\mathbf{x},t), w(\mathbf{x},t)$ of the operator equation* (5.15), *under tha stated assumptions, satisfy the identity*

$$K < \dot{v}(\cdot,\tau), \dot{w}(\cdot,\tau) > - K < \dot{v}(\cdot,0), \dot{w}(\cdot,0) > \\ + < v(\cdot,\tau), Lw(\cdot,\tau) > - < v(\cdot,0), Lw(\cdot,0) > = 0 . \tag{5.56}$$

This is the second identity – additional to (5.28) – and valid in the same context.

Particular importance attaches to the special case arising when

$$v(\mathbf{x},\eta) = u(\mathbf{x},\eta) \quad , \quad w(\mathbf{x},\eta) = -u(\mathbf{x}, 2\tau - \eta) :$$

that is to say

$$K < \dot{u}(\cdot,\tau), \dot{u}(\cdot,\tau) > - < u(\cdot,\tau), Lu(\cdot,\tau) > = \\ = K < \dot{u}(\cdot,0), \dot{u}(\cdot,2\tau) > - < u(\cdot,0), Lu(\cdot,2\tau) > . \tag{5.57}$$

Let us note that the left hand side is often intimately related to the difference between kinetic and potential energies.

We now record the form which this latter identity takes in the context of the wave equation (5.2) subject to (5.3) – where for simplicity, henceforward, we take $K = -1$, and $\alpha = 0, \beta \neq 0$ in (5.3):

$$\frac{1}{2}\int_0^1 \dot{u}^2(x,\tau)dx - \frac{1}{2}\int_0^1 u_x^2(x,\tau)dx =$$
$$= \frac{1}{2}\int_0^1 [\dot{u}(x,0)\dot{u}(x,2\tau) - u_x(x,0)u_x(x,2\tau)]dx\,. \tag{5.58}$$

This may be derived either directly from (5.57), on appropriately identifying the operator L, etc. (cf. Appendix to this chapter) or *ab initio* using the general approach used above. Let us note that the left hand side of (5.58) represents the difference between the kinetic and potential energies at time τ.

The identity (5.58) – rather than (5.13) – can be used as a pivot upon which to mount alternative proofs of two issues previously considered for the system (5.2)–(5.3): the uniqueness issue and the issue of equipartition of energy.

The uniqueness issue may be established – again taking $K = -1$ and $\alpha = 0$, $\beta \neq 0$ in (5.3) for simplicity – on combining the "conservation of energy" (5.49) with (5.58).

With regard to the equipartition of energy, the identities (5.47) or (5.48) may be derived from (5.58) on further integration:

$$2\tau(\overline{K}(\tau) - \overline{V}(\tau)) = \int_0^\tau\int_0^1 [\dot{u}(x,0)\dot{u}(x,2\eta) - u_x(x,0)u_x(x,2\eta)]dxd\eta =$$
$$\frac{1}{2}\int_0^1 \dot{u}(x,0)\{u(x,2\tau) - u(x,0)\}dx + \int_0^\tau\int_0^1 u(x,0)\ddot{u}(x,2\eta)dxd\eta$$

where integration by parts and (5.41) have been used. Further integration with respect to the time variable is readily found to lead to (5.48), and the remainder of the discussion is the same.

5.7 EXERCISES

Exercise 5.1 - The small transverse deflection of a Bernoulli-Euler elastic beam, whose ends are built in horizontally, satisfies

$$u_{tt} + u_{xxxx} = 0\,, \qquad 0 < x < 1 \quad , \quad t > 0\,,$$
$$u(0,t) = u_x(0,t) = u(1,t) = u_x(1,t) = 0\,,$$
$$u(x,0) = u_0(x) \quad , \quad u_t(x,0) = v_0(x)\,.$$

Use the Lagrange identity method

(i) to prove uniqueness of solution;

(ii) to obtain explicit versions of (5.22), (5.28), (5.30);

(iii) to obtain the principle of equipartition of energy appropriate to the context

$$\lim_{\tau \to \infty} \overline{K}(\tau) = \lim_{\tau \to \infty} \overline{V}(\tau)$$

where

$$\overline{K}(\tau) = \frac{1}{\tau}\int_0^\tau \{\int_0^1 \frac{1}{2}\dot{u}^2(x,\eta)dx\}d\eta$$
$$\overline{V}(\tau) = \frac{1}{\tau}\int_0^\tau \{\int_0^1 \frac{1}{2}u_{xx}^2(x,\eta)dx\}d\eta\,.$$

Hint: Let $L\phi = -\phi_{xxxx}$ for the appropriate class of functions ϕ.

Exercise 5.2 - Let D be a bounded, three dimensional domain whose boundary ∂D is sufficiently smooth to admit use of the divergence theorem. Consider classical solutions of the P.D.E.

$$\text{(a)} \qquad K u_{tt} + \nabla^2 u = 0 \quad \text{in } D\,,$$

subject to

$$u = 0 \quad \text{on } \partial D\,,$$
$$\text{(b)} \qquad K u_{tt} - \nabla^4 u = 0 \quad \text{in } D\,,$$

subject to

$$u = \frac{\partial u}{\partial n} = 0 \quad \text{on } \partial D\,,$$

where K is a constant (positive or negative) in both cases, and $\partial/\partial n$ the normal derivative outward from ∂D.

(i) Derive explicit versions of (5.22), (5.28), (5.30) in both cases (i.e. analogues of (5.13), (5.32), (5.34) respectively).

(ii) Consider an initial boundary value problem for each of (a),(b), the initial condition being of the type

$$u(\cdot,0) = \text{specified} \quad , \quad u_t(\cdot,0) = \text{specified}\,.$$

Use the appropiate identities to establish uniqueness.

(iii) Suppose u_{tt} is replaced by u_t in (a),(b). Prove results in such cases which are analogous to those dealt with in (i),(ii) above.

Exercise 5.3 - Suppose $\mathcal{D}$ is as described in the previous question. Let $u(x_1, x_2, x_3, t)$ be a classical solution of the "forward" heat equation

$$u_t = \nabla^2 u \qquad \text{in } \mathcal{D}$$

subject to

$$u = 0 \qquad \text{on } \partial\mathcal{D}\,,$$

while $v(x_1, x_2, x_3, t)$ is a classical solution of the "backward" heat equation

$$v_t + \nabla^2 v = 0 \qquad \text{in } \mathcal{D}$$

subject to

$$v = 0 \qquad \text{on } \partial\mathcal{D}\,,$$

prove that

$$\int_{\mathcal{D}} v(\mathbf{x},t)u(\mathbf{x},t)dA = \int_{\mathcal{D}} v(\mathbf{x},0)u(\mathbf{x},0)dA\,.$$

Exercise 5.4 - Prove (5.31)–(5.34), (5.39), (5.49) which occur in the text.

Exercise 5.5 - Consider the initial boundary value problem

$$Ku_{tt} = u_{xx}\,, \qquad 0 < x < 1 \quad , \quad 0 < t\,,$$

$$u(0,t) = u(1,t) = 0\,,$$

$$u(x,0) = u_0(x) \quad , \quad u_t(x,0) = v_0(x)\,,$$

where K is a constant (positive or negative) and where classical solutions are assumed.

(i) Suppose that $v_0(x) = 0$. Defining

$$G(t) = \int_0^1 u^2(\cdot,t)dx$$

prove that

$$G(\tau) = \frac{1}{2}\int_0^1 u(x,0)u(x,2\tau)dx + \frac{1}{2}\int_0^1 u^2(x,0)dx\,.$$

Use the arithmetic-geometric inequality to prove that

$$G(\tau) \le \frac{3}{4}G(0) + \frac{1}{4}G(2\tau) \le \frac{3}{4}\sum_{i=0}^{N-1}\frac{1}{4^i}G(0) + \frac{1}{4^N}G(2^N\tau)\,.$$

If $\lim_{t\to\infty} G(t)/t^2 = 0$, deduce that $G(t) \le G(0)$.

(ii) Suppose that $u_0(x) = 0$. Defining

$$H(t) = \int_0^1 \dot{u}^2(x,t)dx$$

prove that

$$H(\tau) = \frac{1}{2}H(0) + \frac{1}{2}\int_0^1 \dot{u}(\cdot,2\tau)\dot{u}(\cdot,0)dx \le \frac{3}{4}H(0) + \frac{1}{4}H(2\tau)$$
$$\le \frac{3}{4}\sum_{i=0}^{N-1}\frac{1}{4^i}H(0) + \frac{1}{4^N}H(2^N\tau)\,.$$

[*Hint*: Replace u by $\dot{u}$ in (5.13).]

If $\lim_{t\to\infty} H(t)/t^2 = 0$, deduce that $H(t) \le H(0)$.

Exercise 5.6 - The problem given in exercise 5.5 is well known to be ill-posed when K is negative in that there is no continuity with respect to data in general. Suppose, however, that it is known that

$$|u| \le M$$

where M is a given constant. Prove that (Hölder) continuity with respect to data is recovered by establishing, in the context of and notation of exercise 5.5(i) *supra*:

$$G(\tau) \le \frac{1}{2}M\{G(0)\}^{1/2} + \frac{1}{2}G(0)\,.$$

Obtain an analogous continuous dependence result in the context of and notation of exercise 5.5(ii) *supra*.

Exercise 5.7 - Suppose that $u(x,t)$ is a classical solution of

$$u_t + u_{xx} = F(x,t) \quad , \quad 0 < x < 1 \quad , \quad 0 < t < T$$

where $F(x,t)$ is an assigned 'source' term, subject to the boundary conditions

$$u(0,t) = u(1,t) = 0\,,$$

and to the null initial conditions

$$u(x,0) = 0\,.$$

Use the Lagrange identity approach to prove that

$$\int_0^1 u^2(x,\tau)dx = \int_0^1\int_0^\tau F(\cdot,\eta)u(\cdot,2\tau-\eta)dxd\eta - \int_0^1\int_\tau^{2\tau} F(\cdot,\eta)u(\cdot,2\tau-\eta)dxd\eta$$

$$(\tau < T/2)\,.$$

Assuming that

$$|u(x,t)| < M \qquad \text{for } t \leq T$$

where M is a given constant, establish the estimate

$$\int_0^1 u^2(x,\tau)dx \leq 2^{1/2}M\tau^{1/2}\{\int_0^1\int_0^{2\tau} F^2(\cdot,\eta)dxd\eta\} \quad , \quad \tau < T/2$$

– which exhibits continuous dependence on the source term.
Show how this latter result may be generalized if the initial condition is non-null

$$u(x,0) = u_0(x)\,.$$

Exercise 5.8 - Establish uniqueness of solution for the initial boundary value problem occurring in exercise 5.5 by using (5.57) together with "conservation of energy".
Use an analogous method to establish uniqueness of solution for the initial boundary value problem: $u(x,t)$ is a classical solution of

$$Ku_{tt} = u_{xxxx}\,, \qquad 0 < x < 1 \quad , \quad 0 < t\,,$$

$$u(0,t) = u_{xx}(0,t) = u(1,t) = u_{xx}(1,t) = 0\,,$$

$$u(x,0) = \text{specified} \quad , \quad u_t(x,0) = \text{specified}\,;$$

K is a non-zero constant.

5.8 APPENDIX

The essential properties of scalar products and symmetric operators are sketched here. An example relevant to the text is also given.

(a) The *scalar product* of any two real scalar functions $f(\mathbf{x}), g(\mathbf{x})$ defined in a region R – denoted by $< f, g >$ – is required to satisfy the properties

$$\text{(i)} < f,g >=< g,f > ,$$

$$\text{(ii)} < \alpha f,g >= \alpha < f,g > \qquad \alpha \text{ being a real constant},$$

$$\text{(iii)} < f_1 + f_2, g >=< f_1, g > + < f_2, g > ,$$

where $f_1(\mathbf{x}), f_2(\mathbf{x})$ are any two functions defined in R.

The simplest example of such a product – and the one envisaged in the text – is

$$< f,g >= \int_R f(\mathbf{x})g(\mathbf{x})dx$$

where the integration is carried out over the domain of definition R of the functions. It is readily verified that it has all the requisite properties listed above.

(b) A linear operator L has the linear property

$$L\{\alpha_1 f + \alpha_2 g\} = \alpha_1 L\{f\} + \alpha_2 L\{g\}$$

where α_1, α_2 are any two (real) numbers, and f, g are any two functions from the appropriate vector space to which L relates. It is *symmetric* if

$$< Lf, g >=< f, Lg > .$$

(c) *An example*: Consider the class of real functions $\phi(x,t)$ such that for each t, $\phi \in C^2(0 \leq x \leq 1)$ and satisfies the boundary conditions

$$\alpha_0 \phi_x(0,\cdot) + \beta_0 \phi(0,\cdot) = 0 \qquad , \qquad \alpha_1 \phi_x(1,\cdot) + \beta_1 \phi(1,\cdot) = 0$$

where $\alpha_\cdot\,, \beta_\cdot$ are all real non-zero constants. Plainly this set of functions constitutes a real vector space.

Suppose that an operator L is defined for the aforementioned class of functions by means of

$$L\phi = \phi_{xx} .$$

Such an operator is clearly linear and maps the subspace of C^2 functions satisfying the boundary conditions into the space H of continuous functions. Moreover, the operator is clearly symmetric – using the example of scalar product as given above: for any of two functions ψ, χ belonging to the aforementioned subspace one has

$$\begin{aligned} - < L\psi, \chi > &= -\psi_x \chi]_0^1 + \int_0^1 \psi_x \chi_x dx \\ &= \alpha_1^{-1}\beta_1 \psi(1)\chi(1) - \alpha_0^{-1}\beta_0^{-1}\psi(0)\chi(0) + \int_0^1 \psi_x \chi_x dx \\ &= - < \psi, L\chi > \end{aligned}$$

using integration by parts and the boundary conditions. Note that when the constants α_0, α_1, are both zero, the latter analysis is easily modified.

Chapter six

WEIGHTED INTEGRAL METHODS

6.1 INTRODUCTION

Weighted integrals have been and are so extensively used in the theory of P.D.E.s that their utility can hardly be overestimated. Various types of weight functions – chosen with different aims – have been introduced and so many authors have used them that it is impossible to recall the contributions of each one. We refer the readers to the references [7,34,35,59,72,148-150] which include many papers on the use of weighted integrals. In the following sections, we shall present, by means of examples, some methods of using weighted integrals in the qualitative theory of P.D.E.s. We begin by presenting the weight function approach to uniqueness on bounded (secs. 6.2-6.3) and unbounded domains (sec. 6.4). Successively (sec. 6.5) we introduce the weight function approach to the stability of the Cauchy problem for the Korteweg-Burgers-de Vries equation. After having obtained (sec. 6.6) some weighted Poincaré inequalities for unbounded domains, in the last section (sec. 6.7) we introduce a means of marrying the weight function approach for unbounded domains to the cross-sectional method.

6.2 WEIGHT FUNCTION APPROACH TO UNIQUENESS ON BOUNDED DOMAINS

The weight function approach to uniqueness in bounded domains and, more generally, to continuous dependence involves introducing – instead of the function u at hand – a new function φ linked to u by a weight $f(t)$ by means of $\varphi = f(t)u$. The role of the weight is to generate a more tractable problem. The exponential functions – because of their properties and, in particular, the simplicity of their derivatives – are good candidates for the weight function role. Here we consider the problem of uniqueness for the *displacement problem in linear elastodynamics* {See **1.8** subsection c), eqs. (1.32)–(1.35) with $\partial_2\Omega = \emptyset$}. Because of the linearity, the "difference" displacement $\mathbf{u}$ is governed by

$$\begin{cases} \rho\ddot{\mathbf{u}} = \nabla\cdot\mathbf{T} & \text{on } \Omega\times[0,T]\,, \\ \mathbf{u} = \dot{\mathbf{u}} = 0 & \text{on } \Omega\times\{0\}\,, \\ \mathbf{u} = 0 & \text{on } \partial\Omega\times[0,T] \end{cases} \tag{6.1}$$

where the stress tensor $\mathbf{T}$ is given by (1.36). Following Murray [152], we assume that the elastic coefficients C_{ijkl} satisfy the (non-standard) *skew-symmetric constitutive relationship*

$$C_{ijkl}(x) = -C_{klij}(x)\,. \tag{6.2}$$

Conventional proofs for proving uniqueness – based on sign-definiteness of the strain energy and/or the *symmetric* relationship corresponding to (6.2) above, for example (See Knops and Payne [22]) – are not available when the elastic coefficients satisfy a relationship of the *skew-symmetric* type (6.2). For this reason, one needs to use a different approach, such as the weight function approach described here. Letting α be a positive constant to be chosen suitably later, and setting

$$\begin{cases} \mathbf{u} = g(t)\mathbf{v}\,, \\ g(t) = e^{\alpha t} & (\alpha = \text{const.} > 0)\,, \end{cases} \tag{6.3}$$

it turns out that

$$\ddot{\mathbf{u}} = \left(\alpha^2\mathbf{v} + 2\alpha\dot{\mathbf{v}} + \ddot{\mathbf{v}}\right)g\,. \tag{6.4}$$

The stress tensor $\mathbf{T}$ can be written

$$\mathbf{T} = g\,\mathbf{C}\cdot\mathbf{E}^* \tag{6.5}$$

where $\mathbf{E}^*= \text{sym}\nabla\mathbf{v}$, and $(6.1)_1$ becomes

$$\rho\left(\alpha^2\mathbf{v} + \ddot{\mathbf{v}}\right) + \left[2\alpha\rho\dot{\mathbf{v}} - \nabla\cdot(\mathbf{C}\cdot\mathbf{E}^*)\right] = 0\,. \tag{6.6}$$

Hence, by squaring and disregarding the two "squares" and the positive factor $2\rho(x)$, it follows that

$$\left(\alpha^2\mathbf{v} + \ddot{\mathbf{v}}\right)\cdot\left[2\alpha\rho\dot{\mathbf{v}} - \nabla\cdot(\mathbf{C}\cdot\mathbf{E}^*)\right] \le 0\,. \tag{6.7}$$

But $(6.1)_2$ implies that

$$2\int_0^t dt\int_\Omega \rho\mathbf{v}\cdot\dot{\mathbf{v}}\,dx = \int_\Omega \rho v^2\,dx \quad , \quad 2\int_0^t dt\int_\Omega \rho\dot{\mathbf{v}}\cdot\ddot{\mathbf{v}}\,dx = \int_\Omega \rho\dot{\mathbf{v}}^2\,dx\,. \tag{6.8}$$

From (6.2) it turns out that

$$\nabla\mathbf{v}\cdot\mathbf{C}\cdot\nabla\mathbf{v} = \nabla\dot{\mathbf{v}}\cdot\mathbf{C}\cdot\nabla\dot{\mathbf{v}} = 0 \tag{6.9}$$

which, together with the initial-boundary conditions, imply that

$$\int_\Omega \nabla \overset{\bullet}{\mathbf{v}} \cdot \mathbf{C} \cdot \nabla \mathbf{v} dx = - \int_0^t d\tau \int_\Omega (\alpha^2 \mathbf{v} + \ddot{\mathbf{v}}) \cdot [\nabla \cdot (\mathbf{C} \cdot \mathbf{E}^*)] \, dx \,. \tag{6.10}$$

Noting (6.8)–(6.10), it follows that

$$\alpha \int_\Omega \rho \overset{\circ}{v}{}^2 \, dx + \alpha^3 \int_\Omega \rho v^2 \, dx + \int_\Omega \nabla \overset{\bullet}{\mathbf{v}} \cdot \mathbf{C} \cdot \nabla \mathbf{v} dx \leq 0 \,, \qquad t \in [0,T) \,, \tag{6.11}$$

and hence that

$$\int_\Omega \rho v^2 \, dx \leq -\frac{1}{\alpha^3} \int_\Omega \nabla \overset{\bullet}{\mathbf{v}} \cdot \mathbf{C} \cdot \nabla \mathbf{v} \, dx \qquad t \in [0,T) \,. \tag{6.12}$$

Reintroducing $\mathbf{u}$ and noting that $\nabla \mathbf{u} \cdot \mathbf{C} \cdot \nabla \mathbf{u} = 0$, one has

$$\int_\Omega \rho u^2 dx \leq -\frac{1}{\alpha^3} \int_\Omega \nabla \overset{\bullet}{\mathbf{u}} \cdot \mathbf{C} \cdot \nabla \mathbf{u} \, dx \qquad t \in [0,T) \,. \tag{6.13}$$

Letting $\alpha^3 \to \infty$, one immediately obtains uniqueness of smooth solutions to the displacement problem of linear elastodynamics, under the constitutive assumptions (6.2).

Remark 6.1 - *Continuous dependence on the initial data can be obtained* [152] *from* (6.7) (cfr. exercise 6.4).

6.3 WEIGHT FUNCTION APPROACH TO BACKWARDS UNIQUENESS ON BOUNDED DOMAINS

Now we apply the weight function approach to the uniqueness question for the one-dimensional heat equation, backwards in time. However, we begin by proving – by the standard *energy method* – that if two temperature distributions coincide at time $t = 0$, then the temperatures must coincide at all future times (*forwards in time uniqueness*).

Theorem 6.1 - *There exists, at most, one solution* $\theta \in C^{2,1}\{(0,l) \times (0,T)\} \cap C^1\{(0,l) \times [0,T]\}$ *of the I.B.V.P.*

$$\begin{cases} \theta_t = \theta_{xx} + f(x,t) & \text{in } (0,l) \times (0,T) \,, \\ \theta(x,0) = \theta_0(x) & \text{in } [0,l] \,, \\ \theta(0,t) = \theta_1(t) \quad , \quad \theta(l,t) = \theta_2(t) & \text{in } [0,T] \,, \end{cases} \tag{6.14}$$

where $f \in C\{(0,l) \times (0,T)\}$, $\theta_0 \in C[0,l]$, $\theta_i \in C[0,T]$, $(i = 1,2)$ *are prescribed and* $T > 0$.

Proof. Denoting by $\bar{\theta}$ another solution (if such exists), then it is evident that $u = \theta - \bar{\theta}$ satisfies

$$\begin{cases} u_t = u_{xx} & \text{in } (0,l)\times(0,T)\,, \\ u(x,0) = 0 & \text{in } [0,l]\,, \\ u(0,t) = u(l,t) = 0 & \text{in } [0,T]\,, \end{cases} \tag{6.15}$$

and hence

$$\frac{dE}{dt} = [u\, u_x]_{x=0}^{x=l} - \int_0^l u_x^2\, dx = - \int_0^l u_x^2\, dx \leq 0 \tag{6.16}$$

where

$$E = \frac{1}{2} \int_0^l u^2\, dx \tag{6.17}$$

is the so-called "energy" of the perturbation u. From (6.16) it follows that $E(t) \leq E(0) = 0$ $(\forall t \in [0,T])$, which implies $u \equiv 0$ in $[0,l]\times[0,T]$.

Now we want to show that if two temperature distributions coincide at time $t = 0$, then the temperature must have coincided at all earlier times (*backwards in time uniqueness*). To this end, we indicate by (6.14)* the problem (6.14) when the time intervals $[0,T]$ and $(0,T)$ are replaced by $[-T,0]$ and $(-T,0)$ respectively.

Theorem 6.2 - *There exists, at most, one solution* $\theta \in C^{2,1}\{(0,l)\times(-T,0)\} \cap C^1\{[0,l]\times[-T,0]\}$ *of* (6.14)*.

Proof. We have to show that the problem

$$\begin{cases} u_t = u_{xx} & \text{in } (0,l)\times(-T,0)\,, \\ u(x,0) = 0 & \text{in } [0,l]\,, \\ u(0,t) = u(l,t) = 0 & \text{in } [-T,0]\,, \end{cases} \tag{6.18}$$

has the zero solution only. Introducing the weight g^m such that

$$g(t) = T - t\,, \quad m = \text{const.} > 0\,, \tag{6.19}$$

and setting

$$\varphi(x,t) = g^m\, u\,, \tag{6.20}$$

it turns out that

$$\varphi_t + m g^{-1} \varphi - \varphi_{xx} = 0 \qquad \text{in} \quad (0,l)\times(-T,0)\,, \tag{6.21}$$

and hence

$$(\varphi_t - \varphi_{xx})^2 + m^2 g^{-2} \varphi^2 + 2(\varphi_t - \varphi_{xx}) m g^{-1} \varphi = 0 .$$

Noting that

$$\begin{cases} 2mg^{-1} \varphi\varphi_t = m\dfrac{\partial}{\partial t}(g^{-1}\varphi^2) - mg^{-2}\varphi^2 , \\ \varphi\varphi_{xx} = \dfrac{\partial}{\partial x}(\varphi\varphi_x) - \varphi_x^2 , \end{cases}$$

one has

$$m(m-1)g^{-2}\|\varphi\|^2 + m\frac{d}{dt}(g^{-1}\|\varphi\|^2) + 2mg^{-1}\|\nabla\varphi\|^2 \le 0 , \tag{6.22}$$

with $\|\cdot\| = \|\cdot\|_2$, and hence

$$\begin{aligned} &m(m-1)\int_{-T}^{0} g^{-2}\|\varphi\|^2 dt + \\ &+ m\left[T^{2m-1}\|u(x,0)\|^2 - (2T)^{2m-1}\|u(x,-T)\|^2\right] \le 0. \end{aligned} \tag{6.23}$$

Therefore for $m > 1$, because of $(6.18)_2$, one has the *weighted inequality*

$$\int_{-T}^{0} g^{2(m-1)}\|u\|^2 dt \le \frac{(2T)^{2m-1}}{m(m-1)}\|u(x,-T)\|^2 , \qquad (m>1) . \tag{6.24}$$

Letting $m \to \infty$, (6.24) implies uniqueness in the time interval $[-1/2, 0]$ which is enough to ensure, by a recursive argument, uniqueness backward in time.

Remark 6.2 - *The proof of Theorem* 6.2 *is along the lines of general results of Lees and Protter* [153], *revisited by J.Serrin* [154].

6.4 WEIGHT FUNCTION APPROACH TO UNIQUENESS ON UNBOUNDED DOMAINS

The aim of the weight function approach to uniqueness on unbounded domains is to obtain uniqueness without prescribing, at infinity, conditions on the solutions and on its spatial derivatives which are "too strong". The approach that we present here has been introduced, for "smooth solutions" to the Navier-Stokes equations, by Rionero & Galdi [155,72]. But, for the sake of simplicity and clarity, we here apply the method to the one-dimensional *heat equation*.

Let us consider the case of a semi-infinite bar ($l = \infty$). Then the I.B.V.P. (6.14) becomes

$$\begin{cases} \theta_t = \theta_{xx} + f(x,t) & \text{in } (0,\infty)\times(0,T) , \\ \theta(x,0) = \theta_0(x) & \text{in } [0,\infty) , \\ \theta(0,t) = \theta_1(t) & \text{in } [0,T] , \end{cases} \tag{6.25}$$

and hence (6.15) becomes

$$\begin{cases} u_t = u_{xx} & \text{in } (0,\infty)\times(0,T)\,, \\ u(x,0) = 0 & \text{in } [0,\infty)\,, \\ u(0,t) = 0 & \text{in } [0,T]\,. \end{cases} \tag{6.26}$$

For each $l > 0$, (6.16) then gives

$$\frac{d}{dt}\frac{1}{2}\int_0^l u^2\,dx = u(l,t)u_x(l,t) - \int_0^l u_x^2\,dx\,. \tag{6.27}$$

If one assumes the *decay condition at infinity*

$$\lim_{l\to\infty} u(l,t)\,u_x(l,t) = 0\,, \qquad \forall\, t \in I\!R^+ \tag{6.28}$$

then it follows, once again, that $E(t) = 0\,, \forall\, t \in I\!R^+$, which implies $u \equiv 0$. Therefore, according to the standard energy method, for uniqueness to hold one has to make the "extra" assumption (6.28) *which is not a datum of the problem at hand* (6.26). Of course – as pointed out in Remark 1.9 – restrictions on the solutions are needed which guarantee that they belong to the functional class delimited by counterexamples to uniqueness. In the case at hand – because of the Tikhonov counterexample [Sec.1.8, e)] – one has to require $\|u\| = O(e^{x^m})$ with $m < 2+\sigma$, where $\sigma \in (0,1)$.

Proceeding now to the weight function approach, let us show how one can circumvent the decay condition at infinity (6.28), allowing "*exponential growth*"

$$|\theta|\,,\,|\theta_x| = O(e^{\lambda x}) \quad \text{as} \quad x\to\infty\,, \tag{6.29}$$

for any preassigned constant $\lambda\,(>0)$.

Theorem 6.3 - *There exists, at most, one solution* $\theta \in C^{2,1}\{(0,\infty)\times(0,T)\} \cap C^1\{[0,\infty]\times[0,T]\}$ *of* (6.25) *such that* (6.29) *holds.*

Proof. Denoting any other solution by $\bar\theta$, $u = \theta - \bar\theta$ satisfies (6.26). Instead of (6.17), we introduce now the *weighted energy*

$$\mathcal{E}(t) = \frac{1}{2}\int_0^\infty g\,u^2\,dx\,, \tag{6.30}$$

with the weight

$$g = e^{-\alpha x} \tag{6.31}$$

where α is a constant such that $\alpha > 2\lambda$. This guarantees the finiteness of (6.30), although $|u| = O(e^{\lambda x})$.

Noting that

$$g u u_{xx} = \frac{d}{dx}(g u u_x) - g u_x^2 - g' u u_x\,,$$

it turns out that, for $\forall\, \bar x > 0$,

$$\frac{d}{dt}\frac{1}{2}\int_0^{\bar x} g\,u^2\,dx = g(\bar x)u(\bar x,t)u_x(\bar x,t) - \int_0^{\bar x} g\,u_x^2\,dx + \alpha\int_0^{\bar x} g u u_x\,dx\,.$$

But, for any positive α, Cauchy's inequality gives $\alpha u u_x \le \frac{1}{2}\left(\alpha^2 u^2 + u_x^2\right)$. Hence substituting in the foregoing and letting $\bar{x} \to \infty$, the *weighted energy inequality* follows:

$$\frac{d\,\mathcal{E}}{dt} \le \alpha^2\,\mathcal{E}$$

and, by integration,

$$\mathcal{E}(t) \le \mathcal{E}(0)\, e^{\alpha^2 t} = 0\,. \qquad t \in [0,T]$$

Consequently $u \equiv 0$ in $[0,\infty)\times[0,T]$.

Remark 6.3 - *The quantity* E *defined by* (6.17), *being reminiscent of kinetic energy, may be referred to as "energy", but it has no physical meaning. It is quite natural then to call* $\mathcal{E}$, *given by* (6.30), *"weighted energy".*

6.5 WEIGHT FUNCTION APPROACH TO CAUCHY PROBLEM FOR THE KORTEWEG - DE VRIES - BURGERS EQUATION

The introduction of weight functions depending both on space variables and time, can be very useful. We shall apply this type of weight to the Cauchy problem for the nonlinear Korteweg - de Vries - Burgers (K.d.V.B.) equation [156]

$$\begin{cases} v_t + v v_x + \mu v_{xxx} = \nu v_{xx} + F(x,t) & \text{on } I\!R\times(0,T)\,, \\ v(x,0) = v_0(x) & \text{on } I\!R\,, \end{cases} \tag{6.32}$$

where $\mu \in I\!R$ is the so-called *dispersion constant*, while $\nu =$ const. > 0 is a *coefficient of viscosity*. The function v_0 is prescribed as well as the function $F(x,t)$ – which can represent a forcing action on the physical system, or a control. Denoting by v and $v+u$ two smooth solutions corresponding respectively to the data (v_0, F) and $(v_0 + u_0, F + f)$, one has

$$\begin{cases} u_t = -(v + u)u_x - u v_x - \mu u_{xxx} + \nu u_{xx} + f & \text{on } I\!R\times(0,T)\,, \\ u(x,0) = u_0(x) & \text{on } I\!R\,. \end{cases} \tag{6.33}$$

Denoting by $g(x,t) \ge 0$ a weight function, differentiable as many times as necessary, one has

$$\begin{cases} g\,u\,u_t = \frac{1}{2}(g u^2)_t - \frac{1}{2} g_t u^2\,, \\ (v+u) g\,u\,u_x = \left(\frac{1}{2} g\,v\,u^2 + \frac{1}{3} g\,u^3\right)_x - \frac{1}{2} g_x \left(v + \frac{2}{3} u\right) u^2 - \frac{1}{2} g v_x u^2\,, \\ g\,u\,u_{xx} = \left(g\,u\,u_x - g_x \frac{u^2}{2}\right)_x + g_{xx}\frac{u^2}{2} - g\,u_x^2\,, \\ g\,u\,u_{xxx} = \left(g\,u\,u_{xx} - g_x\,u\,u_x - g\,\frac{u_x^2}{2} + \frac{g_{xx}}{2}\,u^2\right)_x - \frac{g_{xxx}}{2}\,u^2 + \frac{3}{2}\,g_x\,u_x^2\,. \end{cases} \tag{6.34}$$

Choosing

$$g = \exp\left[-\alpha(1+\varepsilon\, x)(t+t_0)^{\beta}\right] \tag{6.35}$$

where the constants α, t_0, β, and the quantity ε satisfy

$$\alpha,\, t_0,\, \beta > 0 \quad , \quad \varepsilon = 1 \quad \text{for} \quad x \geq 0 \quad , \quad \varepsilon = -1 \quad \text{for} \quad x \leq 0\,; \tag{6.36}$$

we introduce the *weighted energy*

$$\mathcal{E} = \int_{I\!R} g\, u^2\, dx\,. \tag{6.37}$$

Then one obtains from (6.33)–(6.34):

Theorem 6.4 - *Let f, v_x, u_{xx} be $O(|x|^k)$, $k=$ const. >0, uniformly with respect to t, as $|x|\to\infty$. Then u satisfies the weighted energy equality*

$$\begin{aligned}\overset{\bullet}{\mathcal{E}} = \int_{I\!R}\Big\{ &[g_t + g_x(v+\frac{2}{3}u) + (\nu g + \mu g_x)_{xx} - g v_x]u^2 + \\ &-(2\nu g + 3\mu g_x)u_x^2 + 2gfu\Big\}\, dx\,.\end{aligned} \tag{6.38}$$

As a consequence of (6.38), one has

Theorem 6.5 - *Let the assumptions of Theorem 6.4 hold with $|v_x|$, $|u_x|=O(1)$, uniformly with respect to t as $|x|\to\infty$. Then*

$$\begin{cases} u_0 \in L^2(I\!R)\,, \\ f \in L^1[0,T; L^2(I\!R)] \end{cases} \Rightarrow \begin{cases} u(t) \in L^2(I\!R)\,, \\ u_x(t) \in L^1[0,T; L^2(I\!R)] \end{cases} \tag{6.39}$$

$\forall\, t \in [0,T]$, and the following L^2 energy inequality holds

$$\|u(t)\|^2 \leq \|u_0\|^2 - \int_0^t\!\!\int_{I\!R} \left(v_x u^2 + 2\nu u_x^2 - 2fu\right)\, dx\, dt\,, \qquad t\in[0,T]\,. \tag{6.40}$$

Proof. Since

$$\begin{cases} g_t = -\alpha\beta(1+|x|)(t+t_0)^{\beta-1}g\,, & g_x = -\alpha\varepsilon(t+t_0)^{\beta}g\,, \\ g_{xx} = \alpha^2(t+t_0)^{2\beta}g\,, & g_{xxx} = -\alpha^3\varepsilon(t+t_0)^{3\beta}g\,, \end{cases} \tag{6.41}$$

for $\beta > \frac{5}{3}M(T+t_0)$, where M is a constant such that $|v|, |u| \leq M(1+|x|)$, $|v_x| \leq M$, one has

$$\begin{cases} (\nu g + \mu g_x)_{xx} \leq \alpha^2(T+t_0)^{2\beta}[\nu + |\mu|\alpha(T+t_0)^{\beta}]\, g\,, \\ g_t + \left(v + \frac{2}{3}u\right) g_x \leq \alpha g(1+|x|)\left(\frac{5}{3}M - \frac{\beta}{T+t_0}\right)(t+t_0)^{\beta} < 0\,. \end{cases} \tag{6.42}$$

Further, for

$$0 < \alpha < \frac{2\nu}{3|\mu|(T+t_0)^\beta}, \tag{6.43}$$

there follows

$$\nu g + \frac{3}{2}\mu g_x \geq \nu - \frac{3}{2}|\mu|\alpha(T+t_0)^\beta > 0. \tag{6.44}$$

Hence, from (6.38), (6.42)–(6.44) and the Cauchy inequality $2fu \leq f^2 + u^2$, it turns out that

$$\overset{\bullet}{\mathcal{E}} \leq A\mathcal{E} + \int_{\mathbb{R}} g(f^2 - 2\overline{\nu}u_x^2)\,dx \tag{6.45}$$

where

$$\begin{cases} 2\overline{\nu} = 2\nu - 3|\mu|\alpha(T+t_0)^\beta \quad , \quad |v_x| \leq M = \text{const.}, \\ \\ A = \alpha^2(T+t_0)^{2\beta}[\nu + |\mu|\alpha(T+t_0)^\beta] + 1 + M. \end{cases} \tag{6.46}$$

By Gronwall's lemma (2.3), one obtains from (6.45) that

$$2\overline{\nu}\int_0^t\int_{\mathbb{R}} g u_x^2\,dxdt + \mathcal{E}(t) \leq \exp(AT)\left[\mathcal{E}(0) + \int_0^T\int_{\mathbb{R}} gf^2\,dxdt\right], \quad t \in [0,T]. \tag{6.47}$$

Since by assumptions (6.39) the right hand side converges to a finite quantity as $\alpha \to 0$, it follows from the monotone convergence theorem [159], on letting $\alpha \to 0$, that

$$\int_0^t\int_{\mathbb{R}} u_x^2\,dxdt + \int_{\mathbb{R}} u^2\,dx < \infty, \qquad \forall t \in [0,T]. \tag{6.48}$$

Let us return now to (6.38). Because of $(6.42)_2$, it follows that

$$\mathcal{E} \leq \mathcal{E}(0) + \int_0^t\int_{\mathbb{R}} g[(\alpha^2 A_1 - v_x)u^2 - 2\overline{\nu}u_x^2 + 2fu]\,dxdt \qquad t \in [0,T] \tag{6.49}$$

where

$$A_1 = (T+t_0)^{2\beta}[\nu + |\mu|\alpha(T+t_0)^\beta]. \tag{6.50}$$

The inequality (6.40) follows from (6.49) on letting $\alpha \to 0$.

We are now in a position to obtain continuous dependence on the data (u_0, f), and stability on the initial data u_0. Denoting by $\|\cdot\|$ the $L^2(\mathbb{R})$-norm, the following theorems hold:

Theorem 6.6 - *Let the assumptions of Theorem* 6.5 *hold. Then*

$$\|u_0\|^2 + \int_0^T \|f\|^2\,dt < \delta \tag{6.51}$$

implies that

$$2\nu\int_0^t \|u_x\|^2\,dt + \|u\|^2 \leq e^{(1+M)T}\delta \qquad \forall t \in [0,T], \tag{6.52}$$

where M is a positive constant such that $|v|, |u| \leq M(1+|x|)$ and $|v_x| \leq M$ on $I\!R \times [0,T]$.

Proof. (6.52) is obtained from (6.46)–(6.47), on letting $\alpha \to 0$.

Theorem 6.7 - *Suppose $f = 0$ and that the assumptions of Theorem 6.5 hold, $\forall T > 0$. Setting*

$$\nu^* = \frac{1}{2} \sup_{t \in I\!R^+} \sup_{w \in \Sigma} \frac{-\int_{I\!R} v_x w^2 \, dx}{\int_{I\!R} (w')^2 \, dx} \tag{6.53}$$

where Σ is the set of once differentiable functions on $I\!R$, then

$$\nu^* < \nu \tag{6.54}$$

implies that

$$||u(t)|| \leq ||u_0|| \quad , \quad \int_0^t ||u_x||^2 \, dt \leq \frac{||u_0||^2}{2(\nu - \nu^*)} \, , \qquad \forall t \geq 0 \, . \tag{6.55}$$

Proof. It follows from (6.40) that

$$||u(t)||^2 + 2(\nu - \nu^*) \int_0^t ||u_x||^2 \, dt \leq ||u_0||^2 \, , \qquad \forall t \geq 0 \, , \tag{6.56}$$

which proves the Theorem.

Remark 6.4 - *Theorems 6.6–6.7 ensure, in particular, continuous dependence and stability in the L^2-norm. We refer to exercises 6.6–6.7, for continuous dependence and stability in the pointwise norm.*

Remark 6.5 - *The weight function (6.35) is more general than those previously used, containing more parameters, all of which have been used in the course of the proof.*

Remark 6.6 - *Instead of the weight function (6.35) – which does not belong to $C^1(I\!R)$ – one can choose the weight function belonging to $C^\infty(I\!R)$:*

$$g^* = \frac{1}{\exp[\alpha x(t+t_0)^\beta] + \exp[-\alpha x(t+t_0)^\beta]} \, , \qquad (\alpha, \, \beta, \, t_0 > 0) \, .$$

6.6 WEIGHTED POINCARÉ INEQUALITIES

In obtaining stability theorems for some parabolic equations in *bounded domains* (intervals) in chapter two, a fundamental role is played by the Poincaré inequalities (2.26) and (2.55). Therefore it is quite natural to investigate if these inequalities continue to hold in unbounded domains. The following theorem holds

Theorem 6.8 - *Let Z be the set of functions $u: I\!R_+ \to I\!R$ such that*

$$\int_{I\!R_+} (u^2 + u'^2)dx < \infty, \qquad \{u(0) = 0,\ u \neq \emptyset\}. \tag{6.57}$$

Then the functional

$$F(u) = \frac{\displaystyle\int_{I\!R_+} u^2 dx}{\displaystyle\int_{I\!R_+} u'^2 dx}, \tag{6.58}$$

is unbounded.

Proof. Denoting by ε and $k(\neq \varepsilon)$ two positive constants and setting

$$f = \frac{1}{1+kx}, \quad g = \frac{1}{1+\varepsilon x}, \quad u = f - g$$

immediately it follows (see exercise 6.10) that $u \in Z$ and

$$\int_{I\!R_+} u^2 dx = \frac{1}{k} + \frac{1}{\varepsilon} - \frac{2}{\varepsilon - k}\int_{I\!R_+}\left(\frac{\varepsilon}{1+\varepsilon x} - \frac{k}{1+kx}\right)dx = \frac{1}{k} + \frac{1}{\varepsilon} - \frac{2}{\varepsilon - k}\log\frac{\varepsilon}{k}$$

$$\int_{I\!R_+} u'^2 dx = \int_{I\!R_+}(f' - g')^2 dx \leq 2\int_{I\!R_+}(f'^2 + g'^2)dx = \frac{2}{3}(k+\varepsilon)$$

$$F(u) \geq \frac{3}{2k\varepsilon} - \frac{3}{\varepsilon^2 - k^2}\log\frac{\varepsilon}{k}.$$

Choosing $\varepsilon = ek$, where "e" is the Napier number, and taking into account that

$$\frac{1}{2e} - \frac{1}{e^2 - 1} > 0.02$$

it turns out that $F(u) \geq \dfrac{3}{k^2}\left(\dfrac{1}{2e} - \dfrac{1}{e^2-1}\right) \geq \dfrac{0.06}{k^2}$, and hence $\lim\limits_{k\to 0^+} F(u) = +\infty$.

As a consequence of Theorem 6.8, the inequality (2.26) does not hold on $I\!R_+$. Now it is well known that $\exists u : I\!R_+ \to I\!R$ which can "grow" faster than their derivatives in such a way that $\{u' \in L^2(I\!R_+), u \notin L^2(I\!R_+)\}$. This happens, for instance, for $u = (1+x)^\alpha$ with $\alpha \in (0, 1/2)$. Theorem 6.8 indicates that on Z, $\|u\|_2$ can "grow" faster than $\|u'\|_2$. Therefore one is led to weaken $\|\cdot\|_2$ by means of a weight $0 \leq g(x) \leq 1$, $\forall x \in I\!R_+$ considering, instead of $\|u\|_2$, the weighted norm $\|g^{1/2}u\|_2$ and requiring that $g(x) \to 0$ as $x \to \infty$. This can be done with different choices of the weight g and several weighted Poincaré inequalities can be obtained [157,72,158,160]. In this section, we confine ourselves to broadening the applicability of the inequality (2.55) to the unbounded interval $\mathbb{R}_+$ and to obtaining generalized weighted Poincaré inequalities for some unbounded domains of $\mathbb{R}^3$. In doing this we use a simple "cut-off" function, introduced in [72], whose properties can be summarized in the following Lemma.

Lemma 6.1 - *Let a be a positive constant and let $\varphi : \mathbb{R}_+ \to \mathbb{R}_+$ denote the "cut-off" function*

$$\varphi(\xi) = \begin{cases} 1 & \xi \in [0,a]\,, \\ 2\left(\dfrac{\xi}{a}\right)^3 - 9\left(\dfrac{\xi}{a}\right)^2 + 12\left(\dfrac{\xi}{a}\right) - 4 & \xi \in [a,2a]\,, \\ 0 & \xi \in [2a,\infty)\,. \end{cases} \tag{6.59}$$

Then it follows that $\varphi \in C^1(\mathbb{R}_+)$ and $(\forall\, \xi \in \mathbb{R}_+)$

$$\varphi(\xi) \geq 0\,,\ \varphi'(\xi) \leq 0\,,\ |\varphi'(\xi)| = O(\frac{1}{a})\,,\ \lim_{a\to\infty} \varphi(\xi) = 1\,,\ \lim_{a\to\infty} \varphi'(\xi) = 0. \tag{6.60}$$

Proof. We leave the immediate proof to the reader.

Theorem 6.9 - *Let Z_k, $(k = const.(>0))$, be the set of functions $\{u : \mathbb{R}_+ \to \mathbb{R}\}$ such that $\{u(0) = 0,\ e^{-kx}u'^2 \in L^1(\mathbb{R}_+)\}$. Then $e^{-kx}u^2 \in L^1(\mathbb{R}_+)$ and the following inequality holds*

$$\int_{\mathbb{R}_+} e^{-kx}u^2 dx \leq \frac{4}{k^2}\int_{\mathbb{R}_+} e^{-kx}u'^2 dx\,. \tag{6.61}$$

Proof. In view of $(6.60)_2$, we have

$$\begin{aligned} ke^{-kx}\varphi u^2 &= e^{-kx}(\varphi' u^2 + 2\varphi u u') - (e^{-kx}\varphi u^2)' \leq \\ &\leq 2e^{-kx}\varphi u u' - (e^{-kx}\varphi u^2)'\,. \end{aligned}$$

Integrating on $[0,2a]$ it turns out that

$$k\int_0^{2a} e^{-kx}\varphi u^2 dx = 2\int_0^{2a} e^{-kx}\varphi u u' dx\,,$$

and hence employing the Cauchy inequality we deduce

$$\int_0^{2a} e^{-kx}\varphi u^2 dx \leq \frac{1}{\varepsilon(k-\varepsilon)}\int_0^{2a} e^{-kx}\varphi u'^2 dx\,, \qquad \varepsilon \in (0,k)\,.$$

Choosing $\varepsilon = k/2$ and letting $a \to \infty$, it follows from the theorem of dominate a convergence that $e^{-kx}u^2 \in L^1(\mathbb{R}_+)$ and (6.61) follows.

Definition 6.1 - *A domain $D \subset \mathbb{R}^n$ is starshaped with respect to a point O if any ray with origin O intersects ∂D in just one point.*

Let $\Omega_0 \subset \mathbb{R}^3$ be a smooth bounded domain. We denote by Ω the domain exterior to Ω_0 (i.e. $\Omega = \mathbb{R}^3 - \Omega_0$) and by $\mathbf{n}_0$ and $\mathbf{n}(= -\mathbf{n}_0)$ the outward unit normal to $\partial\Omega_0$ and $\partial\Omega$ respectively. The following theorem holds [160]

Theorem 6.10 - *Let Ω_0 be starshaped with respect to an inner point O and such that $(P-O)\cdot\mathbf{n}_0 \geq 0, \forall P \in \partial\Omega_0$. Denoting by I the set of vector valued functions $\{\mathbf{u} : \Omega \to \mathbb{R}^3, \mathbf{u} \in L^2(\partial\Omega)\}$ such that*[1]

$$\int_\Omega \left[\frac{u^2}{r^k} + \frac{1}{r^{k-2}}(\nabla\mathbf{u})^2\right] d\mathbf{x} < \infty \qquad k \in [2,3[,\ r = |OP|, \tag{6.62}$$

then, $\forall\,\mathbf{u} \in I$, the following inequality holds

$$\int_\Omega \frac{u^2}{r^k} d\mathbf{x} \leq \left(\frac{2}{3-k}\right)^2 \int_\Omega \frac{1}{r^{k-2}}(\nabla\mathbf{u})^2 d\mathbf{x} - \frac{2}{3-k}\int_{\partial\Omega} \frac{u^2}{r^k}(P-O)\cdot\mathbf{n}_0 d\sigma\,. \tag{6.63}$$

Proof. Let r denote the radial coordinate, and $\mathbf{e}_r$ the unit vector in the radial direction of spherical polar coordinates with origin at O and let φ denote the "*cut-off*" function (6.59) with $r = \xi$. Setting $P - O = \mathbf{x} = r\mathbf{e}_r$, from

$$\nabla\cdot\left[\frac{\varphi(r)u^2(P-O)}{r^k}\right] = \frac{\varphi' u^2}{r^{k-1}} + \frac{2\varphi(P-O)\cdot\nabla\mathbf{u}\cdot\mathbf{u}}{r^k} + (3-k)\frac{\varphi u^2}{r^k} \tag{6.64}$$

and the arithmetic-geometric inequality, it follows that

$$(3-k-\varepsilon)\frac{\varphi u^2}{r^k} \leq \nabla\cdot\left[\frac{\varphi u^2(P-O)}{r^k}\right] - \varphi' r^{1-k}u^2 + \frac{1}{\varepsilon}\varphi r^{2-k}(\nabla\mathbf{u})^2 \tag{6.65}$$

where $\varepsilon < 3-k$ is a positive constant. Denoting by $R = a$ the radius of a sphere $S(R)$ centered at O and containing Ω_0, integrating (6.65) and taking into account $(6.60)_3$ it follows that

$$(3-k-\varepsilon)\int_\Omega \frac{\varphi u^2}{r^k} d\mathbf{x} \leq -\int_{\partial\Omega} \frac{u^2}{r^k}(P-O)\cdot\mathbf{n}_0 d\sigma + \frac{1}{\varepsilon}\int_\Omega \varphi r^{2-k}(\nabla\mathbf{u})^2 d\mathbf{x} + \\ + \frac{C}{R}\int_{B(R)} r^{1-k}u^2 d\mathbf{x} \tag{6.66}$$

---

[1] $(\nabla\mathbf{u})^2 = \nabla\mathbf{u} : \nabla\mathbf{u} \overset{\text{def.}}{=} \sum_{ij}^{1-3}\left(\frac{\partial u_i}{\partial x_j}\right)^2$

where $B(R) = S(2R) - S(R)$ and C is a positive constant associated with $|\varphi'(r)| = O(\frac{1}{R})$. Taking into account (6.60) and (6.62) and letting $R \to \infty$, it turns out that[2]

$$(3-k-\varepsilon)\int_\Omega r^{-k}u^2 d\mathbf{x} \leq \frac{1}{\varepsilon}\int_\Omega r^{2-k}(\nabla \mathbf{u})^2 d\mathbf{x} - \int_{\partial\Omega} r^{-k}u^2 (P-O)\cdot \mathbf{n}_0 d\sigma \tag{6.67}$$

and hence, because the l.u.b. of $[\varepsilon(3-k-\varepsilon)]^{-1}$ is reached for $\varepsilon = (3-k)/2$, (6.63) immediately follows.

Remark 6.7 - *Concerning the inequality* (6.63), *we observe that: i)* $u \in L^2(\partial\Omega)$ *is the sole condition imposed to* u *on* $\partial\Omega$; *ii) on the left hand side of* (6.63) *the presence of a negative term depends on the geometrics assumption* $\mathbf{e}_r \cdot \mathbf{n}_0 \geq 0$ *on* $\partial\Omega$ *and can help in many questions* [158]; *iii) most of the common domains (spheres, ellipsoids, cylinders, cones, parallelepipeds,...) are starshaped.*

Theorem 6.11 - *Let* Ω *be the half-space* $z \geq 0$. *Denoting by* I *the set of vector valued functions* $\{\mathbf{u} : \Omega \to \mathbb{R}^3,\ \mathbf{u}|_{z=0} = 0,\ u^2 \in L^1_{\text{loc}}(\Omega),\ (\nabla \mathbf{u})^2 \in L^1(\Omega)\}$ *then* $\dfrac{u^2}{(1+z)^2} \in L^1(\Omega)$ *and the following inequality holds*

$$\int_\Omega \frac{u^2}{(1+z)^2} d\mathbf{x} \leq 4\int_\Omega (\nabla \mathbf{u})^2 d\mathbf{x}, \qquad \forall \mathbf{u} \in I. \tag{6.68}$$

Proof. Let φ denote the cut-off function (6.59) with $z = \xi$ and $a \equiv l =$ const. > 0. From

$$\frac{\partial}{\partial z}\left(\frac{\varphi(z)u^2}{1+z}\right) = \frac{\varphi' u^2}{1+z} - \frac{\varphi u^2}{(1+z)^2} + 2\varphi\frac{\mathbf{u}}{1+z}\cdot\frac{\partial \mathbf{u}}{\partial z},$$

taking into account that $\{\varphi' \leq 0,\ \mathbf{u}|_{z=0}\}$ and integrating on the cylinder $\{x^2 + y^2 \leq l,\ z \geq 0\}$ it follows that

$$\int_\Omega \varphi\frac{u^2}{(1+z)^2} d\mathbf{x} \leq 2\int_\Omega \varphi\frac{|\mathbf{u}|}{1+z}\left|\frac{\partial \mathbf{u}}{\partial z}\right| d\mathbf{x} \leq$$
$$\leq 2\left(\int_\Omega \varphi\frac{u^2}{(1+z)^2} d\mathbf{x}\right)^{1/2}\left(\int_\Omega \varphi(\nabla \mathbf{u})^2 d\mathbf{x}\right)^{1/2}$$

---

[2] Because of

$$\frac{1}{R}\int_{B(R)} r^{1-k}u^2 d\mathbf{x} < 2\left(\int_{S(2R)} r^{-k}u^2 d\mathbf{x} - \int_{S(R)} r^{-k}u^2 d\mathbf{x}\right),$$

from $r^{-k}u^2 \in L^1(\Omega)$ immediately it follows that

$$\lim_{R\to\infty} R^{-1}\int_{B(R)} r^{1-k}u^2 d\mathbf{x} = 0.$$

and hence

$$\int_\Omega \varphi \frac{u^2}{(1+z)^2} d\mathbf{x} \leq 4 \int_\Omega \varphi(|\nabla \mathbf{u}|^2) dx . \tag{6.69}$$

On letting $l \to \infty$, by the theorem of dominated convergence [159], it follows that $\frac{\mathbf{u}}{1+z} \in L^2(\Omega)$ and (6.68) holds.

6.7 METHOD OF WEIGHTED CROSS-SECTIONAL INTEGRALS

In this section it is shown how to marry the weight function approach (appropriate to infinite regions) to the cross-sectional method – introduced in chapter four – in connection with a simple Dirichlet problem for an unbounded domain. One might term the method the *method of weighted cross-sectional integrals*. It is believed that this is the first example in the literature of such a method. The general aim in this example is to obtain an estimate which implies uniqueness and continuous dependence.

We shall be concerned with the semi-infinite rectangular strip $\{x \in (0,1), y > 0\}$, where (x,y) are rectangular cartesian coordinates. We suppose that $\phi(x,y)$ is a classical solution (i.e. $\phi(x,y) \in C[[0,1] \times \mathbb{R}_+] \cap C^2[(0,1) \times \mathbb{R}_+])$ of

$$\nabla_1^2 \phi = 0 \qquad \text{in } 0 < x < 1, 0 < y < \infty \tag{6.70}$$

subject to

$$\phi(x,0) = 0 \tag{6.71}$$

and

$$\phi(0,y) = f(y) \quad , \quad \phi(1,y) = g(y) \tag{6.72}$$

where f, g are assigned continuous functions; and that, in addition, its asymptotic behaviour as $y \to \infty$ is such that

$$e^{-2\mu y}\phi^2, \quad e^{-2\mu y}\phi_x^2, \quad e^{-2\mu y}\phi_y^2, \quad e^{-2\mu y}\phi_{xx}^2 = O(1) \tag{6.73}$$

where μ is a suitable, positive constant, yet to be chosen.

Consider the weighted cross-sectional integral measure of the solution

$$F(x) = \int_0^\infty g(y)\phi^2 dy \tag{6.74}$$

where $g(y) = e^{-\lambda y}$, λ being a constant such that $\lambda > 2\mu$.

In view of (6.73) it is clear that the integral in (6.74) exists. We proceed to derive a second order differential inequality for $F(x)$, and to use it to obtain an upper estimate for $F(x)$ in terms of data. Uniqueness and continuous dependence follow from this estimate.

Sucessive differentiations (with respect to x) of (6.74) yield

$$F'(x) = \int 2g\phi\phi_x dy , \tag{6.75}$$

$$F''(x) = \int 2g(\phi_x^2 + \phi\phi_{xx})dy\,. \tag{6.76}$$

Let us note that all these integrals – and the integral $\int g\phi_y^2 dy$ arising subsequently – are convergent in view of (6.73). Here and subsequently (generally speaking) the limits of integration are omitted. If one uses (6.70), integration by parts, together with (6.71) and the relation

$$\lim_{y\to\infty} g\phi\phi_y = 0\,, \tag{6.77}$$

one obtains

$$F''(x) = \int 2g(\phi_x^2 + \phi_y^2)dy - \lambda^2 F\,. \tag{6.78}$$

We note that (6.77) is valid in view of (6.73). Identifying ϕ with u and λ with k, one obtains from (6.61) and (6.78) that

$$F'' + \frac{1}{2}\lambda^2 F \geq 2\int g\phi_x^2 dy\,. \tag{6.79}$$

With a view to obtaining an inequality for $F^{1/2}(=\sqrt{F(x)})$ we note that

$$(F^{1/2}(x))'' = \frac{1}{2}F^{-3/2}[FF'' - \frac{1}{2}F'^2]\,,$$

and using this in association with (6.79), (6.75) we obtain

$$\begin{aligned}(F^{1/2})'' &\geq \frac{1}{2}F^{-3/2}[-\frac{1}{2}\lambda^2F^2 + 2\{\int g\phi^2 dy \int g\phi_x^2 dy - (\int g\phi\phi_x dy)^2\}] \\ &\geq -(\lambda^2/4)F^{1/2}\end{aligned}$$

on using Schwarz's inequality in the last step; that is to say,

$$(F^{1/2})'' + (\frac{1}{4}\lambda^2)F^{1/2} \geq 0\,. \tag{6.80}$$

Let us now note the comparison principle embodied in exercise 4.16 – sometimes attributed to Chapyglin – and identifying the F occurring therein with $F^{1/2}$ occurring here, we obtain the estimate embodied in Theorem 6.12 given hereunder. The argument just given implicitly assumes, however, that $F \neq 0$ for any x, $0 < x < 1$. Even if this assumption is not valid, it is easily shown that the Theorem continues to hold; this is left as an exercise to the reader.

Theorem 6.12 - *The weighted cross-sectional integral measure* (6.74) *of the Dirichlet problem defined by* (6.70)–(6.73), *satisfies the estimate*

$$F^{1/2}(x) \leq G(x)$$

where G satisfies

$$G'' + \frac{1}{4}\lambda^2 G = 0$$

subject to

$$G(0) = F(0) \quad , \quad G(1) = F(1),$$

$F(0), F(1)$ *being available in terms of data specified by* (6.72), *provided that*

$$2\mu < \lambda < 2\pi .$$

It is plain that this is a continuous dependence result, and that uniqueness follows on formally putting $F(0) = F(1) = 0$.

Remark 6.8 - *There is a sense in which Theorem* 6.12 *is optimal – so far as uniqueness is concerned: the function*

$$\phi = \sinh \pi y \sin \pi x$$

satisfies (6.70), (6.71), *and* (6.72) *with zero data, but 'barely' violates* (6.73) *i.e. the constants* λ, 2μ *can be as close to one another and to* 2π, *as we like, provided* $2\pi > \lambda > 2\mu$, *but we cannot allow the three numbers to coincide.*

This concludes our discussion of an estimate for a Dirichlet problem in an unbounded domain using the weighted cross-sectional integral method. The aim of the discussion was to exhibit a *methodology*, but it should be noted, of course, that uniqueness and continuous dependence results for the problem may be derived in an arguably more orthodox manner using maximum principles e.g. Protter and Weinberger [131].

6.8 EXERCISES

Exercise 6.1 - Consider the displacement problem of linear elasticity on a bounded domain Ω:

$$\begin{cases} \rho\ddot{\mathbf{u}} = \nabla\cdot\mathbf{T} & \text{on } \Omega\times[0,T)\,, \\ \mathbf{u} = \mathbf{u}_0 \quad , \quad \dot{\mathbf{u}} = \dot{\mathbf{u}}_0 & \text{on } \Omega\times\{0\}\,, \\ \mathbf{u} = 0 & \text{on } \partial\Omega\times[0,T)\,, \end{cases}$$

and establish continuous dependence on the initial data $\mathbf{u}_0$ and $\dot{\mathbf{u}}_0$, in the set of classical solutions such that

$$\left|\int_\Omega \nabla\dot{\mathbf{u}}\cdot\mathbf{C}\cdot\nabla\mathbf{u}\,dx\right|$$

is uniformly bounded and under the constitutive assumption (6.2).

Hint - As in Sect.2 arrive at (6.7) and then at

$$\int_\Omega\left(\alpha\rho\dot{\mathbf{v}}^2 + \alpha^3\rho\mathbf{v}^2 + \nabla\dot{\mathbf{v}}\cdot\mathbf{C}\cdot\nabla\mathbf{v}\right)dx \le \int_\Omega\left(\alpha\rho\dot{\mathbf{v}}_0^2 + \alpha^3\rho\mathbf{v}_0^2 + \nabla\dot{\mathbf{v}}_0\cdot\mathbf{C}\cdot\nabla\mathbf{v}_0\right)dx.$$

Disregard the positive term $\rho\dot{v}^2$ and reintroduce $\mathbf{u}$.

Answer - Cfr [34] pp. 16-17.

Exercise 6.2 - Establish uniqueness for the n-dimensional ($n > 1$) heat equation on a bounded domain, backward in time, using the weight function approach.

Hint - Follow Sect.6.3, noting that

$$\varphi\Delta_2\varphi = \frac{1}{2}\nabla\cdot(\nabla\varphi^2) - (\nabla\varphi)^2\,.$$

Exercise 6.3 - Let Ω be a bounded domain of $I\!R^n$ and let $u(x,t)\in C^{2,1}\{\Omega\times(-T,0)\}\cap C^1\{\overline{\Omega}\times[-T,0]\}$ $(T>0)$ be such that

$$\begin{cases} u(x,0) = 0 & \text{on } \Omega \\ u = 0 & \text{on } \partial\Omega\times[-T,0]\,. \end{cases}$$

Show that, for constant ($m > 1$), one has

$$\int_{-T}^0 g^{2(m-1)}\|u\|^2\,dt \le \frac{(2T)^{2m-1}}{m(m-1)}\|u(x,T)\|^2 + \frac{1}{m(m-1)}\int_{-T}^0 g^{2m}\|Lu\|^2\,dt$$

where

$$Lu = \Delta u - u_t \quad , \quad g = T - t \quad , \quad \|\cdot\| = \|\cdot\|_2\,.$$

Hint - Follow Sect.6.3.

Exercise 6.4 - Let Ω be a bounded domain in $\mathbb{R}^n$ $(n > 1)$, and denote by $\overline{\Omega}$ its closure and by $\widehat{\Omega}$ the exterior of $\overline{\Omega}$. Consider the initial boundary value problem (heat diffusion in n-dimensions):

$$\begin{cases} u_t = \Delta u & \text{in } \widehat{\Omega}\times\mathbb{R}^+ , \\ u = 0 & \text{in } \partial\Omega\times\mathbb{R}^+ , \\ u(x,0) = 0 & \forall\, x \in \widehat{\Omega} , \end{cases}$$

and show that $u \equiv 0$ in the class of solutions growing (a priori) exponentially at infinity:

$$|u| , \left|\frac{\partial u}{\partial r}\right| = O\left(e^{\lambda r}\right) \qquad r = \left(\sum_{i=1}^{n} x_i^2\right)^{1/2}$$

for any preassigned, constant $\lambda > 0$.

Hint - Let $S(O, R)$ be an open ball, centered on a point $O \in \Omega$, with radius R such that $\overline{\Omega} \subset S(O, R)$. Introduce the weighted energy on $\Omega_R = S - \overline{\Omega}$:

$$\frac{1}{2}\int_{\Omega_R} g u^2 \, dx ,$$

with the weight $g = e^{-\alpha r}$ $(\alpha > 2\,\lambda)$, and follow the procedure of Sect.6.4. (All the details can be found in [3], pp. 8-10.)

Exercise 6.5 - In the context of the I.B.V.P. (6.25), suppose that

$$f \equiv 0 \quad , \quad \theta_1 \equiv 0 ,$$
$$\theta_0(x) = \sinh \mu x ,$$

where μ is a constant. Determine a solution of the problem using separable variable methods (i.e. suppose $\theta = \chi(x)T(t)$). Use Theorem 6.3 to prove that the solution so found is unique (under certain assumptions).

Answer -

$$\theta = e^{\mu^2 t} \sinh \mu x .$$

Exercise 6.6 - Let the assumptions of Theorem 6.5 hold. Show that, in the class I of perturbations u where $|u_x|$ is bounded on $\mathbb{R}\times[0, T]$, one has

$$||u_0||^2 + \int_0^t ||f||^2 \, dt \le \delta \Rightarrow \sup_{\mathbb{R}\times[0,T]} |u| < \lambda \delta^{1/2}$$

where $\lambda (> 0)$ is a constant independent of δ {*pointwise continuous stability or Hölder stability*}.

Hint - Take into account (6.52) and that in the aforesaid class of perturbations u, the following pointwise estimate holds (Cfr. Appendix)

$$|u(x)| \leq k \left[\int_x^{x+1} u^2\, dx + \left(\int_x^{x+1} u^2\, dx \right)^{1/2} \right]^{1/2} \qquad x \in \mathbb{R},$$

where $k^2 = \max(1, 2c)$ and $c =$ const. such that $|u_x| \leq c$ on $\mathbb{R} \times [0, T]$, $\forall\, u \in \mathcal{I}$.

Exercise 6.7 - Let the assumptions of Theorem 6.7 hold. Show that, in the class of perturbations u where $|u_x|$ is bounded on $\mathbb{R} \times \mathbb{R}^+$

$$(\|u_0\|^2 < \delta\,,\ \nu^* < \nu) \Rightarrow \sup_{\mathbb{R} \times \mathbb{R}^+} |u| < \lambda \delta^{1/2}$$

where $\lambda (> 0)$ is a constant independent of δ (*pointwise stability*).

Hint - Follow the hint of exercise 6.6.

Exercise 6.8 - Let X be the set of functions $f \in C^1([0,1])$ such that $\{f(0) = 0,\ f \neq \emptyset\}$. Show that the functional

$$F : f \in X \Rightarrow F(f) = \frac{\displaystyle\int_0^1 f^2(x) dx}{\displaystyle\int_0^1 f'^2(x) dx},$$

is bounded.

Hint - See Remark 2.7.

Answer - $|f| = \left| \int_0^1 f'(s) ds \right| \leq \left[\int_0^1 (f')^2 ds \right]^{1/2} \Rightarrow f^2 \leq \int_0^1 (f')^2 ds.$

Hence, integrating on $[0, 1]$, it follows $F(f) \leq 1$.

Exercise 6.9 - Let Y be the set of functions $u \in C^1([0,1])$ such that $u(0) = a =$ const. $\neq 0$. Show that the functional $G : u \in \{Y - \{a\}\} \Rightarrow G(u) = \dfrac{\int_0^1 u^2 dx}{\int_0^1 u'^2 dx}$, is unbounded.

Hint - $u_\lambda = a \cos \sqrt{\lambda} x \in Y$.

Answer - $G(u_\lambda) = \dfrac{\displaystyle\int_0^1 \cos^2 \sqrt{\lambda} x dx}{\displaystyle\int_0^1 \sin^2 \sqrt{\lambda} x dx} = \dfrac{1 + \dfrac{\sin 2\mu}{2\mu}}{1 - \dfrac{\sin 2\mu}{2\mu}}$, with $\mu = \sqrt{\lambda}$. Hence $\lim_{\lambda \to 0^+} G(u_\lambda) = +\infty$.

Exercise 6.10 - Let Y be the set of functions $f : \mathbb{R}_+ \to \mathbb{R}$ such that

$$\begin{cases} \int_{\mathbb{R}_+} (f^2 + f'^2) dx < \infty \\ f(0) = 1 . \end{cases}$$

Show that the functional $F(f) = \int_{\mathbb{R}_+} f^2 \, dx / \int_{\mathbb{R}_+} f'^2 \, dx$, with $f \in \{Y - \{1\}\}$, is unbounded.

Hint - Consider that $\{u_k(x) = \dfrac{1}{1 + kx}, \, x \in \mathbb{R}_+, \, k = \text{const.}(> 0)\}$ belong to Y.

Answer - $\int_{\mathbb{R}_+} u_k^2 dx = \dfrac{1}{k}$, $\int_{\mathbb{R}_+} {u'}_k^2 dx = \dfrac{k}{3}$, $F(u_k) = \dfrac{3}{k^2}$, $\lim_{k \to 0^+} F(u_k) = +\infty$.

Exercise 6.11 - Consider the B.V.P.

$$\begin{cases} g'' + \lambda g = 0 & \qquad x \in [0,1] \\ g(0) = 0 . \end{cases}$$

Verify that, for $\lambda \in \mathbb{R}$, the problem is ill posed.

Answer - For $\lambda \in \mathbb{R}$ exists the zero solution and the solution $g_1 = e^{\mu x} - e^{-\mu x}$, $g_2 = x$, $g_3 = \sin \sqrt{\lambda} x$ respectively for $\lambda = -\mu^2$, $\lambda = 0$ and $\lambda > 0$.

Exercise 6.12 - Consider the operator $L = -\dfrac{d^2}{dx^2}$ and let l be a positive number. Verify that:

i) in the set of functions $C_0^2([0,l])$ the eigenvalues are positive and constitute an increasing sequence $\{\lambda_n\}$ with $\lim_{n \to \infty} \lambda_n = \infty$,

ii) for $l \to \infty$, the eigenvalues move to the left towards zero while the L^2-norm of the eigenfunctions tends to ∞,

iii) in the set I of functions $u = \{u \in C^2(\mathbb{R}_+), \, u(0) = 0\}$ the spectrum of L contains all the real numbers but the eigenfunctions don't belong to $L^2(\mathbb{R}_+)$.

Hint - The eigenvalues are the complex numbers λ for which the equation

$$-\frac{d^2 u}{dx^2} = \lambda u$$

has nontrivial solutions (called eigensolutions associated with λ). Take into account exercises 2.3, 6.11.

Answer - i) $\lambda = \dfrac{\int_0^l u'^2 dx}{\int_0^l u^2 dx} > 0$, λ_n = eigenvalues $= \dfrac{n^2 \pi^2}{l^2}$, u_n = eigenfunctions $= \sin \dfrac{n\pi}{l} x$; ii) $\|u_n\|^2 = \dfrac{l}{2}$; iii) $\forall \lambda > 0, u = \sin \sqrt{\lambda} x \in I$ is a nontrivial solution but does not belong to $L^2(\mathbb{R}_+)$.

Exercise 6.13 - Consider the operator $-(1+z)^2\Delta$ on the set I of vector valued functions $\{\mathbf{u} : \Omega \to \mathbb{R}^3, \mathbf{u}|_{z=0} = 0\}$, where Ω is the half space $z \geq 0$. Verify that on the subset I on which inequality (6.68) holds, the (possible) eigenvalues are positive and not less than $\frac{1}{4}$.

Hint - Consider the B.V.P.

$$\begin{cases} \Delta \mathbf{u} + \lambda \dfrac{\mathbf{u}}{(1+z)^2} = 0 \\ \mathbf{u}|_{z=0} = 0 \,. \end{cases}$$

Multiply by $\varphi \mathbf{u}$, where φ is the cut-off function (6.59), take into account the procedure of Theorem 6.11 and the inequality (6.68).

Answer - $\lambda \int_\Omega \varphi \frac{u^2}{(1+z)^2} d\mathbf{x} = \int_\Omega [\varphi (\nabla \mathbf{u})^2 + \frac{1}{2} \nabla \varphi \cdot \nabla u^2] d\mathbf{x}$. On letting $a \to \infty$ in the cut-off function (6.59), it turns out that

$$\lambda \int_\Omega \frac{u^2}{(1+z)^2} d\mathbf{x} = \int_\Omega (\nabla \mathbf{u})^2 d\mathbf{x} \,,$$

and hence (6.68) implies $\lambda > 1/4$.

Exercises 6.14 - Let Ω be the unbounded region $\Omega = \{(x_1, x_2, x_3) \in \mathbb{R}^3, \, x_1 \geq 0, x_2 \geq 0\}$ and let S be the set of vector valued functions $\mathbf{u} : \Omega \to \mathbb{R}^3$ such that

$$\int_\Omega \left[\sum_{i=1}^{2} \frac{u^2}{(1+x_i)^2} + (\nabla \mathbf{u})^2 \right] d\Omega < \infty \qquad \mathbf{u}|_{\partial\Omega} = 0 \,.$$

Show that the following inequality holds

$$\int_\Omega \frac{u^2}{1+\rho^2} d\Omega \leq 3 \int_\Omega (\nabla \mathbf{u})^2 d\Omega \,, \qquad (\forall \mathbf{u} \in S, \, \rho = (x_1^2 + x_2^2)^{1/2}) \,.$$

Hint - Take into account inequality (6.68) and

$$\frac{1}{(1+x_1)^2} + \frac{1}{(1+x_2)^2} \geq \frac{4}{3} (1 + x_1^2 + x_2^2)^{-1} \,, \qquad \forall \, x_1, x_2 \geq 0 \,.$$

Answer - See [157], Theor. 1.

Exercise 6.15 - Let Ω be the unbounded region $\Omega = \{x_1 \geq 0, x_2 \geq 0, \, x_3 \geq 0\}$ and let S be the set of vector valued functions $\mathbf{u} : \Omega \to \mathbb{R}^3$ such that

$$\int_\Omega \left\{ \sum_{i=1}^{3} \frac{u^2}{(1+x_i)^2} + (\nabla \mathbf{u})^2 \right\} d\Omega < \infty \,, \qquad \mathbf{u}|_{\partial\Omega} = 0 \,.$$

Show that the following inequality holds

$$\int_\Omega \frac{u^2}{1+r^2} d\Omega \leq \frac{16}{9} \int_\Omega (\nabla \mathbf{u})^2 d\Omega, \qquad (\forall \mathbf{u} \in S,\, r = \sqrt{x_1^2 + x_2^2 + x_3^2}).$$

Hint - Take into account inequality (6.68) and

$$\sum_{i=1}^{3} \frac{1}{(1+x_i)^2} \geq \frac{9}{4(1+r^2)}, \qquad \forall x_1, x_2, x_3 \geq 0.$$

Answer - See [157], Theor. 2.

Exercise 6.16 - Let Ω be the unbounded domain contained in the region $\Gamma = \{(x_1, x_2, x_3) \in I\!R^3, x_1 > 0, x_2 > 0\}$ and let S be the set of vector valued functions $\mathbf{u} : \Omega \to I\!R^3$ such that

$$\int_\Omega \left\{ \sum_{i=1}^{2} \frac{u^2}{x_i^2} + (\nabla \mathbf{u})^2 \right\} d\Omega < \infty, \qquad \mathbf{u}|_{\partial\Omega} = 0.$$

Show that the following inequalities hold:

i) $\displaystyle\int_\Omega \frac{u^2}{x_1^2} d\Omega \leq 4 \int_\Omega \left(\frac{\partial u}{\partial x_i} \right)^2 d\Omega;$

ii) $\displaystyle\int_\Omega \frac{u^2}{\rho^2} d\Omega \leq \int_\Omega (\nabla \mathbf{u})^2 d\Omega, \quad (\forall \mathbf{u} \in S,\, \rho = \sqrt{x_1^2 + x_2^2}).$

Hint - In obtaining the inequality i), follow the proof of Theorem 6.11. In obtaining the inequality ii) use the inequality i) and take into account that

$$\sum_{i=1}^{2} \frac{1}{x_i^2} \geq \frac{4}{\rho^2}, \qquad \forall x_1, x_2 > 0.$$

Exercise 6.17 - Let Ω be the unbounded domain contained in the region $\Gamma_1 = \{(x_1, x_2, x_3) \in I\!R^3, x_1 > 0, x_2 > 0, x_3 > 0\}$ and let S be the set of vector valued functions $\mathbf{u} : \Omega \to I\!R^3$ such that

$$\int_\Omega \left\{ \sum_{i=1}^{3} \frac{u^2}{x_i^2} + (\nabla \mathbf{u})^2 \right\} d\Omega < \infty \qquad \mathbf{u}|_{\partial\Omega} = 0.$$

Show that the following inequality holds

$$\int_\Omega \frac{u^2}{\rho^2} d\Omega \leq \frac{4}{9} \int_\Omega (\nabla \mathbf{u})^2 d\Omega, \qquad (\forall \mathbf{u} \in S,\, \rho = \sqrt{x_1^2 + x_2^2 + x_3^2}).$$

Hint - Take into account that inequality i) of exercise 6.16 holds $\forall i \in \{1, 2, 3\}$ and that

$$\sum_{i=1}^{2} \frac{1}{x_i^2} \geq \frac{9}{r^2}, \qquad \forall x_1, x_2, x_3 > 0.$$

Exercise 6.18 - Let Ω be the exterior of the cylinder $C = \{\rho^2 = x^2 + y^2 \leq l, l > 1\}$. Denoting by I the set of the vector valued functions $\{\mathbf{u} : \Omega \to I\!R^3, \mathbf{u}|_{\rho=l} = 0, u^2 \in L^1_{\mathrm{loc}}(\Omega), (\nabla\mathbf{u})^2 \in L^1(\Omega)\}$. Show that $\dfrac{u^2}{\rho^2 \log^2 \rho} \in L^1(\Omega)$ and that the following inequality holds

$$\int_\Omega \frac{u^2}{\rho^2 \log^2 \rho} d\Omega \leq 4 \int_\Omega \|\nabla\mathbf{u}\|^2 d\Omega, \qquad \forall \mathbf{u} \in I.$$

Hint - Consider the $\nabla \cdot \left(\dfrac{\varphi u^2 \mathbf{e}_\rho}{\rho \log \rho}\right)$ where φ is given by (6.59) and take into account the procedure of Theorem 6.10.

Answer - See [160] page 17.

Exercise 6.19 - Let Ω_i $(i = 1,...,n)$ denote n compact domains, Ω a domain exterior to them and S the class of vector valued functions $\{\mathbf{u} : \Omega \to I\!R^3, \mathbf{u}|_{\partial\Omega} = 0\}$ such that

$$\int_\Omega [\frac{u^2}{r^2} + (\nabla\mathbf{u})^2] d\Omega < \infty \quad \text{or} \quad \int_\Omega \log(1+r)\left[\frac{u^2}{r^2} + (\nabla\mathbf{u})^2\right] d\Omega < \infty.$$

Show that, respectively, the following inequalities hold

$$\int_\Omega \frac{u^2}{r^2} d\Omega \leq 4 \int_\Omega (\nabla\mathbf{u})^2 d\Omega,$$

$$\int_\Omega \log(1+r)\frac{u^2}{r^2} d\Omega \leq 4 \int_\Omega \log(1+r)(\nabla\mathbf{u})^2 d\Omega,$$

where $r = |OP|$ with O inner to one of the domains Ω_i.

Hint - See [72] page 36.

Exercise 6.20 - Consider an I.B.V.P. for the simple wave equation, analogous to (6.25): $\theta(x,t)$ is a sufficiently smooth solution of

$$\begin{aligned}
&\theta_{tt} = \theta_{xx} + f(x,t) && \text{in } (0,\infty) \times (0,T) \\
&\theta(x,0) = \theta_0(x),\ \theta_t(x,0) = \psi_0(x) && \text{in } [0,\infty) \\
&\theta(0,t) = \theta_1(t) && \text{in } [0,T].
\end{aligned}$$

Establish that the solution is unique under conditions of exponential growth analogous to (6.29).

Hint - Consider the "difference problem" analogous to (6.26) and prove that the weighted energy therefor

$$\mathcal{E}(t) = \int_0^\infty g(\frac{1}{2}u_t^2 + \frac{1}{2}u_x^2) dx$$

where $g(x) = e^{-\alpha x}$ (for suitable, constant α), satisfies

$$\frac{d\mathcal{E}}{dt} \leq \alpha\mathcal{E}.$$

Exercise 6.21 - Suppose that

$$f \equiv \psi_0 \equiv \theta_1 \equiv 0$$

and that

$$\theta_0(x) = \sinh \nu x, \qquad \psi_0(x) = 0,$$

ν being a positive constant, in the I.B.V.P. of the previous question. Determine the solution and prove that it is unique (under certain assumptions).

Exercise 6.22 - (i) Let $\Phi(y) \in C^1(0 \leq y \leq \infty)$ be such that

$$\int_0^\infty e^{-\lambda y}(\Phi^2 + \dot{\Phi}^2)dy < \infty$$

(where λ is a positive constant) but is an otherwise arbitrary function; a superposed dot denotes differentiation with respect to y throughout. Prove that

$$\int_0^\infty e^{-\lambda y}\Phi^2 dy \leq \frac{1}{\lambda - \varepsilon}\Phi^2(0) + \frac{1}{\varepsilon(\lambda - \varepsilon)}\int_0^\infty e^{-\lambda y}\dot{\Phi}^2 dy \tag{P}$$

where ε is any constant such that $0 < \varepsilon < \lambda$.

Hint - Establish that

$$\begin{aligned} -\Phi^2(0) &= \int_0^\infty (e^{-\lambda y}\Phi^2)^{\bullet} dy \\ &= -\lambda\int_0^\infty e^{-\lambda y}\Phi^2 dy + 2\int_0^\infty e^{-\lambda y}\Phi\dot{\Phi}dy \end{aligned}$$

and apply the Cauchy inequality to the integrand in the last term. [Note that the assumed regularity of Φ here obviates the need for a "cut off" function.]

(ii) Prove that Theorem 6.12 continues to hold if $F(x)$ is redefined thus:

$$F(x) = \int_0^\infty g(y)\phi_y^2 dy,$$

$g(y)$ being unchanged, provided that the smoothness assumptions and the asymptotic conditions are modified as necessary.

Hint - Use the inequality (P) of (i) above, with $\varepsilon = \lambda/2$, suitably identifying Φ.

Show how a pointwise bound can be deduced for ϕ (using Schwarz's inequality) by establishing that

$$\phi^2(x, y) \leq \lambda^{-1}(e^{\lambda y} - 1)F(x).$$

Chapter seven

ELLIPTIC EQUATIONS

7.1 INTRODUCTION

This chapter deals with elliptic equations; exceptionally, reference is made in Section 7.5 to an equation which is not necessarily elliptic. The first part of the chapter deals with linear equations while the remainder is devoted to nonlinear equations. Estimates for integral measures/norms are derived which are of the general type discussed in Part 1.

The part dealing with linear equations commences with some simple, novel estimates for the harmonic equation and then discusses some estimates for the equations of linear elasticity. The elasticity estimates are broadly of two types: the first class deals with cylinders/strips where – usually but not exclusively – zero displacement is prescribed (Dirichlet conditions) on the lateral boundary/ies, while the second deals with such regions where zero traction (Neumann) conditions are prescribed on the lateral boundary. Estimates of the type considered had their origin in the need to justify the celebrated Saint-Venant principle in elasticity (Knowles [99], Toupin [100]). Comprehensive treatments of these latter matters – together with extensive list of references – are available in the review article of Horgan and Knowles [101] and in an update by Horgan [102]. The estimates relevant to linear elasticity which are proved (as distinct from those summarized) in this chapter do not – with one exception – occur in [101-102].

The second part (7.5 and 7.6) dealing with nonlinear equations deals with issues somewhat similar to those dealt with for linear equations.

7.2 LAPLACE'S EQUATION

We commence with some estimates for the harmonic equation in the plane. Whereas it is recognized that this equation has been exhaustively studied for a long time, the justification for including this material is that it is, we believe, appearing for the first time.

We shall, in the first instance, be concerned with a Dirichlet problem in the plane. We consider a simply connected open region R in the x, y (rectangular cartesian) plane which is (cf. 4.4.3)
(i) bounded by straight line segments Γ_0, Γ_L with abscissae $x = 0, x = L$ respectively, but, in limiting cases, Γ_L may degenerate to a point/vanish,
(ii) bounded by two curves C_+ (upper) and C_- (lower); these curves $y = y_+(x)$, $y = y_-(x)$ respectively (say) have the properties:

(α) they are single-valued functions defined in $0 \leq x \leq L$,

(β) they are twice continuously differentiable in $0 < x < L$,

(γ) $y_+(x) \geq y_-(x)$, and the equality sign can only occur therein when Γ_L degenerates/vanishes.

Let us denote by Γ_x the straight line segment contained in R whose points have the abscissa x.

Consider the Dirichlet problem: ψ is a function

(a) three time continuously differentiable in $\overline{R}/(\Gamma_0 \cup \Gamma_L)$,

(b) such that ψ_y is continuously in $\overline{R}$, (or such that $F(x)$ defined below is continuous in $0 \leq x \leq L$) which satisfies

$$\nabla_1^2 \psi = 0 \qquad \text{in } R \tag{7.1}$$

subject to

$$\psi = 0 \qquad \text{on } C_+, C_- \tag{7.2}$$

and

$$\psi = \text{assigned on } \Gamma_0 \text{ and on } \Gamma_L \qquad (\textit{if non-degenerate}). \tag{7.3}$$

Defining the cross-sectional measure of the solution by

$$F(x) = \int_{\Gamma_x} \psi_y^2 dy \tag{7.4}$$

but, as all integrals occurring in this analysis are over Γ_x (or particular cases thereof) there is no need to include the subscript in these integrals henceforward. We proceed to derive a differential inequality for $F(x)$ whence an upper bound thereof may be obtained in terms of data. Use of (4.104)–(4.105) yield

$$F'(x) = 2 \int \psi_y \psi_{xy} dy + \psi_y^2 y'] \, , \tag{7.5}$$

$$F''(x) = 2\int(\psi_{xx}^2 + \psi_{yy}^2)dy - \psi_y^2 y''] \tag{7.6}$$

where the primes and the notation $--]$ are as explained previously (e.g. Lemma 4.3, Chapter four). One may combine (7.5) and (7.6) in a number of different ways to furnish a differential inequality for F; broadly speaking, two approaches are possible which we now describe.

First approach. Applying the arithmetic-geometric inequality to (7.5) we obtain

$$2\int \psi_{xy}^2 dy \geq -2(l\,\alpha)^{-1}F' - 2(l\,\alpha)^{-2}F + 2(l\,\alpha)^{-1}\psi_y^2 y'] \tag{7.7}$$

where $l(x)$ denotes the width of the domain at abscissa x, and α is any positive number. Combining (7.6) and (7.7) gives

$$F'' + 2(l\,\alpha)^{-1}F' \geq 2\int(\psi_{yy}^2 - l^{-2}\alpha^{-2}\psi_y^2)dy + \psi_y^2(-y'' + 2l^{-1}\alpha^{-1}y')]\,. \tag{7.8}$$

It is natural to choose α in such a way that the right hand side of (7.8) is non-negative. Whereas this can be realized in a number of ways the simplest arises through use of the inequality (A2(ii))

$$\int \psi_{yy}^2 dy \geq \pi^2 l^{-2}\int \psi_y^2 dy\,. \tag{7.9}$$

In view of this follows from (7.8), on choosing $\alpha = \pi^{-1}$, that

$$F'' + 2\pi l^{-1}(x)F' \geq 0 \tag{7.10}$$

provided that the lateral boundaries conform to

$$-y_+'' + 2\pi l^{-1}y_+' \geq 0 \quad , \quad y_-'' - 2\pi l^{-1}y_-' \geq 0\,. \tag{7.11}$$

In view of Theorem 4.4

$$F(x) \leq G(x) \tag{7.12}$$

where $G(x)$ satisfies

$$G'' + 2\pi l^{-1}G' = 0$$

subject to

$$G(0) = F(0)\,, \qquad G(L) = F(L)\,,$$

both of which are calculable from data. Integration of the differential equation in (7.12) yields

Theorem 7.1 - *In the context of the Dirichlet problem defined by* (7.1)–(7.3), *inter alia, the cross-sectional measure of solution* (7.4) *satisfies*

$$F(x) \leq F(L) + \{F(0) - F(L)\} \int_x^L \exp\{-\int_0^x 2\pi l^{-1}(x')dx'\}dx \cdot$$
$$\cdot [\int_0^L \exp\{-\int_0^x 2\pi l^{-1}(x')dx'\}dx]^{-1} \qquad (7.13)$$

provided that the lateral boundaries C_+, C_- *satisfy the conditions* (7.11).

Let us note:

(i) As usual pointwise bounds for $|\psi|$, reflecting position are obtainable from (7.13).

(ii) The equality sign occurs in (7.13) in the following circumstances: the region R is a semi-infinite rectangular strip $0 < x < 1, 0 < y < \infty$, the boundary condition on Γ_0 is $\psi = K \sin \pi y$ where K is a constant, while the remaining boundary condition is such that $\psi \to 0$ as $x \to \infty$.

Second approach. We now discuss another way of combining (7.4)–(7.6) to furnish another differential inequality for $F(x)$ whence upper bounds therefor may be deduced. Suppose the region R is contracting in the x direction i.e.

$$y'_+ < 0\,, \qquad y'_- > 0\,. \qquad (7.14)$$

In these circumstances, the following inequality holds (reminiscent of conservation of "energy"):

$$\int \psi_x^2 dy \geq \int \psi_y^2 dy\,; \qquad (7.15)$$

see (4.54)–(4.55).

In order to derive an additional, necessary inequality, we suppose that the lateral boundaries of the region R are also strictly convex

$$y''_+ < 0\,, \qquad y''_- > 0\,. \qquad (7.16)$$

Let $\lambda_1(x)$ be the lowest eigenvalue of

$$\frac{d^2\phi}{dy^2} + \lambda\phi = 0 \qquad (7.17)$$

subject to

$$\frac{d\phi}{dy} - \left(\frac{y''_+}{2y'^2_+}\right)\phi = 0 \quad \text{on } y = l(x)\,, \qquad \frac{d\phi}{dy} - \left(\frac{y''_-}{2y'^2_-}\right)\phi = 0 \quad \text{on } y = 0\,.$$

In view of the fact that

$$\psi_x + \psi_y y' = 0$$

on the lateral boundaries, it follows from inequality A3 that

$$2\int \psi_{xy}^2 dy - \psi_y^2 y''] \geq 2\lambda_1(x)\int \psi_x^2 dy\,. \tag{7.18}$$

In view of (7.4), (7.6), (7.9), (7.15), (7.18), we obtain the differential inequality

$$F'' - 2\{\pi^2 l^{-2}(x) + \lambda_1(x)\}F \geq 0\,. \tag{7.19}$$

In view of this and Theorem 4.4 we have

Theorem 7.2 - *In the context of the Dirichlet problem defined by* (7.1)–(7.3), *inter alia, the cross-sectional measure of solution* (7.4) *satisfies*

$$F(x) \leq G(x) \tag{7.20$_1$}$$

where $G(x)$ *satisfies*

$$G'' - 2\{\pi^2 l^{-2}(x) + \lambda_1(x)\}G \geq 0 \tag{7.20$_2$}$$

subject to

$$G(0) = F(0)\,, \qquad G(L) = F(L)\,, \tag{7.20$_3$}$$

where $\lambda_1(x)$ *is the lowest eigenvalue of* (7.17), *provided that the lateral boundaries of the region satisfy* (7.14) *and* (7.16).

Once again, the equality sign holds in the foregoing in the circumstances described in (ii) following (7.13).

We continue to consider the region R as defined previously (second paragraph of this section), and a harmonic function therein as previously specified except that on the lateral boundaries

$$\psi = \mathcal{F}_+(x) \quad \text{on } C_+\,, \qquad \psi = \mathcal{F}_-(x) \quad \text{on } C_-\,; \tag{7.21}$$

moreover, (7.3) may be ignored. We now consider a cross-sectional measure (cf. Theorem 3.8) generalizing (7.4)

$$F(x) = \int \phi(\psi_y)dy\,, \tag{7.22}$$

where $\phi(u)$ is a convex function of its argument satisfying $\phi(0) = 0$ and we derive some of its properties. In view of the fact – used in sections 3.2.2-3.2.3 – that

$$u\dot{\phi}(u) - \phi(u) \geq 0 \tag{7.23}$$

for such a function, (a dot denoting differentiation with respect to the relevant argument), the following is obtained from Lemma 4.3:

Theorem 7.3 - *In the context of harmonic functions ψ defined on the region R (as previously specified) which satisfy* (7.21) *on the lateral boundaries C_+, C_-, the cross-sectional measure $F(x)$, defined by* (7.22)*, is a convex function of x provided that*

(i) (a) $\phi(u)$ *is a convex function of u such that* $\phi(0) = 0$,

(b) $\dot{\phi}(u) \geq 0$,

(ii) *the lateral boundaries are convex*

$$y''_+(x) \leq 0, \qquad y''_-(x) \geq 0,$$

(iii) *the boundary conditions thereon are such that*

$$\mathcal{F}''_+(x) \geq 0 \qquad \mathcal{F}''_-(x) \leq 0.$$

If, however,

$$\mathcal{F}''_+(x) = \mathcal{F}''_-(x) = 0$$

$F(x)$ continues to be convex provided that (i) (a) *(only) and* (ii) *hold.*

Particular choices of the function $\phi(u)$ lead to interesting conclusions (cf. 3.2.2):

$$\begin{aligned} &(a) \quad \phi(u) = u^{2p}, \\ &(b) \quad \phi(u) = ({}_+u)^{2p}, \end{aligned} \tag{7.24}$$

where

$${}_+u = \begin{cases} u & (u \geq 0), \\ 0 & (u < 0), \end{cases}$$

where p is any positive integer (> 1). We now examine, in succession, conclusions (Theorems 7.3a, 7.3b) which can be drawn from Theorem 7.3 in the case of the above choices of ϕ, and we note interesting corollaries in each case.

Theorem 7.3a - *In the context of a harmonic function ψ defined on the region R (as previously specified) which satisfy* (7.21) *on the lateral boundaries C_+, C_-, where*

$$y''_+(x) \leq 0, \qquad y''_-(x) \geq 0, \tag{7.25$_1$}$$

and

$$\mathcal{F}''_+(x) = \mathcal{F}''_-(x) = 0, \tag{7.25$_2$}$$

the cross-sectional measure

$$F(x) = \int \psi_y^{2p} dy$$

is convex ($F''(x) \geq 0$) *in* $0 < x < L$.

We note the following obvious corollary:

Corollary 7.3a - *Assuming that* ψ_y *is continuous in* $0 \leq x \leq L$ *in addition to assumptions of Theorem 7.3a, one has*

$$\int_{\Gamma_x} \psi_y^{2p} dy \leq \max_{x=0;\, x=L} \int_{\Gamma_x} \psi_y^{2p} dy\,, \tag{7.26$_1$}$$

$$\max_y |\psi_y(x,y)| \leq \max_{x=0;\, x=L} \{\max_y |\psi_y(x,y)|\}\,. \tag{7.26$_2$}$$

Note that (7.26_2) follows from (7.26_1) on noting the well known proposition (cf. 3.3.2)

$$\lim_{p\to\infty} \int_0^l \phi^{2p}(y)dy = \max_y |\phi(y)|$$

for continuous functions defined in $0 \leq y \leq l$.

We obtain similar results using the second choice of ϕ specified in (7.24).

Theorem 7.3b - *In the context of the harmonic function* ψ *defined on the region* R *(as previously specified) which satisfy* (7.21) *on the lateral boundaries* C_+, C_-, *where*

$$y_+''(x) \leq 0\,, \qquad y_-''(x) \geq 0\,, \tag{7.27$_1$}$$

and

$$\mathcal{F}_+''(x) \geq 0\,, \qquad \mathcal{F}_-''(x) \leq 0 \tag{7.27$_2$}$$

the cross-sectional measure

$$F(x) = \int ({}_+\psi_y)^{2p} dy$$

is a convex function in $0 < x < L$.

We note the following obvious corollary which follows in a similar way:

Corollary 7.3b - *Assuming that ψ_y is continuous in $0 \le x \le L$ in addition to the assumptions for Theorem 7.3b, one has*

$$\int_{\Gamma_x} ({}_+\psi_y)^{2p} dy \le \max_{x=0;\, x=L} \int_{\Gamma_x} ({}_+\psi_y)^{2p} dy \tag{7.28$_1$}$$

$$\max_y {}_+\psi_y(x,y) \le \max_{x=0;\, x=L} \{\max_y {}_+\psi_y(x,y)\}\,. \tag{7.28$_2$}$$

Let us make a number of remarks in relation to (7.26_2) and (7.28_2):
(i) They may be viewed as modifications of the weak maximum principle for the contexts in question.
(ii) The assumptions of convex lateral boundaries $(7.25_1), (7.27_1)$ cannot be entirely dispensed with since – as discussed in Part 1 – one may make ψ_y as large as one pleases by suitably indenting the lateral boundary.
(iii) Whereas results such as $(7.26_2), (7.28_2)$ can be derived via maximum principles (Hopf principle) our general objective is to explore properties of P.D.E.s by reference to the variation of integral measures.
This concludes our discussion of harmonic functions.

7.3 DISPLACEMENT-TYPE PROBLEMS IN LINEAR ELASTICITY

7.3.1 Introduction

This section obtains qualitative estimates for cylindrical and other regions whose lateral boundaries are subject to displacement boundary conditions. Firstly, an elastic right cylinder is considered whose lateral boundary is clamped (i.e. subjected to zero displacement) and which is deformed by actions applied to its (plane) ends. Estimates are obtained for the deformation using two methods:
(i) the volume integral method, including Phragmèn-Lindelöf estimates, of the type discussed in Chapter three,
(ii) the cross-sectional method – of the type discussed in Chapter four.
Of course, the aforementioned analysis of the elastic cylinder by means of the cross-sectional method carries over easily to the analogous (two-dimensional) problem for a rectangular elastic strip. The following related matter is considered: we show how a ***weighted cross-sectional method*** (cf. Chapter six) can be used to analyze a semi-infinite elastic strip, on the long (semi-infinite) sides of which (non-null) data are specified, and where the elastic field is allowed to grow rapidly at infinity.

Finally, a two-dimensional elastic region is considered with strictly convex lateral boundary which is subject to *non-zero* displacement conditions. This is analyzed by means of the cross-sectional method.

In summary, ***four analyses are given*** of related elastic problems – broadly of the displacement type – ***which exemplify the full range of relevant techniques introduced*** in Part 1.

We commence by specifying the context for (i), (ii) above. We consider a right cylinder B with plane ends and choose rectangular coordinates x_1, x_2, x_3 such that the x_3 axis is parallel to the generators and such that the origin lies in the plane end $x_3 = 0$. We suppose that the length of the cylinder is L and that $D(x_3) \in \mathbb{R}^2$ represents the cross-section at distance x_3 from the plane end $x_3 = 0$. We distinguish between finite and infinite values of L and thus separately discuss cylinders that are respectively of finite and semi-infinite length. The boundary of the cross-section is assumed sufficiently smooth to admit application of the divergence theorem in the plane of the cross-section.

The cylinder consists of homogeneous, isotropic, linear elastic material and its lateral boundary is displacement-free. We suppose that the rectangular cartesian components of displacement $u_i(x_1, x_2, x_3)$ are sufficiently smooth (classical) solutions of the well known Navier equations

$$u_{i,jj} + \alpha u_{j,ji} = 0 \qquad x_i \in B \tag{7.29$_1$}$$

subject to

$$u_i = 0\,, \quad x_3 \in \partial D(x_3)\,, \qquad x_3 \in (0, L)\,. \tag{7.29$_2$}$$

The usual indicial notation is used throughout: repeated indices imply summation, suffices following commas, differentiation with respect to the appropriate variable, Latin suffices take values 1, 2, 3 and Greek suffices 1, 2 (unless otherwise stated). In the above, α is a constant which, in terms of the Lamé constants, λ, μ, is given by

$$\alpha = (\lambda + \mu)/\mu\,;$$

it shall be supposed throughout that $\alpha > 0$ (although the stronger assumption $\alpha > 1/3$ is common). It is supposed that suitable conditions are specified on the plane ends of the cylinder – for example, that the displacement components are specified thereon: the precise form depends on the context.

7.3.2 Volume Integral Approach

We now show how the volume integral method may be used to furnish qualitative estimates for the elastic field in the cylinder; the analysis is largely based upon [103] which, however, treats a region of varying cross-section.

With a view to using this approach – and in a way that includes finite and semi-infinite cylinders – we introduce the function

$$H(x_3) = -\int_{D(x_3)} (u_i u_{i,3} + \alpha u_{j,j} u_3)\, dA\,. \tag{7.30}$$

This implies

$$-H(x_3+h) = -H(x_3) + \int_{x_3}^{x_3+h}\int_{D(\eta)} (u_{i,j}u_{i,j} + \alpha u_{i,i}u_{j,j})\,dA\,d\eta \qquad (7.31)$$

for $0 \le h \le L - x_3$. The integral forming the second term on the right hand side of (7.31) is akin to the elastic energy contained between the cross-sections $x_3, x_3 + h$; it is the volume integral measure of the deformation between these cross-sections; the quantities $H(x_3), H(x_3 + h)$ are akin to the work done by the tractions acting on the cross-section. Note that when, in particular,

$$u_i(x_\alpha, L) = 0 \qquad (7.32)$$

holds on the plane end $x_3 = L, H(x_3) \ge 0$. In general, however, $H(x_3)$ is nonincreasing and its derivative is plainly given by

$$H'(x_3) = -\int_{D(x_3)} (u_{i,j}u_{i,j} + \alpha u_{i,i}u_{j,j})\,dA\,. \qquad (7.33)$$

We now wish to compute a bound of the form

$$|H(x_3)| \le -n\, H'(x_3) \qquad (7.34)$$

where n is a constant. Once (7.34) is established it leads immediately to the inequalities

$$H(x_3) \ge nH'(x_3) \qquad (7.35_1)$$

and

$$H(x_3) \le -n\, H'(x_3) \qquad (7.35_2)$$

which, on integration, yield the desired information, on the behaviour of the solution. The inequality (7.34) is derived, in the main, using Schwarz's inequality and the variational characterization of the lowest "fixed membrane" eigenvalue λ_1, for the cross-section (see A3(i)(a)): if λ_1 denotes the lowest eigenvalue of

$$\nabla_1^2 v + \lambda v = 0 \qquad \text{in } D$$

$$v = 0 \qquad \text{on } \partial D\,,$$

one has

$$\int_{D(x_3)} \phi_{,\beta}\phi_{,\beta}\,dA \ge \lambda_1 \int_{D(x_3)} \phi^2\,dA \qquad (7.36)$$

where ϕ is an arbitrary C^1 function vanishing on ∂D.

Specifically, Schwarz's inequality combined with (7.36) yields

$$|H(x_3)| = \Big(\int_{D(x_3)} u_i u_i dA \int_{D(x_3)} u_{i,3}u_{i,3}dA\Big)^{1/2}$$
$$+\alpha\Big(\int_{D(x_3)} u_i u_i dA \int_{D(x_3)} u_{i,i}u_{j,j}dA\Big)^{1/2} \le$$
$$\lambda_1^{-1/2}\Big(\int_{D(x_3)} u_{i,\beta}u_{i,\beta}dA\Big)^{1/2}.$$
$$\left\{\Big(\int_{D(x_3)} u_{i,3}u_{i,3}dA\Big)^{1/2} + \Big(\int_{D(x_3)} u_{i,i}u_{j,j}dA\Big)^{1/2}\right\} \le$$
$$\frac{1}{2}\lambda_1^{-1/2}[(C_1+C_2)\int_{D(x_3)} u_{i,\beta}u_{i,\beta}dA$$
$$+C_1^{-1}\int_{D(x_3)} u_{i,3}u_{i,3}dA + \alpha^2 C_2^{-1}\int_{D(x_3)} u_{i,i}u_{j,j}dA] \tag{7.37}$$

where in the last line we have used Cauchy's inequality with arbitrary positive constants C_1, C_2. On setting

$$C_2 = \alpha C_1 = \alpha(1+\alpha)^{-1/2},$$

we find that (7.37) becomes

$$|H(x_3)| \le -\frac{1}{2}\lambda_1^{-1/2}(1+\alpha)^{1/2}H'(x_3) \tag{7.38}$$

which is of the required form (7.34) with

$$n = \frac{1}{2}\{(1+\alpha)\lambda_1^{-1}\}^{1/2}. \tag{7.39}$$

We have already seen that under the conditions (7.32) we must have $H(x_3) \ge 0$, and in view of the non-positivity of $H'(x_3)$ (see (7.33)), it follows that inequality (7.35_2) – with n given by (7.39) – holds. On integration we obtain

Theorem 7.4 - *In the context of classical solutions of the displacement equations (7.29_1) in the right elastic cylinder subject to displacement-free conditions (7.29_2) on the lateral boundary, and (7.32) on the plane end $x_3 = L$, the volume integral measure of the elastic deformation*

$$E(x_3) = \int_{x_3}^{L}\int_{D(\eta)} (u_{i,j}u_{i,j} + \alpha u_{i,i}u_{j,j})dAd\eta \tag{7.40_1}$$

satisfies

$$E(x_3) \leq E(0)\exp\{-2(1+\alpha)^{-1/2}\lambda_1^{1/2}x_3\} \tag{7.40$_2$}$$

where λ_1 *is the lowest fixed membrane eigenvalue for the cross-section.*

In order to make this estimate fully explicit it is, of course, necessary to be able to bound $E(0)$ above in terms of data specified on $x_3 = 0$; the reader is referred to [103] for details.

We now proceed to derive a Phragmèn- Lindelöf version of the Theorem – of the type discussed in Chapter three – appropriate to a semi-infinite cylinder $(L \to \infty)$. In considering the semi-infinite cylinder, we do not impose, *a priori*, a requirement on the asymptotic behaviour of the solution as $x_3 \to \infty$. Suppose that $H(x_3)$ becomes negative at some point $x_3 = t$, $t \in [0, \infty)$, then it follows from (7.35$_1$) that $H(x_3)$ is a negative, decreasing function for $x_3 \geq t$; integrating (7.35$_1$) yields

$$-H(x_3) \geq -H(t)\exp[n^{-1}(x_3 - t)] \tag{7.41}$$

– where n is given by (7.39). On using (7.31), we obtain, for $x_3 > t$,

$$\int_t^{x_3}\int_{D(\eta)}(u_{i,j}u_{i,j} + \alpha u_{i,i}u_{j,j})dAd\eta \geq -H(t)\{\exp[n^{-1}(x_3 - t)] - 1\}. \tag{7.42}$$

By l'Hospital's theorem (concerning the ratio of two functions which tend to infinity as the independent variable tends to a limit), (7.42) yields

$$\lim_{x_3\to\infty} n\int_{D(x_3)}(u_{i,j}u_{i,j} + \alpha u_{i,i}u_{j,j})dA\cdot\{\exp[-n^{-1}(x_3 - t)]\} \geq -H(t). \tag{7.43}$$

If we now impose the asymptotic condition

$$\lim_{x_3\to\infty}\int_{D(x_3)}(u_{i,j}u_{i,j} + \alpha u_{i,i}u_{j,j})dA\cdot\exp(-n^{-1}x_3) = 0 \tag{7.44}$$

we see that it contradicts the earlier assumption that $H(t)$ is negative, i.e.

$$H(x_3) \geq 0, \qquad x_3 \in [0, \infty).$$

Hence on appealing to (7.35$_2$) and integrating we obtain exponential decay for $H(x_3)$ – as defined by (7.30); but as this decay implies that $\lim_{x_3\to\infty} H(x_3) = 0$, (7.31) enables us to express $H(x_3)$ otherwise.

Theorem 7.5 - *In the context of classical solutions of the displacement equations* (7.29$_1$) *subject to the displacement-free conditions on the lateral boundary*

(7.29$_2$), for a semi-infinite, elastic right cylinder ($0 < x_3 < \infty$), in the presence of the asymptotic condition

$$\lim_{x_3 \to \infty} \int_{D(x_3)} (u_{i,j}u_{i,j} + \alpha u_{i,i}u_{j,j})dA \cdot \exp\{-2(1+\alpha)^{-1/2}\lambda_1^{1/2}x_3\} = 0 \quad (7.45_1)$$

the volume integral measure of the elastic deformation

$$E(x_3) = \int_{x_3}^{\infty} \int_{D(\eta)} (u_{i,j}u_{i,j} + \alpha u_{i,i}u_{j,j})dAd\eta \quad (7.45_2)$$

satisfies

$$E(x_3) \leq E(0)\exp\{-2(1+\alpha)^{-1/2}\lambda_1^{1/2}x_3\} \quad (7.45_3)$$

where λ_1 is the lowest fixed membrane eigenvalue for the cross-section.

7.3.3 Cross-sectional Method

We now obtain estimates, by the *cross-sectional method*, of solutions of the elastic displacement equations (7.29$_1$) in the right cylinder subject to null displacement on the lateral boundary (7.29$_2$). This analysis is based upon [136]. For an analogous analysis of a different cross-sectional measure, see [161]; this latter analysis, however, requires that the magnitude of α be restricted.

We define the cross-sectional measure of deformation

$$F(x_3) = \int_{D(x_3)} (u_{\beta,3}u_{\beta,3} + u_{3,\beta}u_{3,\beta})dA. \quad (7.46)$$

Denoting derivatives of $F(x_3)$ by primes, successive differentiations of (7.46) yield

$$F'(x_3) = 2\int_{D(x_3)} (u_{\beta,3}u_{\beta,33} + u_{3,\beta}u_{3,\beta 3})dA, \quad (7.47)$$

$$\begin{aligned} F''(x_3) = &2\int_{D(x_3)} (u_{\beta,33}u_{\beta,33} + u_{3,\beta 3}u_{3,\beta 3})dA \\ &+2\int_{D(x_3)} (u_{\beta,3}u_{\beta,333} + u_{3,\beta}u_{3,\beta 33})dA. \end{aligned} \quad (7.48)$$

We now seek to express the final two terms in (7.48) in an alternative form; the symbol $D(x_3)$ will be dropped in the ensuing arguments and subsequently. Differentiating (7.29$_1$) with respect to x_3 and putting $i = 1$, we obtain

$$\begin{aligned} \int u_{1,3}u_{1,333}dA &= -\int u_{1,3}\{(\alpha+1)u_{1,113} + \alpha(u_{2,213} + u_{3,313}) + u_{1,223}\}dA \\ &= \int \{(\alpha+1)u_{1,13}^2 + u_{1,23}^2 + \alpha(u_{1,32}u_{2,13} + u_{1,13}u_{3,33})\}dA \end{aligned} \quad (7.49)$$

where the divergence theorem together with the boundary conditions (7.29_2) on the lateral surface are used in the last step. Similarly

$$\int u_{\beta,3}u_{\beta,333}\,dA =$$

$$\int\{(\alpha+1)(u_{1,13}^2+u_{2,23}^2)+u_{1,32}^2+u_{2,31}^2+\alpha u_{\beta,\beta 3}u_{3,33}+2\alpha u_{1,32}u_{2,31})\}dA$$

$$= \int(\alpha u_{\beta,\beta 3}u_{\gamma,\gamma 3}+u_{\beta,\gamma 3}u_{\beta,\gamma 3}+\alpha u_{\beta,\beta 3}u_{3,33})dA; \tag{7.50}$$

in the last step we have used the identity

$$\int u_{1,23}u_{2,13}\,dA = \int u_{1,13}u_{2,23}\,dA$$

– readily verifiable using the divergence theorem together with the boundary conditions on the lateral boundary (7.29_2). We now use (7.29_1) (with $i = 3$) to eliminate the term $u_{3,33}$ in (7.50) to obtain

$$\int u_{\beta,3}u_{\beta,333}\,dA = \int[\alpha(\alpha+1)^{-1}u_{\beta,\beta 3}u_{\gamma,\gamma 3}+u_{\beta,\gamma 3}u_{\beta,\gamma 3}]dA$$
$$- \int \alpha(\alpha+1)^{-1}u_{\beta,\beta 3}u_{3,\gamma\gamma}\,dA. \tag{7.51}$$

Now using the divergence theorem together with the boundary conditions on the lateral surface and using (7.29_1) (with $i = 3$), we obtain

$$\int u_{3,\beta}u_{3,\beta 33}\,dA = -\int u_{3,\beta\beta}u_{3,33}\,dA = \int(\alpha+1)^{-1}u_{3,\beta\beta}u_{3,\gamma\gamma}\,dA$$
$$+ \int \alpha(\alpha+1)^{-1}u_{\beta,\beta 3}u_{3,\gamma\gamma}\,dA. \tag{7.52}$$

On combining (7.48), (7.51)–(7.52), we obtain

$$F''(x_3) = \int\{u_{\beta,33}u_{\beta,33}+u_{3,\beta 3}u_{3,\beta 3}+u_{\beta,\gamma 3}u_{\beta,\gamma 3}+\alpha(\alpha+1)^{-1}u_{\beta,\beta 3}u_{\gamma,\gamma 3}$$
$$+ (\alpha+1)^{-1}u_{3,\beta\beta}u_{3,\gamma\gamma}\}dA. \tag{7.53}$$

Noting that each of the terms in the integrand is non-negative we obtain the following

Theorem 7.6 - *The cross-sectional measure $F(x_3)$ (defined by (7.46)) of the deformation in the elastic cylinder, subjected to imposed zero displacement*

(7.29$_2$) *on its lateral surface, is a convex function of* x_3.

Remark 7.1 - *Let us note that Theorem 7.6 continues to hold if the boundary conditions on the lateral boundary (7.29$_2$) are generalized to*

$$u_\beta = f_\beta(x_1, x_2), \qquad u_3 = g(x_1, x_2) + x_3 h(x_1, x_2) \tag{7.54}$$

where f_β, g, h *are arbitrary smooth functions. The proof is left as an exercise.*

We now show how to generalize Theorem 7.6 replacing convexity by generalized convexity. In view of the inequalities A4(i)(a), A4(ii),

$$\int_D u_{\beta,3\gamma} u_{\beta,3\gamma} dA \geq \lambda_1 \int_D u_{\beta,3} u_{\beta,3} dA, \tag{7.55}$$

$$\int_D u_{3,\beta\beta} u_{3,\gamma\gamma} dA \geq \lambda_1 \int_D u_{3,\beta} u_{3,\beta} dA, \tag{7.56}$$

λ_1 being the "fixed membrane" eigenvalue for the cylinder cross-section. Using (7.46), (7.53), (7.55)–(7.56), we obtain the inequality

$$F''(x_3) \geq 2 \int (u_{\beta,33} u_{\beta,33} + u_{3,\beta 3} u_{3,\beta 3}) dA + 2\lambda_1 (\alpha + 1)^{-1} F. \tag{7.57}$$

Writing the positive square root of F as $F^{1/2}$, we obtain

$$\begin{aligned}
(F^{1/2})'' &= \frac{1}{2} F^{-3/2} (F F'' - \frac{1}{2} F'^2) \\
&\geq \lambda_1 (\alpha + 1)^{-1} F^{1/2} + F^{3/2} [\int (u_{\beta,3} u_{\beta,3} + u_{3,\beta} u_{3,\beta}) dA \cdot \\
&\int (u_{\beta,33} u_{\beta,33} + u_{3,\beta 3} u_{3,\beta 3}) dA \\
&- \{\int (u_{\beta,3} u_{\beta,33} + u_{3,\beta} u_{3,\beta 3}) dA\}^2]
\end{aligned} \tag{7.58}$$

where (7.46)–(7.47) and (7.57) have been used. We notice that the quantity within the square brackets in (7.58) is non-negative by virtue of Schwarz's inequality. We thus obtain the following

Theorem 7.7 - *The cross-sectional measure* $F(x_3)$ *(defined by (7.46)) of the deformation in the elastic cylinder, with null-displacement boundary conditions on its lateral surface, satisfies the (generalized convexity) inequality*

$$(F^{1/2})'' - \lambda_1 (\alpha + 1)^{-1} F^{1/2} \geq 0 \tag{7.59}$$

where λ_1 denotes the lowest eigenvalue of the fixed membrane coinciding with the cylinder cross-section.

We now discuss three further (related) propositions which are consequences of the foregoing. First, suppose that the ends of the cylinder are situated at $x_3 = 0, x_3 = L$. It follows from Theorem 4.4 that

$$F^{1/2} \leq G\,, \tag{7.60$_1$}$$

where G satisfies

$$G'' - \lambda_1(\alpha+1)^{-1}G = 0\,,$$
$$G(0) = F^{1/2}(0)\,, \qquad G(L) = F^{1/2}(L)\,. \tag{7.60$_2$}$$

One obtains from (7.60_1) and from the explicit solution of (7.60_2) the following

Theorem 7.8 - *The cross-sectional measure $F(x_3)$ of the deformation in the elastic cylinder $0 \leq x_3 \leq L$, with null-displacement boundary conditions on its lateral surface, satisfies*

$$F^{1/2}(x_3) \leq \{F^{1/2}(0)\sinh\{\lambda^{1/2}(1+\alpha)^{-1/2}(L-x_3)\}$$
$$+ F^{1/2}(L)\sinh\{\lambda^{1/2}(1+\alpha)^{-1/2}x_3\}\}/\sinh\{\lambda^{1/2}(1+\alpha)^{-1/2}L\}\,. \tag{7.61}$$

A weaker result, but one which more transparently exhibits decay of effects away from the ends, is easily deduced therefrom.

Theorem 7.8a - *The cross-sectional measure $F(x_3)$ of the deformation in the elastic cylinder $0 \leq x_3 \leq L$, with null-displacement boundary conditions on its lateral surface, satisfies*

$$F^{1/2}(x_3) \leq F^{1/2}(0)\exp\{-\lambda_1^{1/2}(1+\alpha)^{-1/2}x_3\}$$
$$+ F^{1/2}(L)\exp\{-\lambda_1^{1/2}(1+\alpha)^{-1/2}(L-x_3)\}\,. \tag{7.62}$$

Contemplating a semi-infinite cylinder $L \to \infty$ we have the following.

Theorem 7.8b - *The cross-sectional measure $F(x_3)$ of the deformation in elastic cylinder $0 \leq x_3 \leq \infty$, with null boundary conditions on its lateral surface satisfies*

$$F(x_3) \leq F(0)\exp\{-2\lambda_1^{1/2}(1+\alpha)^{-1/2}x_3\} \tag{7.63}$$

provided either

$$F(x_3) \to 0 \quad \text{as} \quad x_3 \to \infty \qquad \text{(case 1)},$$

or

$$\lim_{L\to\infty} \{m^{-1}(F^{1/2})' + F^{1/2}\}_L e^{-mL} \to 0 \qquad \text{(case 2)}$$

where

$$m = \lambda_1^{1/2}(1+\alpha)^{-1/2},$$

and where the bracketed quantity with the subscript L means that it is evaluated at $x_3 = L$.

That (7.63) holds in case 1 is a simple consequence of (7.62). That it holds in case 2 is left as an exercise (See Exercises). Theorem 7.8b, case 2, may be termed a Phragmèn-Lindelöff result.

Remark 7.2 - *The decay exponent occurring in* (7.63) *is the same as that occurring in* (7.45_3).

Let us note that $F(0)$, and $F(L)$ when appropriate, may be related to conventional data in the circumstances (a), (b) discussed hereunder.

(a) Suppose that the normal component of displacement components u_3 and the complementary shear stress components $\tau_{3\alpha} = \mu(u_{3,\alpha} + u_{\alpha,3})$ are specified on the plane ends $x_3 = 0, L$. Then $F(0), F(L)$ are calculable therefrom. A similar remark applies to a semi-infinite cylinder (dealt with in Theorem 7.8b) *mutatis mutandis.*

(b) Suppose that a semi-infinite cylinder $L \to \infty$ is envisaged; that $u_{i,j} \to 0$ as $x_3 \to \infty$; and that the displacement components u_i are specified on the end $x_3 = 0$. In view of the behaviour of $u_{i,j}$ at infinity it follows that the constant E occurring in the "conservation law"

$$\int \{(\alpha+1)u_{3,3}^2 - \alpha u_{\beta,\beta}u_{\gamma,\gamma} - u_{\beta,\gamma}u_{\beta,\gamma} - u_{3,\gamma}u_{3,\gamma} + u_{\gamma,3}u_{\gamma,3}\}dA = E \quad (7.64)$$

is zero, and there follows

$$\int (\alpha u_{\beta,\beta}u_{\gamma,\gamma} + u_{\beta,\gamma}u_{\beta,\gamma} + u_{3,\beta}u_{3,\beta})dA \geq \int u_{\beta,3}u_{\beta,3}dA. \qquad (7.65)$$

[See Exercises for an indication of how the "conservation law" (7.64) is proved.] Thus an explicit upper bound for the quantity $\int u_{\beta,3}u_{\beta,3}dA$ occurring in $F(0)$ is available in terms of the stated data. One thus obtains the explicit result appropriate to this case:

$$F(x_3) \leq [\int_{D(0)} (\alpha u_{\beta,\beta}u_{\gamma,\gamma} + u_{\beta,\gamma}u_{\beta,\gamma} + 2u_{3,\beta}u_{3,\beta})dA] \cdot \exp\{-2\lambda_1^{1/2}(1+\alpha)^{-1/2}x_3\}. \qquad (7.66)$$

This completes the discussion of the right cylinder.

Similar results are obtainable for a homogeneous isotropic strip $0 < x_2 < 1$ in a state of plane strain. Essentially all equations and formulae continue to hold with italic indices taking the values 1, 2, and Greek indices having the value 1.

Envisaging displacement components u_1, u_2 as functions of x_1, x_2 satisfying

$$\begin{aligned}(\alpha+1)u_{1,11} + u_{1,22} + \alpha u_{2,21} &= 0\\ \alpha u_{1,12} + u_{2,11} + (\alpha+1)u_{2,22} &= 0\end{aligned} \tag{7.67}$$

subject to

$$u_1 = u_2 = 0 \qquad \text{on the edges} \quad x_2 = 0, 1$$

we define

$$F(x_1) = \int_0^1 (u_{1,2}^2 + u_{2,1}^2)dx_2\,. \tag{7.68}$$

One has the analogue of (7.59)

$$(F^{1/2})'' - \pi^2(1+\alpha)^{-1}F^{1/2} \geq 0 \tag{7.69}$$

where null-displacement conditions obtain on the lateral edges $x_1 = 0, x_2 = 1$ in the latter case. These yield consequences similar to their analogues in the three-dimensional case.

We note, *in this case*, that pointwise upper estimates for the displacement component u_3 – reflecting position and valid up to the boundary – are deducible from the upper estimates for $F(x_3)$ on using (see A5 of Appendix)

$$|u_1(x_1, x_2)| \leq \{(1-x_2)x_2F(x_1)\}^{1/2}\,. \tag{7.70}$$

7.3.4 Weighted Cross-sectional Method

We now prove an analogue of (7.69) for a modified version of (7.68), appropriate to a body which extends to infinity in the x_2 direction, and draw some conclusions therefrom. This is a further example of the method of *weighted cross-sectional integrals*, described in Chapter six; we are unaware of any other example, or any other comparable analysis, in the literature. There are, of course, a considerable number of analyses of elastostatic problems in infinite regions based upon the "weighted energy" approach: for elastic (and other) problems with asymptotic growth conditions at infinity, one obtains inequality estimates for measures of the solution which may be envisaged as weighted energies – that is weighted/mollified so that the relevant integral(s) is (are) convergent in the presence of posited growth conditions at infinity; see [72, 162, 163] for example. We emphasize that what is novel in the ensuing analysis is the marriage between the weight function approach and the cross-sectional method.

We consider a semi-infinite rectangular region

$$0 < x_1 < 1, \qquad 0 < x_2 < \infty$$

occupied by elastic material in plane strain, as described in the previous section, the plane boundary $x_2 = 0$ being displacement-free:

$$u_1(x_1, 0) = u_2(x_1, 0) = 0. \tag{7.71}$$

It is supposed that the boundaries $x_1 = 0, x_1 = 1$ are subjected to specified normal displacement and complementary shear stress:

$$u_1 = \text{specified}, \quad u_{1,2} + u_{2,1} = \text{specified}, \tag{7.72}$$

on both $x_1 = 0, x_1 = 1 (0 < x_2 < \infty)$. It is evident – provided that u_1 is sufficiently smooth – that these latter boundary conditions are equivalent to specifying $u_{1,2}$ and $u_{2,1}$ on the relevant boundaries. For a start, we consider classical slutions of (7.67) (in $0 < x_1 < 1$) satisfying (7.71) together with the following prescribed asymptotic conditions:

$$u_{1,2},\, u_{2,1},\, u_{1,11},\, u_{1,21},\, u_{2,11},\, u_{2,12},\, u_{1,211},\, u_{2,111},\, u_{2.221} = O(e^{\mu x_2}) \tag{7.73}$$

where μ is a positive number which is subsequently restricted. It will be noted that $u_{1,22}$ has similar asymptotic behaviour in view of (7.67).

We consider the weighted cross-sectional measure of deformation

$$F(x_1) = \int_0^\infty g(x_2)(u_{1,2}^2 + u_{2,1}^2)dx_2 \tag{7.74$_1$}$$

where

$$g(x_2) = e^{-\lambda x_2} \tag{7.74$_2$}$$

λ being a positive constant such that

$$\lambda > 2\mu, \tag{7.74$_3$}$$

μ being the constant arising in (7.73). It is evident that the prescribed conditions ensure the existence of the integral in (7.74$_1$), and a similar comment applies to all integrals which arise subsequently. In what follows, the limits of all integrals arising shall be omitted.

Successive differentiations yield

$$F'(x_1) = \int 2g(u_{1,2}u_{1,21} + u_{2,1}u_{2,11})dx_2, \tag{7.75}$$

$$F''(x_1) = \int 2g(u_{1,21}^2 + u_{2,11}^2 + u_{1,2}u_{1,211} + u_{2,1}u_{2,111})dx_2\,. \tag{7.76}$$

Use of (7.67), integration by parts using (7.71), (7.73), yields

$$\begin{aligned}&\int g(u_{1,2}u_{1,211} + u_{2,1}u_{2,111})dx_2 = \\ &\int (1+\alpha)^{-1}(gu_{1,22}^2 - \frac{1}{2}\lambda^2 gu_{1,2}^2)dx_2 \\ &+ \int (2\alpha+1)(1+\alpha)^{-1}(gu_{2,12}^2 - \frac{1}{2}\lambda^2 gu_{2,1}^2)dx_2 + \frac{1}{2}(1+\alpha)^{-1}\lambda u_{1,2}^2(x_1,0) \\ &- \alpha(1+\alpha)^{-1}\lambda \int g(u_{1,2}u_{2,21} - u_{2,1}u_{1,22})dx_2.\end{aligned} \tag{7.77}$$

Now we suppose that θ, ψ are arbitrary positive numbers, and β an arbitrary number such that $0 \leq \beta < 1$. On using the arithmetic-geometric inequality applied to the last group of terms in (7.77) etc., we obtain

$$\begin{aligned}&\int 2g(u_{1,2}u_{1,211} + u_{2,1}u_{2,111})dx_2 \geq \\ &(1+\alpha)^{-1} \int g[2u_{1,22}^2 + (2\alpha+1)\{2(1-\beta)u_{2,12}^2 + 2\beta u_{2,12}^2 - \lambda^2 u_{2,1}^2\} \\ &-\alpha(\theta^{-1}\lambda^2 u_{1,2}^2 + \theta u_{2,12}^2 + \psi^{-1}\lambda^2 u_{2,1}^2 + \psi u_{1,22}^2) - \lambda^2 u_{1,2}^2]dx_2\,.\end{aligned} \tag{7.78}$$

Using Theorem 6.9 as applied to $u_{2,1}$, we obtain

$$\begin{aligned}&\int 2g(u_{1,2}u_{1,211} + u_{2,1}u_{2,111})dx_2 \geq \\ &(1+\alpha)^{-1}\{2(2\alpha+1)(1-\beta) - \theta\alpha\} \int gu_{2,12}^2 dx_2 \\ &+ (1+\alpha)^{-1}(2-\alpha\psi) \int gu_{1,22}^2 dx_2 \\ &- \lambda^2(1+\alpha)^{-1} \int g(1+\theta^{-1}\alpha)u_{1,2}^2 dx_2 \\ &-\lambda^2(1+\alpha)^{-1} \int g\{(2\alpha+1)(1-\frac{1}{2}\beta) + \alpha\psi^{-1}\}u_{2,1}^2 dx_2\,.\end{aligned} \tag{7.79}$$

Choosing ϕ, ψ to make the coefficients of the first two terms vanish gives

$$\theta^{-1} = \frac{1}{2}\alpha(2\alpha+1)^{-1}(1-\beta)^{-1}\,, \qquad \psi^{-1} = \frac{1}{2}\alpha$$

and thus we obtain

$$\int 2g(u_{1,2}u_{1,211} + u_{2,1}u_{2,111})dx_2 \geq -\lambda^2(1+\alpha)^{-1}mF \tag{7.80}$$

where

$$m = \max[1 + \frac{1}{2}\alpha^2(2\alpha+1)^{-1}(1-\beta)^{-1},\ \frac{1}{2}\alpha^2 + (2\alpha+1)(1-\frac{1}{2}\beta)], \tag{7.81}$$

the latter meaning the maximum of the two terms. It is obviously desirable to minimize this maximum with respect to β $(0 \leq \beta < 1)$.

In view of $(7.74_1), (7.76), (7.80)$ we obtain

$$F'' + \lambda^2(1+\alpha)^{-1}mF \geq 0. \tag{7.82}$$

A simple use of Schwarz's inequality, bearing (7.75) in mind, yields the enhanced inequality

$$(F^{1/2})'' + \frac{1}{2}\lambda^2(1+\alpha)^{-1}mF^{1/2} \geq 0. \tag{7.83}$$

This leads to (cf. proof of Theorem 6.12):

Theorem 7.9 - *The weighted cross-sectional measure of deformation given by* (7.74), *for classical solutions of* (7.67) *which satisfy* (7.71) – (7.73) *satisfies*

$$F^{1/2}(x_1) \leq G(x_1) \tag{7.84$_1$}$$

where $G(x_1)$ *satisfies*

$$G'' + \frac{1}{2}\lambda^2(1+\alpha)^{-1}mG = 0 \tag{7.84$_2$}$$

subject to

$$G(0) = F(0) \quad , \quad G(1) = F(1), \tag{7.84$_3$}$$

where $F(0), F(1)$ *are available from* (7.72), *provided that*

$$2\mu < \lambda < \sqrt{2}\pi(1+\alpha)^{1/2}m^{-1/2}, \tag{7.84$_4$}$$

m *being defined by* (7.81).

Let us note that a uniqueness theorem is deducible from Theorem 7.9. If the data on the edges $x_1 = 0, x_1 = 1$ are identically zero (zero normal displacement and zero complementary shear stress), it follows that $F \equiv 0$ implying $u_{1,2} \equiv u_{2,1} \equiv 0$. This, in turn, implies that $u_1 = 0, u_2 = sx_2$ (s being any constant) – a qualified uniqueness. To recover uniqueness proper, one needs to impose a further condition such as

$$u_2 \to 0 \quad \text{as} \quad x_2 \to \infty.$$

Note that pointwise bounds for u_1 in terms of data (reflecting position) may be deduced from the estimate given in Theorem 7.9 via Schwarz's inequality:

$$u_1^2(x_1,x_1) = \{\int_0^{x_2} u_{1,2}dx_2\}^2 \leq \{\int_0^{x_2} e^{\lambda x_2}dx_2\}\{\int_0^{x_2} e^{-\lambda x_2}u_{1,2}^2 dx_2\}$$
$$\leq \lambda^{-1}(e^{\lambda x_2}-1)F(x_1). \tag{7.85}$$

It is evident from this that u_1 exhibits pointwise continuous dependence with respect to data.

7.3.5 Non-null Conditions on Lateral Boundary

We now apply the cross-sectional method to a two-dimensional elastic problem with non-null conditions on its lateral boundary and our analysis is based upon [164]. We consider a region R bounded by lateral boundaries $\mathcal{C}_+, \mathcal{C}_-$ and by, possibly degenerate, straight line segments Γ_0, Γ_L at $x_1 = 0, x_2 = L$ – as specified prior to Lemma 4.3; we shall, however – consistent with the other analyses of this section – use indicial notation here: x_1, x_2 replace x, y respectively, and the lateral boundaries $\mathcal{C}_+, \mathcal{C}_-$ shall be denoted by $x_2 = x_2^+(x_1), x_2 = x_2^-(x_1)$ respectively. The lateral boundaries here shall be supposed to be *strictly* convex i.e.

$$x_2^{+''}(x_1) < 0 \quad , \quad x_2^{-''}(x_1) > 0\,.$$

We suppose that the region R is occupied by elastic material in a state of plane strain due to the applications of displacements on its boundaries. We shall be concerned with classical solutions $u_\beta(x_1,x_2)$ of the P.D.E.s (7.67) – with indices running over $1,2$ – subject to

$$u_\beta = \mathcal{F}_\beta^+(x_1) \quad \text{on } \mathcal{C}_+\,, \quad u_\beta = \mathcal{F}_\beta^-(x_1) \quad \text{on } \mathcal{C}_-\,, \tag{7.86}$$

and subject to specified displacements on Γ_0, Γ_L – assumed non-degenerate.

We define the cross-sectional measure of deformation

$$F(x_1) = \int_{\Gamma_{x_2}} (k_1 u_{1,2}^2 + k_2 u_{2,2}^2)dx_2 \tag{7.87}$$

where k_1, k_2 are positive constants yet to be chosen: henceforward, the suffix attached to the integral is dropped. Using Lemma 4.3b, we obtain

$$F''(x_1) \geq 2\int\{(k_1u_{1,12}^2 + k_2u_{2,12}^2) - (k_1u_{1,11}u_{1,22} + k_2u_{2,11}u_{2,22})\}dx_2$$
$$+(k_1\mathcal{F}_1^{''2} + k_2\mathcal{F}_2^{''2})/(x_2'')] \tag{7.88}$$

(recall that the notation $--]$, occurring at the end of (7.88) and subsequently, has been explained in Chapter four, for example).

Substitution of (7.67) in (7.88) yields

$$F''(x_1) \geq 2\int\{k_1u^2_{1,22} + k_1u^2_{2,12})dx$$

$$+2\int\{(k_1u_{1,22}[\alpha(1+\alpha)^{-1}u_{2,21} + (1+\alpha)^{-1}u_{1,22}]$$

$$+k_2u_{2,22}[(1+\alpha)u_{2,22} + \alpha u_{1,12}]\}dx_2 + \{(k_1\mathcal{F}_1''^2 + k_2\mathcal{F}_2''^2)/(x_2'')\}]$$

$$\geq [2(1+\alpha)^{-1} - \alpha(1+\alpha)^{-1}C_1]\int k_1u^2_{1,22}dx_2 + [2(1+\alpha) - \alpha C_2]\int k_2u^2_{2,22}dx_2$$

$$+[2k_2 - k_1\alpha(1+\alpha)^{-1}C_1^{-1})]\int u^2_{2,21}dx_2 + [2k_1 - k_2\alpha C_2^{-1})]\int u^2_{1,12}dx_2$$

$$+\{(k_1\mathcal{F}_1''^2 + k_2\mathcal{F}_2''^2)/(x_2'')\}] \tag{7.89}$$

where C_1, C_2 are arbitrary positive constants, and where the arithmetic - geometric inequality has been used. Let us choose the constants C_1, C_2 so that the coefficients of the third and fourth terms in (7.89) vanish:

$$C_2 = \alpha k_2/(2k_1), \qquad C_1 = \alpha(\alpha+1)^{-1}k_1/(2k_2). \tag{7.90}$$

Thus

$$F''(x_1) \geq [2(\alpha+1)^{-1} - \alpha^2(\alpha+1)^{-2}k_1/(2k_2)]\int k_1u^2_{1,22}dx_2$$

$$+[2(\alpha+1) - \alpha^2(k_2/2k_1)]\int k_2u^2_{2,22}dx_2 + \{(k_1\mathcal{F}_1''^2 + k_2\mathcal{F}_2''^2)/(x_2'')\}]. \tag{7.91}$$

The constants k_1, k_2 are chosen so that the coefficients of the first two terms in (7.91) are non-negative:

$$2(\alpha+1)^{-1} \geq \alpha^2(\alpha+1)^{-2}k_1/(2k_2),$$

$$(2\alpha+2) \geq \alpha^2k_2/(2k_1),$$

or

$$4(\alpha+1)\alpha^{-2} \geq k_2/k_1 \geq \alpha^2(\alpha+1)^{-1}/4. \tag{7.92}$$

For consistency one requires

$$4(\alpha+1) \geq \alpha^2; \tag{7.93$_1$}$$

this is equivalent to

$$\alpha \leq 2\sqrt{2}+2 \tag{7.93$_2$}$$

or, in terms of Poisson's ratio σ,

$$\sigma \leq \frac{1}{4}(3-\sqrt{2}) \simeq 0.396\,. \tag{7.93$_3$}$$

It is plain that $k_2/k_1 = 1$ if $\alpha = 2\sqrt{2}+2$, while k_2/k_1 has a range of values including 1 if $\alpha < 2\sqrt{2}+2$. Note that the range of α embraces a wide spectrum of materials; furthermore, it includes all materials if the context is that of generalized plane stress (see [165] for example). The restrictions on the elastic constants arising in (7.93) and in the ensuing propositions appear to be dictated by mathematical considerations rather than by intrinsic considerations.

The above considerations, together with Lemma 4.4, lead to the following:

Theorem 7.10 - *If $\alpha \leq 2\sqrt{2}+2$, if the lateral boundaries C_+, C_- are* **strictly** *convex, and if k_1, k_2 are positive constants satisfying* (7.92) *then the cross-sectional measure*

$$F(x_1) = 2\int (k_1 u_{1,2}^2 + k_2 u_{2,2}^2)dx_2$$

in the context of classical solutions of (7.67) *subject to the boundary conditions* (7.86) *on the lateral boundaries, satisfies*

$$\{k_1(\mathcal{F}_1])^2 + k_2(\mathcal{F}_2])^2\}l^{-1}(x_1) \leq F(x_1) \leq G(x_1) \tag{7.94}$$

where $G(x_1)$ satisfies

$$G'' = \{(k_1\mathcal{F}_1{''}^2 + k_2\mathcal{F}_2{''}^2)/(x_2'')\}]$$

subject to

$$G(0) = F(0) \quad , \quad G(L) = F(L)\,,$$

and where $l(x_1)$ is the width of the region at abscissa x_1.

The inequality sign occurs in (7.94) *when the boundary conditions are compatible with an arbitrary homogeneous displacement (i.e. $u_\beta = A_\beta x_1 + B_\beta x_2 + C_\beta$ where $A., B., C.$ are constants).*

It is left as an exercise to prove the last statement.

The interest of the foregoing result is that estimates of the cross-sectional or of the "volume" type appear to be virtually non-existent in the case of non-null conditions on the lateral boundaries – at least in the context of elasticity.

Let us note that it is possible to improve the upper bound for $F(x_1)$ by choosing k_1, k_2 appropriately; it involves the use of a Poincaré inequality. For details, see [164].

7.4 TRACTION-TYPE PROBLEMS IN LINEAR ELASTICITY

7.4.1 Two dimensional stress systems: Saint-Venant issue

We consider a rectangular region R: $0 < x < L, 0 < y < 1$, consisting of homogeneous, isotropic linear elastic material, in a state of plane strain; occasionally we consider a semi-infinite rectangle $L \to \infty$. It is supposed to be subjected to (in-plane) tractions applied to its edges. Stress components $\tau_{xx}, \tau_{xy}, \tau_{yy}$, functions of rectangular cartesian coordinates x, y, may be expressed in terms of the Airy stress function $\phi(x, y)$ as follows:

$$\tau_{xx} = \phi_{yy} \quad \tau_{yy} = \phi_{xx}\,, \quad \tau_{xy} = -\phi_{xy} \tag{7.95}$$

where ϕ satisfies the P.D.E.

$$(\nabla_1^4 \phi =) \phi_{xxxx} + 2\phi_{xxyy} + \phi_{yyyy} = 0 \tag{7.96}$$

in R; see [165], for example. We suppose that $\phi \in C^4$ throughout – although less stringent requirements often suffice. Suppose that the three edges $y = 0, 1$, $x = L$ are traction-free and that (necessarily) the edge $x = 0$ is subjected to tractions which are equipollent to zero (i.e. have zero resultant and zero moment). The issue of Saint Venant's Principle (e.g. [101]) is that of establishing that there is, in a suitable sense, rapid decay of effects away from the loaded end. This issue was resolved by Knowles [99] by a method which we have already called "volume integral" method. This together with the analysis of an elastic right cylinder by Toupin [100] – both of which appeared about the same time – provided a vastly superior theoretical underpinning for Saint Venant's principle compared to anything existing up to that time.

In the circumstances described in the last paragraph, one may, without loss of generality, assume that the arbitrariness inherent in ϕ (an arbitrary linear term) may be exploited to express the boundary conditions thus:

$$\begin{aligned} &\phi = \phi_y = 0 && \text{on edges } y = 0, 1\,, \\ &\phi = \phi_x = 0 && \text{on edge } x = L\,, \\ &\phi, \phi_x && \text{specified on edge } x = 0\,. \end{aligned} \tag{7.97}$$

7.4.2 Knowles' Analysis ("Volume Integral" Method)

We denote by R_z the set of points (x, y) in R for which $x \geq z$, and by L_z the set of points of R which have abscissa z. Knowles' result, with a slightly improved decay constant [166], is expressible as follows:

Theorem 7.11 - *The "volume measure" of stress (or "energy")*

$$E(z) = \iint_{R_z} (\phi_{xx}^2 + 2\phi_{xy}^2 + \phi_{yy}^2)dA \tag{7.98}$$

satisfies

$$E(z) \leq 2E(0)\exp(-\pi\sqrt{2}z) \tag{7.99}$$

where $\phi(x, y) \in C^4(\overline{R})$ *satisfies* (7.96), (7.97).

Proof. The result is established by obtaining a differential inequality for a function related to $E(z)$. The differential inequality is based on an identity satisfied by the integral

$$\int_z^L E(x)dx$$

and the derivative $E'(z)$ of $E(z)$.

We first derive a suitable representation for the integral quantity just mentioned. For $0 < z < L$, one may write

$$\begin{aligned} E(z) = &\iint_{R_z} (\phi_x\phi_{xx} - \phi\phi_{xxx} + 2\phi_y\phi_{xy})_x dA \\ + &\iint_{R_z} (\phi_y\phi_{yy} - \phi\phi_{yyy} - 2\phi\phi_{xxy})_y dA, \end{aligned} \tag{7.100$_1$}$$

using (7.96) and (7.98) Green's Theorem and the boundary conditions (7.97) as applied thereto furnish:

$$E(z) = -\int_{L_z} (\phi_x\phi_{xx} - \phi\phi_{xxx} + 2\phi_y\phi_{xy})dy. \tag{7.100$_2$}$$

By differentiation, (7.100$_2$) may be expressed as

$$E(z) = \frac{d}{dz}\iint_{R_z} (\phi_x\phi_{xx} - \phi\phi_{xxx} + 2\phi_y\phi_{xy})dA \tag{7.101$_1$}$$

whence it follows that

$$\int_z^L E(x)dx = -\iint_{R_z} (\phi_x\phi_{xx} - \phi\phi_{xxx} + 2\phi_y\phi_{xy})dA \tag{7.101$_2$}$$

since the integral on the right vanishes when $z = L$. A further use of Green's theorem and the boundary conditions furnishes

$$\begin{aligned} \int_z^L E(x)dx &= -\iint_{R_z} \frac{\partial}{\partial x}(\phi_x^2 + \phi_y^2 - \phi\phi_{xx})dA \\ &= \int_{L_z} (\phi_x^2 + \phi_y^2 - \phi\phi_{xx})dy. \end{aligned} \tag{7.102}$$

Furthermore, differentiation of (7.98) gives

$$E'(z) = -\int_{L_z} (\phi_{xx}^2 + 2\phi_{xy}^2 + \phi_{yy}^2)dy\,. \tag{7.103}$$

Now for any real constant K, (7.102)–(7.103) furnish

$$\begin{aligned} E'(z)+4K^2 &\int_z^L E(x)dx \\ &= -\int_{L_z} [(\phi_{xx}^2 + \phi_{yy}^2 + 2\phi_{xy}^2 - 4K^2(\phi_x^2 + \phi_y^2 - \phi\phi_{xx})]dy\,. \end{aligned} \tag{7.104}$$

We now define a function $F(z)$ such that

$$F(z) = E(z) + 2K\int_z^L E(x)dx\,. \tag{7.105}$$

In terms of F, (7.104) may be rewritten as

$$\begin{aligned} F'(z) + 2KF(z) = -\int_{L_z} &[(\phi_{xx} + 2K^2\phi)^2 + (2\phi_{xy}^2 - 4K^2\phi_x^2) \\ &+ (\phi_{yy}^2 - 4K^2\phi_y^2 - 4K^4\phi^2)]dy\,. \end{aligned}$$

An obvious inequality may be derived therefrom:

$$\begin{aligned} F'(z) + 2KF(z) \le -\int_{L_z} &[(2\phi_{xy}^2 - 4K^2\phi_x^2)+ \\ &(\phi_{yy}^2 - 4K^2\phi_y^2 - 4K^4\phi^2)]dy\,. \end{aligned} \tag{7.106}$$

We now aim to determine a positive value of K for which the right hand side of (7.106) is nonpositive. In view of inequality A1(i)(a), we have, bearing in mind the boundary conditions $\phi = 0$ on $y = 0, 1$,

$$\int_{L_z} \phi_y^2 dy \ge \pi^2 \int_{L_z} \phi^2 dy\,, \tag{7.107$_1$}$$

$$\int_{L_z} \phi_{xy}^2 dy \ge \pi^2 \int_{L_z} \phi_x^2 dy\,. \tag{7.107$_2$}$$

In view of the inequality A2(i) and the boundary conditions $\phi = \phi_y = 0$ on $y = 0, 1$, we have

$$\int_{L_z} \phi_{yy}^2 dy \ge 4\pi^2 \int_{L_z} \phi_y^2 dy\,. \tag{7.108}$$

In view of (7.107) – (7.108), (7.106) furnishes

$$F'(z)+2KF(z) \leq -\int_{L_z} [2(\pi^2-2K^2)\phi_x^2+(1-K^2/\pi^2-K^4/\pi^4)\phi_{yy}^2]dy\,. \quad (7.109)$$

It is easily checked that if one chooses

$$K = \pi/\sqrt{2} \quad (7.110)$$

one has

$$1 - K^2/\pi^2 - K^4/\pi^4 > 0\,.$$

Thus for K, given by (7.110), one has

$$F'(z) + 2KF(z) \leq 0\,. \quad (7.111)$$

Integration of this latter inequality gives

$$F(z) \leq F(0)e^{-2Kz}$$

and, in view of (7.105), this implies

$$E(z) \leq F(0)e^{-2Kz}\,. \quad (7.112)$$

It remains to compute a suitable upper bound for $F(0)$. Using (7.105) in (7.112), we easily obtain

$$-\frac{d}{dz}[e^{-2Kz}\int_z^L E(x)dx] \leq F(0)e^{-4Kz}$$

and integration between $z = 0, L$, furnishes

$$\begin{aligned}\int_0^L E(x)dx &\leq \frac{F(0)}{4K}(1 - e^{-4KL}) \\ &= \frac{1}{4K}[E(0) + 2K\int_0^L E(x)dx](1 - e^{-4KL})\,.\end{aligned}$$

Solving this inequality, we obtain

$$2K\int_0^L E(x)dx \leq \frac{1 - e^{-4KL}}{1 + e^{-4KL}}E(0) \leq E(0)\,. \quad (7.113)$$

Now inserting (7.113) in (7.105) with $z = 0$, we obtain

$$F(0) \leq 2E(0)\,. \quad (7.114)$$

Thus (7.112)–(7.114) together with (7.110) establish the theorem.

To make the analysis complete, one needs to bound $E(0)$ above in terms of data (ϕ, ϕ_x) specified on the edge $x = 0$ (cf. Theorem 3.2). The following may be established:
If $\overline{\phi}$ is any C^2 function satisfying the same boundary conditions as ϕ, one has

$$\int_R [\overline{\phi}_x^2 + 2\overline{\phi}_{xy}^2 + \overline{\phi}_{yy}^2] dA \geq E(0)\,;$$

the equality sign holds iff $\overline{\phi} = \phi$ (cf.[99]).

Let us note the following:
(i) The above analysis is applicable, with minimal changes, to a non-rectangular region, the context for which Knowles, in fact, derived the result.
(ii) The decay constant occurring in (7.99) underestimates the exact value (got from an eigenfunction analysis) by a factor of about 2.
(iii) Knowles [167] has applied to a *semi-infinite strip*, in the same context, a modified "volume integral" method involving a "second order energy" in addition to (7.98), obtaining an estimate which considerably improves the decay constant in (7.99).
(iv) Vafeades and Horgan [168] have shown how to replace the "amplitude" term $2E(0)$, occurring in (7.99), by $E(0)$.
(v) An alternative analysis of this and related issues is given in [169-170].

7.4.3 Cross-sectional Method

We now apply the cross-sectional method to the problem considered above but we shall here limit our attention to a semi-infinite rectangular region $0 < x < \infty, 0 < y < 1$. Two preliminary lemmas are first proved, the first being a conservation principle analogous to that derived in Chapter four for the harmonic equation.

Lemma 7.1 - *Solutions of the biharmonic equation* (7.96) *in the strip* $0 < x < 1$ *subject to boundary conditions on the lateral edges*

$$\phi_{xy} = \phi_{xx} = 0 \qquad on\ y = 0, y = 1 \tag{7.115}$$

satisfy the conservation law

$$\int_0^1 (\phi_{yy}^2 - \phi_{xx}^2 - 2\phi_{xy}^2 + 2\phi_x \phi_{xxx}) dy = E(a\ constant)\,. \tag{7.116}$$

Proof. We observe that

$$\int_0^1 \phi_x \phi_{xxxx}\, dy = \frac{d}{dx}\int_0^1 (\phi_x\phi_{xxx} - \frac{1}{2}\phi_{xx}^2)dy\,,$$
$$\int_0^1 2\phi_x\phi_{xxyy}\, dy = -\int_0^1 2\phi_{xy}\phi_{xxy}\, dy = -\frac{d}{dx}\int_0^1 \phi_{xy}^2\, dy\,,$$
$$\int_0^1 \phi_x\phi_{yyyy}\, dy = \int_0^1 \phi_{yy}\phi_{yyx}\, dy = \frac{d}{dx}\int_0^1 \frac{1}{2}\phi_{yy}^2\, dy\,, \tag{7.117}$$

where integration by parts and the boundary conditions (7.115) have been used. Use of (7.117) together with (7.96) yields (7.116) on integration.

We now proceed to the second lemma (which implies the convexity of a certain cross-sectional measure):

Lemma 7.2 - *The cross-sectional measure*

$$F(x) = \int_0^1 (\phi_{xx}^2 + \phi_{yy}^2)dy \tag{7.118}$$

satisfies

$$F''(x) = 2\int_0^1 (\phi_{xxx}^2 + \phi_{xyy}^2 + 2\phi_{xxy}^2)dy \tag{7.119}$$

for solutions of the biharmonic equation (7.96) *which satisfy the boundary conditions* (7.115).

Proof. Twice differentiation of (7.118) yields

$$F''(x) = 2\int_0^1 [\phi_{xxx}^2 + \phi_{xyy}^2 + \phi_{xx}\phi_{xxxx} + \phi_{yy}\phi_{yyxx}]dy\,. \tag{7.120}$$

Using integration by parts, the differential equation (7.96) and the boundary conditions (7.115), we obtain

$$\int_0^1 \phi_{xx}\phi_{xxxx}\, dy = \int_0^1 (\phi_{xxy}\phi_{yyy} + 2\phi_{xxy}^2)dy\,, \tag{7.121$_1$}$$

$$\int_0^1 \phi_{yy}\phi_{yyxx}\, dy = -\int_0^1 \phi_{yyy}\phi_{yxx}\, dy\,. \tag{7.121$_2$}$$

Combining (7.120), (7.121) we obtain the required result (7.119).

We now proceed to the principal issue. The analysis is a slightly simplified version of that of Flavin and Knops [134].

Theorem 7.12 - *In the context of a semi-infinite strip* $0 < x < \infty$, $0 < y < 1$, *classical* (C^4) *solutions of* (7.96) *satisfying* (7.97$_1$), (7.97$_3$) *together with the (enhanced) asymptotic conditions*

$$\phi_x \to 0\,, \quad \phi_{xx} \to 0\,, \quad \phi_{xy} \to 0\,, \quad \phi_{yy} \to 0\,, \quad \phi_{xxx} = O(1) \qquad (7.122_1)$$

uniformly as $x \to \infty$, *are such that the cross-sectional (stress) measure*

$$F(x) = \int_0^1 (\phi_{xx}^2 + \phi_{yy}^2)dy \qquad (7.122_2)$$

satisfies

$$F(x) \leq F(0)e^{-\pi\sqrt{2}x}\,. \qquad (7.122_3)$$

Proof. In view of (7.122$_1$), the constant E occurring in (7.116) is zero. Applying the arithmetic-geometric to the resulting equation yields

$$\int_0^1 \phi_{xx}^2 dy \geq \int_0^1 (\phi_{yy}^2 - 2\phi_{xy}^2 - \theta\phi_x^2 - \theta^{-1}\phi_{xxx}^2)dy \qquad (7.123)$$

for any positive constant θ.

In view of inequality A1(i)(a) we see that

$$\int_0^1 \phi_{xxy}^2 dy \geq \pi^2 \int_0^1 \phi_{xx}^2 dy \qquad (7.124)$$

since $\phi_x = 0$ on $y = 0, 1$. Thus (7.119) implies that

$$F''(x) \geq 2\int_0^1 (\phi_{xxx}^2 + \phi_{yyx}^2 + 2\pi^2\phi_{xx}^2)dy$$

and applying (7.123) to this yields

$$\begin{aligned} F''(x) \geq \int_0^1 [&(4\pi^2 - \delta)\phi_{xx}^2 + \delta\phi_{yy}^2 + (2 - \delta\theta^{-1})\phi_{xxx}^2 \\ &+ 2\phi_{yyx}^2 - \delta(2\phi_{xy}^2 + \theta\phi_x^2)]dy \end{aligned} \qquad (7.125)$$

where δ is an arbitrary positive number. Now we choose

$$\delta = 2\pi^2\,, \quad \theta = \delta/2 = \pi^2$$

and obtain

$$F''(x) - 2\pi^2 F(x) \geq 2\int_0^1 [\phi_{yyx}^2 - \pi^2(2\phi_{xy}^2 + \pi^2\phi_x^2)]dy\,. \tag{7.126}$$

In view of inequalities A2(i), A1(i)(a), together with the boundary conditions (7.97_1), we have

$$\int_0^1 \phi_{yyx}^2 dy \geq 4\pi^2 \int_0^1 \phi_{xy}^2 dy\,,$$

$$\int_0^1 \phi_{xy}^2 dy \geq \pi^2 \int_0^1 \phi_x^2 dy\,,$$

and it follows that

$$\int_0^1 [\phi_{yyx}^2 - \pi^2(2\phi_{xy}^2 + \pi^2\phi_x^2]dy \geq 0\,.$$

In view of the latter inequality, (7.126) implies the "generalized convexity" of $F(x)$

$$F'' - 2\pi^2 F \geq 0\,. \tag{7.127}$$

In view of (7.127), Theorem 4.4, and the fact that $\phi_{xx} \to 0, \phi_{yy} \to 0$, as $x \to \infty$ (i.e. $F \to 0$ as $x \to \infty$), we obtain Theorem 7.12.

The amplitude term $E(0)$ may be computed explicitly in terms of conventional data in the following circumstances:
(i) If the load applied to the edge $x = 0$ is normal i.e. $\tau_{xy} = 0$ thereon; in these circumstances, one may use (7.116) to establish that

$$F(0) = 2\int_0^1 \tau_{xx}^2(0, y)dy\,. \tag{7.128}$$

(ii) If the normal traction and the complementary/tangential displacement component (assumed continuously differentiable) are known on the edge $x = 0$, it is possible to compute τ_{yy} thereon, and, hence, to obtain $F(0)$.

7.4.4 Analogous Issues in Three Dimensions

The analogous issues in three dimensions are considerably less tractable. Toupin's seminal work [100], already mentioned, relates to a right (linear) elastic cylinder, one plane end of which is subjected to a self-equilibrated load, the remainder of its boundary being (traction) free; it establishes exponential decay of effects away from the loaded end. Being a "volume integral" method the proof has much in common with that of Knowles. This result was subsequently generalized by Fichera [171] who considered a right (elastic) cylinder with free lateral boundary, each of whose (two) plane ends is subjected to self equilibrated loads. This was, in turn, generalized by Oleinik and Yosifian [172] to cater, *inter alia*, for a region whose cross-section varies. We now quote some of this results, but omit proofs.

To set the context for the results of Toupin, and of Oleinik and Yosifian, we consider a linear elastic material for which the (rectangular cartesian) components of displacement u_i, and of stress τ_{ij}, are related by

$$\tau_{ij} = c_{ijkl} u_{k,l} \tag{7.129}$$

where the elastic moduli c_{ijkl} are such that

$$c_{ijkl} = c_{jikl}\ , \quad c_{ijkl} = c_{ijlk}\ , \quad c_{ijkl} = c_{klij}\ ; \tag{7.130}$$

commas here and throughout, denote partial differentiation with respect to the appropriate variable, and the usual summation convention is adopted. For any real symmetric matrix $\eta = \{\eta_{ij}\}$, there exist positive constants K_1, K_2 such that

$$K_1 \eta_{ih} \eta_{ih} \leq c_{ijkl} \eta_{ij} \eta_{kl} \leq K_2 \eta_{ih} \eta_{ih}\ . \tag{7.131}$$

The displacement components, in the absence of body force, satisfy

$$\tau_{ij,j} = 0\ . \tag{7.132}$$

The boundary condition corresponding to a (traction-) free boundary is

$$\tau_{ij} \nu_j = 0 \tag{7.133}$$

where ν_i denote the (rectangular cartesian) components of the unit outward normal.

Toupin's result is concerned with a homogeneous (i.e. constant c_{ijkl}) elastic right cylinder B. The origin of rectangular cartesian coordinates is taken in one of the plane ends and the x_3 axis is parallel to the generators of the cylinder. The plane end $x_3 = 0$ is subjected to a self-equilibrated load i.e.

$$\int \tau_{3i}\, dA = \int e_{ijk} x_j \tau_{3k}\, dA = 0 \tag{7.134}$$

thereon, and the remainder of the boundary of the cylinder is traction-free (i.e. satisfies (7.133) thereon). Let us denote the (positive definite) elastic energy contained in $B(s) = B \cap x_3 \geq s$ by

$$U(s) = \int_{B(s)} \frac{1}{2} c_{ijkl} u_{i,j} u_{k,l}\, dV\ . \tag{7.135}$$

Toupin's result is embodied in

Theorem 7.13 - *A right cylinder consisting of homogeneous, linear elastic material is loaded on one plane end with an arbitrary self-equilibrated load, and the remainder of its boundary is free. Then the elastic energy $U(s)$ contained in the cylinder beyond a distance s from the loaded end satisfies the inequality*

$$U(s) \leq U(0)\exp[-(s-l)/s_c(l)] \tag{7.136$_1$}$$

where
(i) the characteristic decay length $s_c(l)$ is given by

$$s_c(l) = \sqrt{\frac{\mu^*}{\rho\omega_0^2(l)}}\,, \tag{7.136$_2$}$$

(ii) μ^* is a constant dependent on the elastic moduli,
(iii) ρ is the mass density,
(iv) $\omega_0(l)/(2\pi)$ *is the lowest frequency of free vibration of a section of the cylinder of length l.*

The "amplitude" term $U(0)$ in (7.136$_1$) may be bounded above in terms of data using standard "energy" principles; see [100].

We now turn to the fundamental result of Oleinik and Yosifian [172]. Let Ω be a (suitably restricted) domain of the half-space $x_3 \geq 0$, x_3 being a rectangular cartesian coordinate. Suppose that for every (relevant) $t \geq 0$, $S_t = \Omega\{x : x_3 = t\}$ is not-empty and bounded (where $x = (x_1, x_2, x_3)$). We introduce the notation

$$\Omega(t_1,t_2) = \Omega \cap \{x : t_1 < x_3 < t_2\}\,,$$
$$\Gamma(t_1,t_2) = \partial\Omega \cap \{x : t_1 < x_3 < t_2\}$$

for all relevant $t_1, t_2 : t_2 > t_1$. Denote by V the linear space of all rigid displacements in $\mathbb{R}^3$, i.e. the set of all vector-valued functions of the form $a + Ax$ where (a_1, a_2, a_3) is a constant vector, A is a skew symmetric matrix with constant real elements. By η^γ $(\gamma = 1, 2, ...6)$ we denote a basis for V. Finally, let us denote the elastic energy density (per unit volume) by

$$E(u) = \frac{1}{2}c_{ijkl}u_{i,j}u_{k,l}\,. \tag{7.137}$$

We are now in a position to quote the fundamental result of Oleinik and Yosifian [172]; it is relevant to an elastic material which is not necessarily homogeneous.

Theorem 7.14 - *Let s, h be integers such that $s > h > 0$, and let $u_i(x)$ be (suitably defined, weak) solutions of* (7.132) *(in association with* (7.129) *in* $\Omega(s-h, s+1+h)$, *subject to boundary conditions* (7.133) *on* $\Gamma(s-h, s+1+h)$. *Suppose that*

$$\int_{S_{s+1}} c_{i3jk} u_{j,k} \eta_i^\gamma \, dx = 0 \tag{7.138}$$

$(\gamma = 1, 2, ..., 6)$. *Then the elastic energy contained in* $\Omega(s, s+1)$ *satisfies the decay law*

$$\int_{\Omega(s,s+1)} E(u)dx \leq e^{-Ah} \int_{\Omega(s-h,s+1+h)} E(u)dx \tag{7.139}$$

where A is a constant independent of u, s, h (A depends only on Ω and the coefficients c_{ijkl}).

We make two comments:
(i) The conditions (7.138), in effect, specify that the tractions on each plane perpendicular to the x_3 axis are self-equilibrated.
(ii) To appreciate the physical meaning of the theorem, suppose that the region considered is bounded by two plane ends $x_3 = s - h$, $x_3 = s + 1 + h$, each of which is subjected to a self-equilibrated load. The above theorem implies that the elastic field – as measured by the elastic energy – is (relatively) small in subdomains of the elastic region remote from *both* plane ends. Fichera's result [171] also refers to this type of context, but for a right cylinder.

Oleikin and Yosifian also consider more general results as well as considering displacement and mixed problems [173-174]. As to other work on Saint-Venant's principle in a three-dimensional context, we mention the work of Berdichevskii [175]; this might be envisaged as a more explicit version of Toupin's result, and Horgan and Knowles [101] have shown how it can be made more explicit still. We also refer to the work of Knops, Payne and Rionero [176]. For more comprehensive lists of references, both in two and three dimensions, one should consult [101] and [102].

7.5 SEMILINEAR EQUATION

Both the cross-sectional approach and the volume integral approach are used to derive estimates for a semilinear equation in a right cylinder in the presence of zero boundary conditions on the lateral boundary. The analysis given here is largely based on that of [177]. See also [105] for a similar analysis of somewhat different elliptic equations. The problem of elliptic equations (including semilinear equations) on either the half strip or the semi-infinite cylinder arises in many applications including the elastic membrane, combustion theory and phase transitions, and has been the subject of many studies (e.g. [178-180]).

We consider a right cylinder, and select rectangular cartesian coordinates such that the base of the cylinder lies in the (x_1, x_2) coordinate plane and contains the origin. Suppose $D(x_3)$ represents the plane cross-section at a distance x_3 from the base, and that the boundary ∂D is sufficiently smooth to admit application of the divergence theorem in the plane of the cross-section. Depending on the context, both a finite cylinder $(0 < x_3 < L)$ and an infinite cylinder $(0 < x_3 < \infty)$ are considered.

For convenience, let us write

$$g(\sigma) = |\sigma|^n$$

where n is a positive number. We consider classical solutions of

$$\nabla^2 u - ug(u^2) = 0 \qquad \text{in } D(x_3) \times (0, L) \tag{7.140}$$

subject to the boundary conditions

$$u = 0 \qquad \text{on } \partial D(x_3) \times (0, L) \tag{7.141}$$

and we shall suppose that the values of u are prescribed on the ends $x_3 = 0, L$ (finite cylinder).

Defining the L^{2p} cross-sectional measure of the solution

$$F(x_3) = \int u^{2p} dA, \tag{7.142}$$

p being a positive integer, where integration over the cross-section is understood here and subsequently. Differentiating successively we obtain

$$F'(x_3) = \int [2pu^{2p-1} u_{,3}] dA \tag{7.143}$$

$$F''(x_3) = \int [2p(2p-1)u^{2p-2}u_{,3}^2 + 2pu^{2p-1}u_{,33}] dA. \tag{7.144}$$

On using (7.140), integration by parts using (7.141), together with the inequality A4(i)(a), we obtain

$$\begin{aligned} \int u^{2p-1} u_{,33} dA &= p^{-2}(2p-1) \int (\nabla_1 u^p)^2 dA + \int u^{2p} g(u^2) dA \\ &\geq p^{-2}(2p-1)\lambda_1 F + \int u^{2p} g(u^2) dA, \end{aligned} \tag{7.145}$$

where λ_1 is the lowest "fixed membrane" eigenvalue for the cross-section. Using Hölder's inequality we obtain

$$\int u^{2p} g(u^2) dA \geq F g\{(F/A)^{1/p}\} \tag{7.146}$$

where A is the area of the cross-section. Combining (7.144) – (7.146) leads to

$$F'' \geq 2(2-p^{-1})\lambda_1 F + 2pFg\{(F/A)^{1/p}\} + 2p(2p-1)\int u^{2p-2}u_{,3}^2 dA\,. \tag{7.147}$$

If we define a modified cross-sectional (pth mean)

$$\psi_p(x_3) = F^{1/(2p)} = \left\{\int u^{2p} dA\right\}^{1/(2p)} \tag{7.148}$$

and proceed to derive an analogous inequality for it. Thus, using the foregoing, we have

$$\begin{aligned} \psi_p'' &= (2p)^{-1} F^{1/(2p)-2}[FF'' - (1-\frac{1}{2p})F'^2] \\ &\geq p^{-2}(2p-1)\lambda_1 F^{1/(2p)} + F^{1/(2p)} g\{F/A\}^{1/p}\} \qquad (7.149) \\ &+(2p-1)F^{1/(2p)-2}\left[\int u^{2p} dA \int u^{2p-2}u_{,3}^2 dA - \{\int u^{2p-1}u_{,3} dA\}^2\right]. \end{aligned}$$

Since the last term in brackets is non-negative in view of Schwarz's inequality, we obtain, on recalling (7.148),

$$\psi_p'' \geq p^{-2}(2p-1)\lambda_1\psi_p + \psi_p g[\psi_p^2/A^{1/p}]\,. \tag{7.150}$$

In view of Theorem 4.5 this gives

Theorem 7.15 - *In the context of classical solutions of* (7.140), (7.141), *the cross-sectional mean (defined by* (7.148)*) satisfies*

$$\psi_p(x_3) \leq G_p(x_3)$$

where $G_p(x_3)$ *is any smooth function satisfying*

$$G_p'' \leq p^{-2}(2p-1)\lambda_1 G_p + G_p g(G_p^2/A^{1/p}) \tag{7.151$_1$}$$

such that

$$G_p(0) \geq \psi_p(0) \quad , \quad G_p(L) \geq \psi_p(L)\,. \tag{7.151$_2$}$$

Let us note that
(i) the best G_p arising above occurs when equality signs occur in both (7.151_1), (7.151_2);
(ii) the result continues to hold if the second term on the right hand side of (7.151_1) is dropped.

Henceforward we consider a semi-infinite cylinder ($L \to \infty$) and suppose that no a *priori* conditions are specified at infinity. We proceed to prove two theorems, one a decay theorem for the pth cross-sectional mean the other an "energy" decay result.

For convenience we write

$$m_p = \{p^{-2}(2p-1)\lambda_1\}^{1/2} \tag{7.152}$$

in what follows. The following is a Phragmèn-Lindelöf type result:

Theorem 7.16 - *When a smooth solution of* (7.140), (7.141) *exists globally – in the context of a semi-infinite cylinder* $0 < x_3 < \infty$ *– one has*

$$\psi_p(x_3) \le \psi_p(0)e^{-m_p x_3}\,, \tag{7.153}$$

where m_p *is defined by* (7.152).

Proof. This result is proved by first establishing that if

$$\int u^{2p-1}u_{,3}\,dA > 0 \tag{7.154}$$

for any value of x_3, then no global solution can exist. We now establish this.

Suppose that (7.154) holds for $x_3 = x_0$ (a non-negative constant). Then it follows from the definition of ψ_p that

$$\psi_p'(x_0) > 0\,. \tag{7.155}$$

In view of (7.150) it follows that

$$\psi_p'(x_3) > 0 \tag{7.156}$$

for all $x_3 \ge x_0$. Multiplying (7.150) by ψ_p' and integrating between x_0, x_3, we obtain

$$\begin{aligned}\psi_p'^{\,2}(x_3) \ge{}& m_p^2\{\psi_p^2(x_3) - \psi_p^2(x_0)\} + \psi_p'^{\,2}(x_0)\\ &+ 2\int_{\psi_p(x_0)}^{\psi_p(x_3)} \sigma g\{\sigma^2/A^{1/p}\}\,d\sigma\end{aligned} \tag{7.157}$$

in view of (7.156). Taking the square root (bearing (7.156) in mind) we obtain

$$\int_{\psi_p(x_0)}^{\psi_p(x_3)} \frac{d\eta}{\{2\int_{\psi_p(x_0)}^{\eta} \sigma g(\sigma^2/A^{1/p})d\sigma\}^{1/2}} \geq x_3 - x_0 \tag{7.158}$$

(for $x_3 \geq x_0$). If we let $x_3 \to \infty$, we have a contradiction as the left hand side is finite. Thus we have established that

$$\psi_p' \leq 0. \tag{7.159}$$

On the other hand (7.150) implies that

$$\psi_p'' \geq m_p^2 \psi_p. \tag{7.160}$$

In view of (7.159), (7.160) it is readily established that

$$\psi_p \to 0, \quad \psi_p' \to 0 \qquad \text{as } x_3 \to \infty. \tag{7.161}$$

Multiplying (7.160) by ψ_p' (which is non-positive) and integrating from x_3 to ∞, we obtain

$$\psi_p'^{\,2} \geq m_p^2 \psi_p^2 \tag{7.162}$$

using (7.161). In view of (7.159) and the non-negativity of ψ_p it follows that

$$-\psi_p' \geq m_p \psi_p. \tag{7.163}$$

The result (7.153) follows from this.

We now establish exponential decay for an energy-type measure. The context continues to be that of a semi-infinite cylinder $0 < x_3 < \infty$.

Theorem 7.17 - *If a smooth solution of* (7.140) – (7.141) *exists globally, then the pth-order energy defined by*

$$\chi_p(x_3) = \int_{x_3}^{\infty} d\eta \int_{D(\eta)} [p^{-2}(2p-1)(\nabla u^p)^2 + u^{2p} g(u^2)] dA \tag{7.164}$$

(where p is an arbitrary positive integer) satisfies the inequality

$$\chi_p(x_3) \leq \chi_p(0) \exp\{-2(2-p^{-1})\sqrt{\lambda_1} x_3\}. \tag{7.165}$$

Proof. Let us define

$$\overline{\chi}_p(x_3) = -\int_{D(x_3)} u^{2p-1} u_{,3} dA; \tag{7.166}$$

a non-negative quantity in view of (7.159). Using the divergence theorem together with (7.140) – (7.141) we obtain

$$\overline{\chi}_p(x_3) = \overline{\chi}_p(x_0) - \int_{x_0}^{x_3} \int_{D(\eta)} [p^{-2}(2p-1)(\nabla u^p)^2 + u^{2p} g(u^2)] dA \tag{7.167}$$

for any positive x_0. On differentiating we obtain

$$\overline{\chi}_p'(x_3) = -\int_{D(x_3)} [p^{-2}(2p-1)(\nabla u^p)^2 + u^{2p} g(u^2)] dA. \tag{7.168}$$

It follows from (7.166) that

$$\begin{aligned} \overline{\chi}_p(x_3) &= -p^{-1} \int_{D(x_3)} u^p (u^p)_{,3} dA \\ &\leq p^{-1} \left[\lambda_1^{-1} \int (\nabla_1 u^p)^2 dA \int (u^p_{,3})^2 dA \right]^{1/2} \end{aligned} \tag{7.169}$$

$$\begin{aligned} &\leq (2\sqrt{\lambda_1})^{-1} p (2p-1)^{-1} \int_{D(x_3)} [p^{-2}(2p-1)(\nabla u^p)^2 + u^{2p} g(u^2)] dA \\ &= -[2\sqrt{\lambda_1}(2-p^{-1})]^{-1} \{-\overline{\chi}_p'(x_3)\}, \end{aligned} \tag{7.170}$$

where we have used Schwarz's inequality, the arithmetic-geometric inequality together with (7.168) and A4(i)(a). Integration of this latter inequality yields

$$\overline{\chi}_p(x_3) \leq \overline{\chi}_p(x_0) \exp\{-2(2-p^{-1})\sqrt{\lambda_1} x_3\}. \tag{7.171}$$

But since

$$\overline{\chi}_p(x_3) \to 0 \qquad \text{as } x_3 \to \infty \tag{7.172}$$

it is apparent that $\overline{\chi}_p(x_3)$ and $\chi_p(x_3)$ are synonymus. The theorem is thus proved.

We note that a bound for $\int u^{2p} dA$ could have been obtained by integrating (7.165) from x_3 to infinity.

To make the estimate (7.165) fully explicit one needs an upper bound, in terms of data on the end $x_3 = 0$, for $\chi_p(0)$. Assuming that $u = f$ (a continuously differentiable function of x_1, x_2) on $D(0)$, we note that

$$\chi_p(0) = -\int_{D(0)} f^{2p-1} u_{,3}\, dA\,. \tag{7.173}$$

Schwarz's inequality yields

$$\chi_p(0) \leq \{\int_{D(0)} f^{4p-2} dA\}^{1/2} \{\int_{D(0)} u_{,3}^2 dA\}^{1/2}\,. \tag{7.174}$$

We consider

$$\int_0^\infty \int_{D(x_3)} u_{,3}[\nabla^2 u - u g(u^2)] dA dx_3 = 0 \tag{7.175}$$

or

$$\int_0^\infty \int_{D(x_3)} [(\frac{1}{2} u_{,3}^2)_{,3} + \nabla_1 \cdot (u_{,3} \nabla_1 u) - \frac{1}{2}\{(\nabla_1 u)^2\}_{,3} - \frac{1}{2}(u^2)_{,3} g(u^2)] dA dx_3 = 0. \tag{7.176}$$

Integration, including use of the divergence theorem together with (7.141), yields

$$\int_{D(0)} u_{,3}^2 dA = \int_{D(0)} (\nabla_1 f)^2 dA + \int_{D(0)} \{\int_0^{f^2} g(\eta) d\eta\} dA \tag{7.177}$$

bearing in mind the behaviour at infinity implied by the finiteness of $\chi_p(0)$. We thus have

Theorem 7.17a - *An explicit upper bound for the quantity $\chi_p(0)$, occurring in Theorem 7.17, is given by (7.174) in association with (7.177).*

7.6 QUASILINEAR P.D.E.

There is a very important sequence of papers by Payne and Horgan [181, 104, 182-183] concerning decay type estimates for a class of quasi-linear P.D.E.s. Much of the motivation for these papers lies in the need to provide a version of Saint-Venant's principle appropriate to nonlinear elastic strips and beams. The essential idea is to prove that in the context of "long" regions with homogeneous boundary conditions on the long/lateral sides, solutions of the complete equations, approach (in a suitable sense) – away from the ends – solutions of simplified equations (of lower dimensionality). We summarize one result [182] which may be regarded as representative.

Here we are concerned with second-order quasilinear equations in two dependent variables of the form

$$[\rho(q^2)u_{,\alpha}]_{,\alpha} + 2k = 0\,, \qquad (q^2 \equiv u_{,\beta}u_{,\beta}) \tag{7.178}$$

where $k > 0$ is a constant. In (7.178) the usual summation convention is employed with subscripts preceded by a comma denoting partial differentiation with respect to the corresponding Cartesian coordinate. Equations of the form (7.178) occur in geometry as well as in problems of nonlinear continuum mechanics.

Consider a Dirichlet problem for (7.178) on the rectangular region $R^* = \{(x_1, x_2) : 0 < x_1 < 2l,\, 0 < x_2 < h\}$, where $l >> h$. On the long sides of the rectangle, we have homogeneous boundary conditions so that

$$u(x_1, 0) = u(x_1, h) = 0\,, \qquad 0 \leq x_1 \leq 2l \tag{7.179}$$

while the data at the ends $x_1 = 0, x_1 = 2l$ are assumed to be symmetrically distributed so that

$$u(0, x_2) = f(x_2)\,, \qquad u(2l, x_2) = f(x_2)\,, \qquad 0 \leq x \leq h\,, \tag{7.180}$$

where the prescribed function f is sufficiently smooth and satisfies $f(0) = f(h) = 0$. In view of the form of the differential equation (7.178) and the symmetry of the data in (7.180), we conclude that if (7.178) is elliptic, then the solution $u(x_1, x_2)$ is such that $u(x_1, x_2) = u(2l - x_1, x_2)$ so that

$$u_{,1} = 0 \quad \text{on } x_1 = l \qquad 0 \leq x_2 \leq h\,. \tag{7.181}$$

Furthermore, sufficiently far away from the ends $x_1 = 0$, $x_1 = 2l$, one might expect $u(x_1, x_2)$ to be well approximated by the solution $v(x_2)$ of the one-dimensional problem

$$\frac{d}{dx_2}\left[\rho(p^2)\frac{dv}{dx_2}\right] + 2k = 0\,, \qquad 0 \leq x_2 \leq h\,, \tag{7.182}$$

$$v(0) = 0\,, \qquad v(h) = 0\,, \tag{7.183}$$

where $p^2 \equiv (dv/dx_2)^2$. To investigate this issue, it is sufficient to confine attention to the half-rectangle $R = \{(x_1, x_2) : 0 < x_1 < l\,,\; 0 < x_2 < h\}$ and so we consider solutions of (7.178) on R subject to the boundary conditions

$$u(x_1, 0) = u(x_1, h) = 0\,, \qquad 0 \leq x_1 \leq l\,, \tag{7.184}$$

$$u_{,1} = 0 \quad \text{on } x_1 = l\,, \qquad 0 \leq x_2 \leq h\,, \tag{7.185}$$

$$u(0, x_2) = f(x_2)\,, \qquad 0 \leq x_2 \leq h\,, \tag{7.186}$$

where $l >> h$. We shall examine the spatial evolution of $u(x_1, x_2)$ to the solution $v(x_2)$ of the one-dimensional problem (7.182) – (7.183). We assume the existence of classical solutions $u \in C^2(R) \cap C^1(\overline{R})$ of the problem (7.178), (7.184) – (7.186).

Payne and Horgan establish an energy-decay estimate for solutions of (7.178), (7.184) – (7.186). They show that the energy measure ("volume integral" measure)

$$E(z) = \int_{R_z} \rho(q^2) w_{,\alpha} w_{,\alpha} \, dA \,, \qquad (q^2 = u_{,\alpha} u_{,\alpha}) \tag{7.187}$$

contained in the subdomain $R_z = \{(x_1, x_2) : 0 \le z \le x_1 < l \,,\ 0 < x_2 < h\}$ has exponential decay in z, where the function $w(x_1, x_2)$ is defined by

$$w(x_1, x_2) = u(x_1, x_2) - v(x_2) \tag{7.188}$$

and v is the solution to the one-dimensional problem (7.182) – (7.183). Thus $E(z)$ can be viewed as a "weighted energy" associated with the difference (7.188) between $u(x_1, x_2)$ and $v(x_2)$.

It is assumed that there exist positive constants m, c, d such that for all values of the arguments s, t we have

$$\rho^{-1} \ge m > 0 \tag{7.189}$$

$$|\rho^{-1}(s^2) - \rho^{-1}(t^2)| \, |t| \, \rho(t^2) \le c|t - s| \,, \qquad 0 < c < 1 \tag{7.190}$$

$$|\rho^{-1}(s^2) - \rho^{-1}(t^2)| \, [\rho(t^2) + \rho(s^2)] \le d\rho(t^2)\rho(s^2)|s^2 - t^2| \,. \tag{7.191}$$

In applications, the argument t is taken to be p, where $p = |dv/dx_2|$, and c, d will, in general, depend on the maximum value of p on $[0, h]$. It should be noted that (7.189) – (7.191) do *not* require that equation (7.178) be elliptic, that is, $\rho + 2\rho' q^2 \ge 0$ ($\rho' \equiv d\rho/dq^2$) for all solutions u at all points of R, even though the issues of concern here are of interest primarily for elliptic equations.

Horgan and Payne [182] establish the following "volume integral" type estimate

Theorem 7.18 - *The energy measure* (7.187)

$$E(z) = \int_{R_z} \rho(q^2) w_{,\alpha} w_{,\alpha} \, dA$$

where

$$w(x_1, x_2) = u(x_1, x_2) - v(x_2)$$

in which
(*a*) $u(x_1, x_2)$ *is a classical solution of* (7.178), (7.184) – (7.186),

(b) $v(x_1,x_2)$ is the solution of the one-dimensional problem (7.182) – (7.183), satisfies the exponential decay estimate

$$E(z) \leq Ke^{-2(\pi/h)\nu z}, \qquad z \geq 0,$$

where

$$\nu = \nu(p) = \frac{m(1-c)}{F(\alpha,p^2,\gamma)}$$

and

$$K = E_0 \exp\left[\frac{2m\pi E_0\alpha(1-c)(1+\gamma)}{F^2(\alpha,p^2,\gamma)h^2}\right],$$

wherein
(i) m, c, d are the constants appearing in (7.189) – (7.191), and $\gamma > 0$ is an arbitrary parameter;
(ii)

$$F(\alpha,p^2,\gamma) = \frac{1}{h}\int_0^h \left[\frac{\alpha p^2}{\gamma}\rho(p^2) + \rho^{-1}(p^2)\right] dx_2\,;$$

(iii) the total energy $E_0 = E(0)$ is boundable above, in terms of data, thus:

$$\begin{aligned} E_0 \leq\, & 2(1-c)^{-1}[-2k(1-e^{-l})\int_0^h (f-v)dx_2 \\ & + \{(1-c)4m\}^{-1}\int_0^h (f_{,2}-v_{,2})^2 dx_2 \\ & + \{(1-c)4m\}^{-1}\int_0^h (f-v)^2 dx_2 \\ & + (4m)^{-1}\int_0^h f_{,2}^2 dx_2] \end{aligned}$$

They also establish similar estimates under hypotheses other then those contained in (7.189) – (7.191).

7.7 EXERCISES

Exercise 7.1 - Consider a two dimensional region R of the type described in the second paragraph of Section 7.2. Consider classical solutions of (assuming $r(x)$ positive)

$$\{r(x)\psi_x\}_x + r(x)\psi_{yy} = 0$$

therein, subject to $\psi = 0$ on the lateral boundaries C_+, C_- and on the plane end Γ_L (i.e. $x = L$), and such that $\psi =$ specified on Γ_0 (i.e. $x = 0$). Prove that the cross-sectional measure defined by

$$F(x) = \int \psi^2 \, dy$$

satisfies the inequality

$$F'' + (2\lambda + r^{-1}r')F' \geq 0$$

where

$$\lambda = \pi l^{-1} ,$$

$l = l(x)$ being the width of the strip at abscissa x. Hence prove that

$$F(x) \leq F(0)\{\int_x^L r^{-1}e^{-2\lambda x}dx\}\{\int_0^L r^{-1}e^{-2\lambda x}dx\}^{-1} .$$

Investigate particular circumstances in which the equality sign holds in this inequality.

Exercise 7.2 - Referring to smooth solutions in the elastic right cylinder of the displacement equations (7.29_1) subject to the homogeneous boundary conditions (7.29_2), prove the conservation law

$$\int_{D(x_3)} \{(\alpha+1)u_{3,3}^2 - \alpha u_{\beta,\beta}u_{\gamma,\gamma} - u_{\beta,\gamma}u_{\beta,\gamma} - u_{3,\gamma}u_{3,\gamma} + u_{\gamma,3}u_{\gamma,3}\}dA$$
$$= E(\text{constant})$$

(where summation over repeated Greek suffices – which take the values 1, 2 – is understood) by considering

$$\int_{D(x_3)} u_{i,3}(u_{i,jj} + \alpha u_{j,ji})dA = 0 ,$$

integrating by parts, and proving that the resulting quantity can be expressed as a derivative with respect to x_3 of the conserved quantity. Prove that the conservation law continues to hold if the boundary conditions (7.29_2) on the lateral boundary are replaced by the traction-free (Neumann) conditions

$$(\alpha - 1)u_{j,j}n_i + (u_{i,j} + u_{j,i})n_j = 0$$

where $(n_1, n_2, 0)$ denote the rectangular cartesian components of the unit outward normal to the boundary surface.
Derive analogous results in a two dimensional context.

Exercise 7.3 - Prove a generalization of the conservation law (given in the previous question) for an *inhomogeneous* elastic right cylinder, as follows: consider smooth solutions, in the cylinder, of the equations

$$\tau_{ij,j} = 0$$

where

$$\tau_{ij} = \lambda u_{k,k}\delta_{ij} + \mu(u_{i,j} + u_{j,i}),$$

λ, μ being smooth functions of x_1, x_2, subject to the boundary conditions

$$u_i = 0 \qquad \text{or} \qquad \tau_{ij}n_j = 0$$

on the lateral surface.

Exercise 7.4 - Prove Theorem 7.8b (case 2) by establishing, in the first instance, that the inequality

$$F'' - m^2 F \geq 0 \qquad \text{in } x_3 \in \{0, L\}$$

successively implies

$$\{e^{-mx_3}(F' + mF)\}' \geq 0,$$
$$F' + mF \leq e^{-m(L-x_3)}(F' + mF)_L,$$
$$[e^{mx_3}F - \frac{1}{2}e^{-m(L-2x_3)}(m^{-1}F' + F)_L]' \leq 0$$

where the subscript L means that the relevant quantity is evaluated at $x_3 = L$.

Exercise 7.5 - Consider a rectangular region $0 < x_2 < 1$ of an incompressible, inhomogeneous, isotropic linear elastic material, in plane strain, the boundaries $x_2 = 0, 1$ of which are subject to zero displacement conditions $u_1 = u_2 = 0$. The relevant differential equations are

$$-p_{,1} + \mu u_{1,\alpha\alpha} + \dot{\mu}(u_{1,2} + u_{2,1}) = 0,$$

$$-p_{,2} + \mu u_{2,\alpha\alpha} + 2\dot{\mu} u_{2,2} = 0\,,$$
$$u_{\alpha,\alpha} = 0\,,$$

where u_α ($\alpha = 1,2$) are the displacement components, p the "pressure", and where the elastic coefficient (shear modulus) μ is a smooth (positive) function of x_2; differentiation with respect to x_2 is denoted by a superposed dot.
Prove that the cross-sectional measure

$$F(x_1) = \int \mu(u_{1,2}^2 + u_{2,1}^2)dx_2$$

satisfies

$$F''(x_1) = 2\int \mu(u_{2,11}^2 + u_{1,21}^2 + 2u_{2,12}^2)dx_2 + \int \ddot{\mu}\, u_{2,1}^2 dx_2\,.$$

Hence prove that $F^{1/2}(x_1)$ is a convex function provided that μ is a convex function of x_2. [This latter is assumed henceforward.]
Suppose that the normal displacement and the complementary (i.e. tangential) shear stress are prescribed on the remaining edges $x_1 = 0, x_1 = L$. Determine an upper bound for $F^{1/2}(x_1)$ in terms of data.
Show how a pointwise pper bound for u_1, reflecting position, may be derived in these circumstances, by proving (with the help of Schwarz's inequality) that

$$u_1^2(x_1, x_2) \le \left\{ \frac{\displaystyle\int_0^{x_2} \mu^{-1} dx_2 \int_{x_2}^1 \mu^{-1} dx_2}{\displaystyle\int_0^1 \mu^{-1} dx_2} \right\} F(x_1)\,.$$

Exercise 7.6 - Consider a right cylinder (with x_3 axis parallel to the generators) consisting of material of the type discussed in the previous section. The appropriate P.D.E. are (Latin indices taking the values $1,2,3$ and rectangular cartesian coordinates being understood)

$$-p_{,1} + \mu u_{1,jj} + \dot{\mu}\,(u_{1,2} + u_{2,1}) = 0\,,$$
$$-p_{,2} + \mu u_{2,jj} + 2\dot{\mu} u_{2,2} = 0\,,$$
$$-p_{,3} + \mu u_{3,jj} + \dot{\mu}\,(u_{2,3} + u_{3,2}) = 0\,,$$

$$u_{i,i} = 0 .$$

Assuming that the conditions $u_i = 0$ again obtain on the lateral boundary, prove that the cross-sectional measure

$$F(x_3) = \int \mu(u_{\alpha,3}u_{\alpha,3} + u_{3,\alpha}u_{3,\alpha})dA,$$

satisfies

$$F''(x_3) = 2\int \mu[u_{3,\alpha 3}u_{3,\alpha 3} + u_{\alpha,33}u_{\alpha,33} + u_{\beta,3\alpha}u_{\beta,3\alpha} + u_{3,33}^2]dA + \int \ddot{\mu}\, u_{2,3}^2 dA .$$

Hence prove that $F^{1/2}(x_3)$ is a convex function provided that $\mu(x_2)$ is a convex function. Show how to deduce an upper bound for $F^{1/2}(x_3)$ if the normal displacement component u_3 and the complementary shear stress components $\tau_{3\alpha} = \mu(u_{3,\alpha} + u_{3,\alpha})$ are prescribed on the plane ends, $x_3 = 0, L$ of the cylinder. [In the above, Greek indices take the values 1,2.]

Exercise 7.7 - Obtain conservation laws in the contexts described in the previous two questions.

Exercise 7.8 - In obtaining Theorem 7.9, the term involving $\int g u_{2,12}^2 dx_2$ arising in (7.77) was "split" into two parts. Can an improvement of Theorem 7.9 be obtained by, also, similarly "splitting" the term $\int g u_{1,22}^2 dx_2$ arising in (7.77) and treating it similarly?

Exercise 7.9 - Consider a linear, homogeneous isotropic elastic material which sustains an axisymmetric deformation: the r, z components of displacement $u(r,z), w(r,z)$ satisfy the P.D.E.s

$$(\alpha+1)(u_{rr} + r^{-1}u_r - r^{-2}u^2) + u_{zz} + \alpha w_{rz} = 0,$$

$$\alpha(u_r + r^{-1}u)_z + (w_{rr} + r^{-1}w_r) + (\alpha+1)w_{zz} = 0,$$

where α is a positive constant. The material is bounded by planes $z = 0, 1$ and by surfaces $r = r_0(> 0), r = r_1(> 0)$. Assume that

$$u = w = 0 \qquad \text{on } z = 0, 1 .$$

Defining the cross-sectional measure deformation by

$$F(r) = \int_0^1 [w_r^2 + u_z^2]dz,$$

derive second order differential inequalities for $F(r)$ of the convexity/generalized convexity type. If u_z and w_r (or their equivalent) is specified on the surfaces $r = r_0, r = r_1$, shown how $F(r)$ may be bounded above in terms of data. How can one deduce pointwise bounds for $|u(r,z)|$ in terms of data?

Exercise 7.10 - Consider again the plane strain context of Question 5 above where now, however, the plane boundaries are subject to the traction free boundary conditions

$$\tau_{22} = \tau_{21} = 0$$

on $x_2 = 0, 1$, and the material is assumed incompressible. The stress components are given in terms of the Airy stress function $\phi(x_1, x_2)$ by

$$\tau_{11} = \phi_{,22} \qquad \tau_{22} = \phi_{,11} \qquad \tau_{12} = -\phi_{,12}$$

where ϕ satisfies

$$(\varepsilon\phi_{,11})_{,11} + 2(\varepsilon\phi_{,12})_{,12} + (\varepsilon\phi_{,22})_{,22} - \ddot{\varepsilon}\,\phi_{,11} = 0$$

with $\varepsilon = \varepsilon(x_2)$ being an elastic modulus.
Prove that the cross-sectional measure

$$F(x) = \int_0^1 \varepsilon(\tau_{11}^2 + \tau_{22}^2)dx_2$$

is such that $F^{1/2}(x)$ is convex, assuming $\varepsilon(x_2)$ is a positive, convex function. Also prove that the cross-sectional measure

$$F(x_1) = \int_0^1 \varepsilon(\tau_{11}^2 + \tau_{22}^2 + \alpha\tau_{12}^2)dx_2$$

where α is an arbitrary number such that $0 \le \alpha < 2$, is a convex function provided that $\varepsilon(x_2)$ is as in the previous paragraph.

Exercise 7.11 - Referring to solutions of the P.D.E. mentioned in the last question (subject to the boundary conditions $\phi_{,11} = \phi_{,12} = 0$ on $x_2 = 0, 1$) prove the conservation law

$$\int \{\varepsilon(2\phi_{,1}\phi_{,111} - \phi_{,11}^2 + \phi_{,22}^2 - 2\phi_{,12}^2) - \ddot{\varepsilon}\,\phi_{,1}^2\}dx_2 = E\,(\text{constant}).$$

Exercise 7.12 - Seek a generalization of Theorem 7.12 relevant to the following: ϕ is a sufficiently smooth solution of

$$a\phi_{xxxx} + 2b\phi_{xxyy} + c\phi_{yyyy} = 0$$

in a semi-infinite strip $0 < x < \infty$, $0 < y < 1$, subject to

$$\phi = \phi_y = 0 \qquad \text{on } y = 0, 1$$

and asymptotic conditions analogous to (7.122_1).
The quantities a, b, c are constants such that

$$a > 0, \quad c > 0, \quad ac - b^2 > 0\,.$$

Chapter eight

PARABOLIC EQUATIONS

8.1 INTRODUCTION

Difffusion phenomena of the real world (heat diffusion, diffusion of epidemics in populations, diffusion of a pollutant, fluid motions,...) are modelled by parabolic equations. Therefore it is perfectly understandable that some of the most celebrated P.D.E.s are of parabolic type (heat conduction equations, Navier-Stokes equations,...). We here present some typical aspects of the qualitative analysis of parabolic equations. We begin (sect.8.2) by showing how the boundedness of solutions can be obtained together with its importance for the existence of global solutions. Next (sect.8.3) we perform a stability analysis of a parabolic P.D.E. arising in anisotropic magnetohydrodynamics, representing an inhomogeneous diffusion. After a short introduction (subsect.8.4.1), the long time behaviour of the solution is considered (subsect.8.4.2), some instability results are presented (subsect.8.4.3) and travelling wave solutions are introduced (subsect.8.4.4). Sections 8.5-8.6 are devoted to the energy stability of viscous flows in bounded and unbounded regions respectively. The last section (sect.8.7) introduces the general problem of the onset of natural convection (Bènard problem) and studies the case where the horizontal layer is rotating about a vertical axis (Bènard problem with rotation).

8.2 BOUNDEDNESS AND GLOBAL EXISTENCE OF SOLUTIONS TO A PARABOLIC EQUATION IN GRADIENT FORM

For the sake of simplicity, let us begin (sects.8.2-8.3) by considering parabolic P.D.E.s in one dimension. Consider the I.B.V.P.

$$u_t + \frac{\partial}{\partial x} F'(u) = D u_{xx} \qquad (x,t) \in]0,1[\times \mathbb{R}_+ \,, \tag{8.1}$$

$$u(x,0) = u_0(x) \qquad x \in [0,1] \,, \tag{8.2}$$

$$u(0,t) = u(1,t) = 0 \qquad t \geq 0 \,, \tag{8.3}$$

where D is a positive constant and $F : R \to R$, $u_0 : [0,1] \to R$ is a smooth function such that $F(0) = u_0(0) = u_0(1) = 0$. The following theorem holds.

Theorem 8.1 - *If*

$$1 + \max_{|\xi| \leq m} |F''(\xi)| \leq \alpha m^\beta \qquad \forall m > 0 \,, \tag{8.4}$$

where α and $\beta(< 2)$ are positive constants, then there exists a positive number m_0, independent of t, such that

$$|u(x,t)| \leq m_0 \,, \tag{8.5}$$

on any rectangle $[0,1] \times [0,T]$ on which there exists a strong solution to the I.B.V.P. (8.1) – (8.3).

Proof. Setting $\|\cdot\| = \|\cdot\|_2$, from (8.1) – (8.3), it easily follows that

$$\frac{1}{2}\frac{d}{dt}\|u_x\|^2 = \int_0^1 u_x u_{xt}\, dx = -\int_0^1 u_t u_{xx}\, dx =$$

$$= -D\|u_{xx}\|^2 + \int_0^1 F''(u) u_x u_{xx}\, dx \,. \tag{8.6}$$

But

$$F''(u) u_x u_{xx} \leq \frac{\varepsilon}{2} u_{xx}^2 + \frac{1}{2\varepsilon} |F''|^2 u_{xx}^2 \,, \qquad \varepsilon > 0 \,, \tag{8.7}$$

hence

$$\frac{1}{dt}\frac{d}{dt}\|u_x\|^2 \leq -\left(D - \frac{\varepsilon}{2}\right) \|u_{xx}\|^2 + \frac{1}{2\varepsilon} \int_0^1 |F''|^2 u_x^2 dx \,. \tag{8.8}$$

Setting

$$m = \max_{\mathbb{R}_T} |u(x,t)| \,, \qquad k = \max_{|\xi| \leq m} |F''(\xi)| \tag{8.9}$$

and choosing $\varepsilon = D/2$, we have

$$\|u_x(t)\|^2 \le \|u_{0,x}\|^2 + \frac{k^2}{D}\int_0^t \|u_x\|^2 d\tau\,, \qquad t \in [0,T]\,. \tag{8.10}$$

On the other hand, it turns out that

$$\frac{1}{2}\frac{d}{dt}\|u\|^2 = \int_0^1 u(Du_x - F')_x dx = -D\|u_x\|^2 + \int_0^1 F' u_x dx =$$

$$= -D\|u_x\|^2 + \int_0^1 \frac{\partial}{\partial x}F(u)dx = -D\|u_x\|^2\,, \tag{8.11}$$

and hence

$$\|u(t)\|^2 + 2D\int_0^t \|u_x\|^2 d\tau = \|u_0\|^2\,, \qquad t \in [0,T]\,. \tag{8.12}$$

Coupling (8.12) with (8.10) it follows that

$$\|u_x(t)\|^2 \le \|u_{0,x}\|^2 + \frac{k^2}{2D^2}\|u_0\|^2\,. \tag{8.13}$$

Since $u(0,t) = 0$, $\forall\, t \ge 0$, implies

$$|u(x,t)| = |\int_0^x u_x(s)ds| \le (\int_0^x ds)^{1/2} \cdot (\int_0^x u_x^2(s)ds)\,, \tag{8.14}$$

we have

$$u^2(x,t) \le \|u_x(0)\|^2 + \frac{k^2}{2D^2}\|u_0\|\,, \qquad \forall\,(x,t) \in I\!R_T\,. \tag{8.15}$$

Therefore taking into account (8.4) and introducing the positive constant γ, independent of t,

$$\gamma = \max\{\|u_x(0)\|^2, \frac{1}{2D^2}\|u_0\|^2\} \tag{8.16}$$

it follows that

$$m \le m_0 = (\alpha\gamma)^{1/\beta} \tag{8.17}$$

which implies (8.5).

Remark 8.1 - *The interest in proving that the solution is a priori bounded in the L_∞-norm on $[0,T]$, $T < \infty$, is due to the fact that this ensures global existence in time. For this matter and for the proof of a theorem analogous to theorem* 8.1, *concerning a quasi-linear parabolic system, we refer to chapters* 14 *and 21 of [5]. Let us add that following the procedures of 2.7; we easily obtain that $||u||_\infty \to 0$ as $t \to \infty$. In fact (8.13) and Remark 2.8 imply*

$$\sup_{x\in[0,1]} |u| \leq \sqrt{2}m^{1/4}||u||_2^{1/2}$$

with m given by the R.H.S. of (8.13). On the other hand, (8.11) and (2.26), imply $||u||_2^2 \leq ||u_0||_2^2 \cdot \exp(-\alpha t)$, with $\alpha = -2D\pi^2$.

8.3 STABILITY OF THE MAGNETIC FIELD OF COUETTE-POISEUILLE FLOWS IN ANISOTROPIC MAGNETO-HYDRO-DYNAMICS

We perform here a stability analysis for an equation of the type considered in {2.7, g)} which is encountered in anisotropic magnetohydrodynamics (M.H.D.).

Suppose that the fluid occupies the horizontal layer $-1 \leq z \leq 1$, let $\mathbf{e}$ denote the horizontal unit vector, while $\mathbf{v}$ and $\mathbf{H}$ respectively denote the velocity and the magnetic field. The boundary conditions

$$\mathbf{v}(-1) = 0, \qquad \mathbf{v}(1) = V\mathbf{e}$$
$$\mathbf{H}(-1) = H_1\mathbf{e}, \qquad \mathbf{H}(1) = H_2\mathbf{e}$$

where V, H_1 and H_2 are prescribed constants, are required to be satisfied by the laminar flow

$$\mathbf{v} = v(z)\mathbf{e}, \qquad \mathbf{H} = H(z)\mathbf{e},$$

– called Couette-Poiseuille flow.

It may be shown that the velocity field of a basic Couette-Poiseuille flow is given by

$$v(z) = z^2 + az + b$$

with a and b being determined by the boundary conditions, while the magnetic field is the unique real solution of the algebraic equation

$$\beta_I H^3 + 3\eta_e H + c_1 z + c_2 = 0$$

where $c_i (i = 1, 2)$ are given by

$$\begin{cases} c_1 = \dfrac{1}{2}[\beta_I(H_1^3 - H_2^3) + 3\eta_e(H_1 - H_2)], \\ c_2 = -\dfrac{1}{2}[\beta_I(H_1^3 + H_2^3) + 3\eta_e(H_1 + H_2)], \end{cases}$$

and β_I (*ion-slip coefficient*) and η_e (*magnetic viscosity*) are positive constants.

If one considers the stability of the Couette-Poiseuille flow with respect to laminar perturbations, the following I.B.V.P. for the perturbations $\mathbf{h} = h(z)\mathbf{e}$ to the magnetic field $\mathbf{H}$ arises [90]

$$h_t = [(1 + RH^2)h + RHh^2 + \frac{1}{3}Rh^3]_{zz}\,, \qquad z \in (-1,1),\, t \geq 0\,, \tag{8.18}$$

$$h(\pm 1, t) = 0\,, \tag{8.19}$$

$$h(z,0) = h_0(z)\,, \tag{8.20}$$

where R is a positive constant.[1]

Setting

$$F = (1 + RH^2)h + RHh^2 + \frac{1}{3}Rh^3\,, \tag{8.21}$$

$$G(h) = \int_0^h F(H,s)ds\,, \tag{8.22}$$

it follows that

$$G(h) = \frac{h^2}{12}[6 + 2RH^2 + R(2H + h)^2] \tag{8.23}$$

which implies

$$G(0) = 0, \qquad G(h) > 0, \qquad h \neq 0\,. \tag{8.24}$$

On introducing the Liapunov function (2.70), it follows that

$$\mathcal{E} = \int_{-1}^1 G[H(z), h(z,t)]dz \geq \frac{1}{2}\|h\|_2^2\,, \tag{8.25}$$

$$\dot{\mathcal{E}} = -\int_{-1}^1 (F_z)^2 dz\,. \tag{8.26}$$

From (8.19) it follows that

$$z = \pm 1 \rightarrow F = 0\,,$$

hence the Poincaré inequality (2.26) implies

$$\dot{\mathcal{E}} \leq -\pi^2 \int_{-1}^1 F^2 dz\,. \tag{8.27}$$

---

[1] The constant R is given by $R = \dfrac{\beta_I}{\eta_e}\tilde{H}^2$, $\tilde{H}$ being a reference magnetic field.

But it is an easy matter to verify that

$$F = \frac{h}{3}[3 + 3RH(h+H) + Rh^2] = \frac{h}{6}[6 + 6RH(h+H) + 2Rh^2]$$
$$= \frac{h}{6}\{6 + RH^2 + R[(H+h)^2 + (2H+h)^2]\},$$

$$F^2 = \frac{h^2}{36}\{6 + RH^2 + R[(H+h)^2 + (2H+h)^2]\}^2 \geq$$
$$\geq \frac{h^2}{36}\{6 + RH^2 + R(2H+h)^2\}^2$$
$$\geq \frac{h^2}{36\sqrt{2}}\{6 + 2RH^2 + R(2H+h)^2\}^2,$$

and immediately it follows that

$$F^2 \geq \sqrt{2}G\,. \tag{8.28}$$

Therefore from (8.27) we obtain

$$\dot{\mathcal{E}} \leq -\sqrt{2}\pi^2\mathcal{E}$$

i.e.

$$\mathcal{E}(t) \leq \mathcal{E}(0)e^{-\sqrt{2}\pi^2 t}\,, \qquad \forall t > 0\,. \tag{8.29}$$

In view of (8.25) it follows that

$$\|h\|_2^2 \leq 2\mathcal{E}(0)e^{-\sqrt{2}\pi^2 t}\,, \tag{8.30}$$

which ensures the asymptotic stability in the L_2-norm of the basic magnetic field **H** with respect to laminar perturbations. One may similarly establish the stability of the basic Couette-Poiseuille velocity field with respect to such perturbations. For this and further results we refer the interested reader to [90].

8.4 ASYMPTOTIC BEHAVIOUR OF REACTION-DIFFUSION SYSTEMS

8.4.1 Introduction

Let Ω be an open bounded connected subset of $\mathbb{R}^m$, $m \in \mathbb{N}_+$, with a sufficiently smooth boundary and D a positive (constant) diagonal matrix. As remarked in $\{1.12, c)\}$, the equation

$$\mathbf{u}_t = D\Delta\mathbf{u} + \mathbf{f}(\mathbf{u}) \qquad \text{in } \Omega \times \mathbb{R}_+ \tag{8.31}$$

i.e. the system of semilinear parabolic equations ($D_i = \text{const.} > 0$)

$$\frac{\partial u_i}{\partial t} = D_i \Delta u_i + f_i(u_1, u_2, ..., u_n)\,, \qquad i \in \{1, 2, .., n\}\,, \tag{8.32}$$

models the diffusion of many phenomena. To (8.32) we append the I.B.C.

$$\mathbf{u}(\mathbf{x}, 0) = \mathbf{u}_0(\mathbf{x}) \qquad \text{in } \overline{\Omega}\,, \tag{8.33}$$

$$\mathbf{u}(\mathbf{x}, t) = \mathbf{u}_1(\mathbf{x}, t) \qquad \text{on } \Sigma_1 \times \mathbb{R}_+\,, \tag{8.34}$$

$$\frac{d\mathbf{u}}{d\mathbf{n}} = \mathbf{u}_2 \qquad \text{on } \Sigma_2 \times \mathbb{R}_+\,, \tag{8.35}$$

where $\partial\Omega = \Sigma_1 \cap \Sigma_2, \Sigma_1 \cup \Sigma_2 = \emptyset$, $\mathbf{n} =$ outward unit normal$\}$.

Theorems of (local in time) existence and uniqueness of generalized and smooth solutions to the I.B.V.P. (8.32) – (8.35) are well known in the literature {see, for instance, [87] chapter V, [5] chapters XI and XIV and [184]}. Further, when the solutions are a-priori bounded {[5] Theor. 14.4}, global existence can be obtained. In the section 8.2, theorem 8.1, represents an example of conditions ensuring the boundedness of the solutions of the I.B.V.P. (8.1) – (8.3). Our aim, in this section, is to:

(i) compare the solutions of the I.B.V.P. (8.32) – (8.35) to the solution of the *kinetic equation*

$$\frac{d\mathbf{u}}{dt} = \mathbf{f}(\mathbf{u})\,; \tag{8.36}$$

(ii) present some cases of instabilities;

(iii) introduce, as an example of travelling wave solutions for the reaction diffusion systems, the travelling wave solutions for the Fisher equation.

8.4.2 Asymptotic homogenization

We consider here the case of homogeneous Neumann boundary conditions:

$$\Sigma_1 = \emptyset, \quad \Sigma_2 = \partial\Omega, \quad \mathbf{u}_2 = 0\,. \tag{8.37}$$

Our aim is to recall some conditions ensuring that $\mathbf{u}$, as $t \to \infty$, tends asymptotically – in some norm – to its spatial average ($|\Omega|$= measure of Ω)

$$\mathbf{U}(t) = \frac{1}{|\Omega|}\int_{\Omega} \mathbf{u}(\mathbf{x},t)d\mathbf{x}, \tag{8.38}$$

and to link $\mathbf{U}$ to the solutions of the kinetic equation (8.36).

Denoting by d the quantity $\inf(D_1, D_2, .., D_n)$ and by λ the first eigenvalue of $-\Delta$ on Ω with homogeneous Neumann boundary conditions, the following theorem holds

Theorem 8.2 - *Let I be a bounded, invariant region, and let*

$$m = \max_{I} |\nabla_{\mathbf{u}} f|. \tag{8.39}$$

If

$$\mathbf{u}_0 \in I, \qquad \lambda d - m > 0, \tag{8.40}$$

then there exists a positive constant c such that $||\nabla\mathbf{u}||_{L^2(\Omega)}$ and $||\mathbf{u}-\mathbf{U}||_{L_\infty(\Omega)}$ does not exceed $ce^{-(\lambda d - m)t}$. Further $\mathbf{U}$ satisfies

$$\frac{d\mathbf{U}}{dt} = \mathbf{f}(\mathbf{U}) + g(t), \qquad \mathbf{U}(0) = \frac{1}{|\Omega|}\int_{\Omega} \mathbf{u}_0(\mathbf{x})d\mathbf{x}, \tag{8.41}$$

with

$$g(t) \le ce^{-(\lambda d - m)t}. \tag{8.42}$$

Proof. We refer to [5] pp.223-225 (see also exercises 8.2-8.3.)

Remark 8.2 - *Let the assumptions of theorem 8.2 hold. Then it follows that i) there cannot exist a nonconstant steady solution of the (elliptic) system*

$$\begin{cases} D\Delta\mathbf{u} + \mathbf{f}(\mathbf{u}) = 0 & \text{in } \Omega, \\ \dfrac{d\mathbf{u}}{d\mathbf{n}} = 0 & \text{on } \partial\Omega; \end{cases} \tag{8.43}$$

ii) the solutions of the I.B.V.P.

$$\begin{cases} \mathbf{u}_t = D\Delta\mathbf{u} + \mathbf{f}(\mathbf{u}) & \text{in } \Omega \times I\!R_+, \\ \mathbf{u}(\mathbf{x},0) = \mathbf{u}_0(\mathbf{x}) & \text{in } \Omega, \\ \dfrac{d\mathbf{u}}{d\mathbf{n}} = 0 & \text{on } \partial\Omega, \end{cases} \tag{8.44}$$

and the solution of the kinetic equation (8.36), *have the same* ω*-limit sets.*

The property i) is an immediate consequence of the fact that solutions of the B.V.P. (8.43) *depending on* $\mathbf{x}$ *cannot tend, as* $t \to \infty$, *to solutions independent of* $\mathbf{x}$. *For the proof of ii) we refer to* [185].

8.4.3 Instability of the steady nonconstant solutions of reaction - diffusion equations with homogeneous Neumann boundary conditions

Let us consider the scalar reaction-diffusion equation

$$u_t = \Delta u + f(u) \qquad (\mathbf{x}, t) \in \Omega \times \mathbb{R}_+ \tag{8.45}$$

where Ω is a smooth domain in $\mathbb{R}_n$ and $f \in C^1(\mathbb{R})$. Our aim is to introduce – as a complement to theorem 8.2 – conditions ensuring that nonconstant steady solutions of (8.45) together with the homogeneous boundary conditions

$$\frac{du}{d\mathbf{n}} = 0 \qquad \text{on } \partial\Omega \times \mathbb{R}_+ \tag{8.46}$$

are unstable. Denoting by I the class of scalar field $\varphi : \Omega \to \mathbb{R}$ such that $\{\varphi \in W_1^2(\Omega), \dfrac{d\varphi}{d\mathbf{n}} = 0 \text{ on } \partial\Omega\}$, the following theorem holds

Theorem 8.3 - *Let* v *be a steady nonconstant solution to* (8.45) – (8.46). *If*

$$\overline{\lambda} = \max_{\varphi \in I} \frac{\displaystyle\int_\Omega [-|\nabla\varphi|^2 + f'(v)\varphi^2]d\Omega}{\|\varphi\|_2^2} > 0, \tag{8.47}$$

then v *is unstable.*

Proof. Setting $u = v + w$, from

$$\begin{cases} \Delta v + f(v) = 0 & \text{in } \Omega, \\ \dfrac{dv}{d\mathbf{n}} = 0 & \text{on } \partial\Omega, \end{cases} \tag{8.48}$$

and (8.45) – (8.46), it follows that

$$\begin{cases} w_t = \Delta w + f(v + w) - f(v) & \text{in } \Omega \times \mathbb{R}_+, \\ \dfrac{dw}{d\mathbf{n}} = 0 & \text{on } \partial\Omega. \end{cases} \tag{8.49}$$

We show that v is linearly unstable. Linearizing f in v, we obtain

$$\begin{cases} w_t = \Delta w + f'(v)w & \text{in } \Omega \times \mathbb{R}_+, \\ \dfrac{dw}{d\mathbf{n}} = 0 & \text{on } \partial\Omega. \end{cases} \tag{8.50}$$

Then, looking for solutions of the form $w = e^{\lambda t}\psi(\mathbf{x})$, there follows the eigenvalue problem

$$\begin{cases} \lambda\psi = \Delta\psi + f'(v)\psi & \text{in } \Omega, \\ \dfrac{d\psi}{d\mathbf{n}} = 0 & \text{on } \partial\Omega, \end{cases} \tag{8.51}$$

and hence

$$\lambda\|\psi\|_2^2 = \int_\Omega [\psi\Delta\psi + f'(v)\psi^2]d\Omega. \tag{8.52}$$

But, in view of $(8.51)_2$, it turns out that

$$\int_\Omega \psi\Delta\psi d\Omega = -\int_\Omega |\nabla\psi|^2 d\Omega, \tag{8.53}$$

and hence it follows that

$$\lambda = \frac{\int_\Omega [-|\nabla\psi|^2 + f'(v)\psi^2]d\Omega}{\|\psi\|_2^2}. \tag{8.54}$$

But it is an easy matter to verify that (8.51) are the Euler-Lagrange equations of the variational problem (8.47). Hence denoting the maximizing function by $\overline{\psi}$, (8.50) admits the solution $e^{\overline{\lambda} t}\overline{\psi}(\mathbf{x})$, which increases exponentially as $t \to \infty$.

Remark 8.3 - *As an immediate consequence of theorem* 8.3 *it follows that*

$$\int_\Omega f'[v(\mathbf{x})]d\Omega > 0$$

ensures instability. In fact, for any $\varphi = const. \neq 0$, *the ratio on R.H.S. of* (8.47) *is positive.*

For the proof of the following instability theorem, we refer to [5].

Theorem 8.4 - *Let* $\Omega \in \mathbb{R}^n$, $n \geq 1$, *and let* $v(x)$ *be a nonconstant steady solution of* (8.45) – (8.46). *If one of the following restrictions holds:*

i) Ω *is a convex subset of* $\mathbb{R}^n$;

ii) f'' *is of one sign on the range of* v,

then v *is unstable.*

8.4.4 Travelling wave solutions of reaction-diffusion equations

Let c denote a real number and consider the generic P.D.E. (1.1). Solutions of the form $\mathbf{u} = \mathbf{u}(\mathbf{x} - ct)$, if such exist, are called *travelling waves,* and $|c|$ is the *speed* of the wave. As is well known, many physical phenomena are modelled by travelling wave solutions of hyperbolic equations. But also many biological and chemical phenomena (insect dispersal, waves of an epidemic in interacting populations, waves of chemical concentrations,...), modelled by parabolic reaction-diffusion equations, are represented by travelling waves. We present here a brief introduction to this matter, referring to [5,47,50] for a deeper and extensive study.

For the sake of simplicity, we begin by considering the prototype of reaction-diffusion equations admitting travelling wave solutions, i.e. the Fisher equation (1.150) in one dimension. Rescaling (1.150) by setting

$$u = \frac{\rho}{s}, \qquad t^* = rt, \qquad x^* = x\sqrt{\frac{r}{k}} \tag{8.55}$$

and omitting the asterisks, it follows that

$$u_t = u(1-u) + u_{xx}\,. \tag{8.56}$$

In particular we look for travelling wave solutions $u = u(\xi)$, $\xi = x - ct$ with $0 \le u \le 1$. We notice that $u = 0$ and $u = 1$ are steady solutions of (8.56) and that they are respectively unstable and stable (see exercise 8.6). Substituting $u(x - ct) = U(\xi)$ into (8.56) it follows that

$$\frac{d^2U}{d\xi^2} + c\frac{dU}{d\xi} + U(1-U) = 0\,. \tag{8.57}$$

To (8.57) we append the boundary conditions

$$\lim_{\xi\to\infty} U(\xi) = 0\,, \qquad \lim_{\xi\to-\infty} U(\xi) = 1 \tag{8.58}$$

i.e. we look for travelling wave solutions connecting the steady states. Introducing the phase plane (U,V), where

$$\begin{cases} \dfrac{dU}{d\xi} = V, \\ \dfrac{dV}{d\xi} = -cV - U(1-U)\,, \end{cases} \tag{8.59}$$

any travelling wave trajectory connects the points $(0,0)$ and $(1,0)$ of the phase plane and belongs to the strip $0 \le U \le 1$. The following theorem holds:

Theorem 8.5 - *Between the positive solutions of*

$$\frac{dV}{dU} + \frac{U(1-U)}{V} = -c\,, \qquad U \in (0,1)\,, \tag{8.60}$$

satisfying the boundary conditions

$$V(0) = V(1) = 0 \tag{8.61}$$

and the travelling wave solutions there is a one-one correspondence (modulo shifts in ξ *).*

Proof. In order to give a sketch of the existence proof, for some c, of a trajectory connecting the critical points $(0,0), (0,1)$ of (8.59), we perform a linear stability analysis of these points. Linearizing (8.59) about $(0,0)$ and setting

$$\begin{cases} W = (U,V), \\ A = \begin{Vmatrix} 0 & 1 \\ -1 & -c \end{Vmatrix}, \end{cases}$$

it follows that

$$\frac{dW}{d\xi} = AW .$$

The eigenvalues of A are

$$\lambda = \frac{1}{2}(-c \pm \sqrt{c^2 - 4}),$$

and hence for $c < 0$, $c^2 \geq 4$, $(0,0)$ is a *stable node* and all the orbits near $(0,0)$ must emanate from that point.

Linearizing (8.59) about $(1,0)$, it follows that

$$\frac{dW}{d\xi} = BW$$

where

$$B = \begin{Vmatrix} 0 & 1 \\ 1 & -c \end{Vmatrix} .$$

The eigenvalues of B are

$$\lambda = \frac{1}{2}(-c \pm \sqrt{c^2 + 4}),$$

i.e. $(1,0)$ is a *saddle point* and all the orbits near $(1,0)$ must converge to that point. By continuity arguments one can deduce that, for each $c < 0$ with $|c| \geq 2$, there exists a trajectory in the phase plane emanating from $(0,0)$ and reaching $(1,0)$ from the left and from above (we refer to [50], pp. 101-103, for the details). In this way the existence of travelling waves solutions is proved.

We notice now that, at least for $c = 5/\sqrt{6}$, there exists an exact analytical travelling wave solution. In fact, looking for solutions of the form

$$U(\xi) = (1 + e^{\alpha\xi})^{-2}, \tag{8.62}$$

where α is a positive constant to be found, (8.58) are automatically satisfied. Taking into account that

$$U' = \frac{-2\alpha e^{\alpha\xi}}{(1+e^{\alpha\xi})^3}, \quad U'' = \frac{-2\alpha^2 e^{\alpha\xi}}{(1+e^{\alpha\xi})^3} + \frac{6\alpha^2 e^{2\alpha\xi}}{(1+e^{\alpha\xi})^4},$$

and substituting in (8.57), it follows that

$$(4\alpha^2 - 2\alpha c + 1)e^{2\alpha\xi} + 2(1 - \alpha^2 - \alpha c)e^{\alpha\xi} = 0,$$

which implies

$$\begin{cases} 4\alpha^2 - 2\alpha c + 1 = 0, \\ \alpha^2 + \alpha c - 1, \end{cases}$$

and hence

$$\alpha = \frac{1}{\sqrt{6}}, \qquad c = \frac{5}{\sqrt{6}}. \tag{8.63}$$

Remark 8.4 - *By setting* $\varepsilon = 1/c^2$, $z = \xi/c$, $f(z) = U(\xi, \varepsilon)$, (8.57) – (8.58), *with* $|c| \leq 2$, *become*

$$\begin{cases} \varepsilon \dfrac{d^2 f}{dz^2} + \dfrac{df}{dz} + f(1-f) = 0, \\ f(-\infty) = 1, \quad f(\infty) = 0, \end{cases}$$

with $0 < \varepsilon \leq 0.25$. *Looking for asymptotic solutions considering* ε *a small parameter, it is found that any travelling waves, are of the form* (8.62). *Specifically it is found that* [47]

$$\begin{aligned} U(\xi,\varepsilon) =& (1+e^{\xi/c})^{-1} + c^{-2}e^{\xi/c}(1+e^{\xi/c})^{-2}\ln[4e^{\xi/c}(1+e^{\xi/c})^{-2}] + \\ &+ O(c^{-4}). \end{aligned} \tag{8.64}$$

There exist many reaction-diffusion equations admitting exact travelling wave solutions (see exercises). We confine ourselves to considering the following Fisher equation with density dependent diffusion:

$$u_t = u(1-u) + \frac{\partial}{\partial x}(uu_x). \tag{8.65}$$

This equation represents an example of an equation of the form (1.154) in which migration due to crowding is considered. In fact, comparing (8.65) to (1.54), with $\{\rho = u,\ \sigma = u(1-u)\}$, it follows that $|v| = |u|$, i.e. the diffusion velocity grows with the density, and the population dispersion from a region A to a region B of lower density increases more rapidly as A becomes more crowed. Looking for travelling wave solutions $u(x-ct) = U(\xi)$, from (8.65) it follows that

$$\frac{1}{2}\frac{d^2}{d\xi^2}U^2 + c\frac{dU}{d\xi} + U(1-U) = 0 . \tag{8.66}$$

For the equation (8.66) with the boundary conditions (8.58) the following exact solution has been found [187], [188]

$$U(\xi) = \begin{cases} 1 - e^{(\xi - a)/\sqrt{2}} & \xi \le a, \\ 0 & \xi \ge a \end{cases} \tag{8.67}$$

where a is an arbitrary constant.

8.5 STABILITY OF VISCOUS FLOWS IN BOUNDED REGIONS

8.5.1 Nonlinear energy stability of steady viscous incompressible fluid motions in bounded regions: Serrin's theorem

The modern version of the method of nonlinear stability, in the L_2-norm ("energy method"), of isothermal, viscous incompressible fluid motions in regions which are bounded at least in one direction, was introduced originally by Serrin in 1959 [189]. Since then several authors have exploited, generalized and developed this method [59,71,72]. The energy method is now well established within the theory of the Liapunov direct method, and can give results which are useful in practice. In order to present the method, let $m_0 = (\mathbf{v}, p)$ and $m_0' = (\mathbf{v}+\mathbf{u}, p+\pi)$ be two solutions, corresponding to the same boundary data and force $\mathbf{F}$, of the I.B.V.P.

$$\begin{cases} \mathbf{V}_t + \mathbf{V}\cdot\nabla\mathbf{V} = -\nabla P + \nu\Delta\mathbf{V} + \mathbf{F}, \\ \nabla\cdot\mathbf{V} = 0, \end{cases} \quad \text{in } \Omega \tag{8.68}$$

$$\begin{cases} \mathbf{V}(\mathbf{x},0) = \mathbf{V}_0(\mathbf{x}) & \text{in } \Omega \\ \mathbf{V}(\mathbf{x},t) = \mathbf{V}^*(\mathbf{x},t) & \text{on } \partial\Omega . \end{cases} \tag{8.69}$$

Then the *perturbation* $(\mathbf{u}, \pi)$ to the *basic motion* m_0 satisfies the I.B.V.P.

$$\begin{cases} \mathbf{u}_t + (\mathbf{u} + \mathbf{v}) \cdot \nabla \mathbf{u} + \mathbf{u} \cdot \nabla \mathbf{v} = -\nabla \pi + \nu \Delta \mathbf{u}, \\ \qquad\qquad\qquad\qquad\qquad\qquad\qquad\qquad \text{in } \Omega \\ \nabla \cdot \mathbf{u} = 0, \end{cases} \tag{8.70}$$

$$\begin{cases} \mathbf{u}(\mathbf{x}, 0) = \mathbf{u}_0(\mathbf{x}) & \text{in } \Omega, \\ \mathbf{u}(\mathbf{x}, t) = 0 & \text{on } \partial\Omega. \end{cases} \tag{8.71}$$

On taking into account $(1.106)_1-(1.107)$ and $(8.70)_2-(8.71)$, there immediately follows the *energy equality* ($\|\cdot\| = L_2(\Omega)$-norm)

$$\frac{1}{2}\frac{d}{dt}\|\mathbf{u}\|^2 = -\int_\Omega \mathbf{u} \cdot \mathbf{D} \cdot \mathbf{u}\, d\Omega - \nu \|\nabla \mathbf{u}\|^2, \tag{8.72}$$

where $\mathbf{D} \equiv \mathrm{Sym}(\nabla \mathbf{v})$ is the deformation rate tensor of the basic motion m_0.

In order to put (8.72) in dimensionless form, we set

$$\begin{cases} \mathbf{u} = U\mathbf{u}^*, \; x_i = dx_i^*, \; t = \dfrac{d^2}{\nu} t^*, \; \mathbf{v} = U\mathbf{v}^*, \\ \nabla^* = d\nabla, \; R = \dfrac{Ud}{\nu} \; \text{(Reynolds number)} \end{cases} \tag{8.73}$$

where U is a reference velocity (generally a typical velocity for the problem) and d is a length scale (linked generally to the diameter of Ω). Then, omitting the asterisks for notational simplicity, (8.72) becomes in dimensionless form

$$\frac{1}{2}\frac{d}{dt}\|\mathbf{u}\|^2 = (R\mathcal{F} - 1)\|\nabla \mathbf{u}\|^2 \tag{8.74}$$

where $\mathcal{F}$ is the functional

$$\mathcal{F}(\mathbf{u}, \mathbf{v}, t) = \frac{-\displaystyle\int_\Omega \mathbf{u} \cdot \mathbf{D} \cdot \mathbf{u}\, d\Omega}{\|\nabla \mathbf{u}\|^2} \tag{8.75}$$

– depending explicitly on t ***iff*** the basic motion is unsteady.

Denoting by $\mathcal{I}$ the class of ***kinematically admissible perturbations***, i.e.

$$\mathcal{I} \overset{\text{def.}}{=} \{\mathbf{u} : \nabla \cdot \mathbf{u} = 0, \mathbf{u} \in H_0^1(\Omega)\}, \tag{8.76}$$

we are in a position to prove the following theorem of Serrin [189]

Theorem 8.6 - *Let the basic motion $m_0 = (\mathbf{v}, p)$ be steady, and let*

$$\frac{1}{R_E} = \max_{\mathbf{u} \in \mathcal{I}} \mathcal{F} \tag{8.77}$$

exist. If

$$R < R_E \,, \tag{8.78}$$

then the basic motion m_0 is unconditionally asymptotically stable, according to

$$\|\mathbf{u}(t)\|^2 \leq \|\mathbf{u}_0\|^2 \exp\left(\frac{2(R - R_E)\lambda}{R_E}\right) t \,, \tag{8.79}$$

where λ is the first eigenvalue of $-\Delta$ on Ω, with homogeneous Dirichlet boundary conditions.

Proof. In fact, on taking into account the Poincaré inequality (see appendix)

$$\|\mathbf{u}\|^2 \leq \lambda \|\nabla \mathbf{u}\|^2 \,, \tag{8.80}$$

from (8.74) and (8.77) it turns out that

$$\frac{1}{2}\frac{d}{dt}\|\mathbf{u}\|^2 \leq \lambda \left(\frac{R}{R_E} - 1\right) \|\mathbf{u}\|^2 \,. \tag{8.81}$$

Remark 8.5 - *Let us remark that*

i) $R < R_E$ ensures the exponential asymptotic stability of the basic motion for all initial perturbations $\mathbf{u}_0$;

ii) the existence of the maximum (8.77) is crucial, and the effective evaluation of R_E requires the solution of the Euler-Lagrange equations of the variational problem (8.77) and is the core of the method;

iii) the existence of regular solutions to the I.B.V.P. (8.68) – (8.69) when Ω is a three-dimensional domain without symmetry is not yet established, and the existence of generalized solutions has been proved in $H_0^1(\Omega)$ only [160];[2]

iv) assuming that $m = (\mathbf{v}, p)$ is a regular solution of (8.68) – (8.69), Theorem 8.4 holds at least for regular solutions $(\mathbf{u}, \pi)$ of the I.B.V.P. (8.70) – (8.71) and for (generalized) perturbations satisfying (8.72).

We end by recalling that the existence of the maximum (8.77) has been proved by Rionero in [192], and that by deriving an upper bound for the functional (8.75), restrictive estimates $R_E^*(< R_E)$ of R_E can easily be obtained, such that $R < R_E^*$ represents a stability condition which holds quite independently of the form of the basic flow and of the geometry of the bounded domain Ω (*universal stability*). (See [189], [72] and exercise 8.9.)

---

[2]Let us point out that many classical solutions are known [191].

8.5.2 Connection between nonlinear energy stability and linear stability of steady flows

Linearizing the I.B.V.P. (8.70) – (8.71) it follows that

$$\begin{cases} \mathbf{u}_t + \mathbf{v}\cdot\nabla\mathbf{u} + \mathbf{u}\cdot\nabla\mathbf{v} = -\nabla\pi + \nu\Delta\mathbf{u}, \\ \\ \nabla\cdot\mathbf{u} = 0 \end{cases} \qquad \text{in } \Omega \tag{8.82}$$

$$\begin{cases} \mathbf{u}(\mathbf{x},0) = \mathbf{u}_0(\mathbf{x}) & \text{in } \Omega \\ \mathbf{u}(\mathbf{x},t) = 0, & \text{on } \partial\Omega\,. \end{cases} \tag{8.83}$$

Because (8.82) – (8.83) is autonomous and linear, setting

$$\mathbf{u} = \mathbf{q}(\mathbf{x})e^{-\sigma t}, \qquad \pi(\mathbf{x},t) = \pi_0(\mathbf{x})e^{-\sigma t} \tag{8.84}$$

where σ is a-priori a complex parameter, and substituting into (8.82) – (8.83), it turns out that

$$\begin{cases} \sigma\mathbf{q} = \mathbf{v}\cdot\nabla\mathbf{q} + \mathbf{q}\cdot\nabla\mathbf{v} + \nabla\pi_0 - \nu\Delta\mathbf{q} & \text{in } \Omega, \\ \nabla\cdot\mathbf{q} = 0 & \text{in } \Omega, \\ \mathbf{q}(\mathbf{x}) = 0 & \text{on } \partial\Omega\,. \end{cases} \tag{8.85}$$

The linear stability method consists in the previous procedure [10], and the basic motion m_0 is ***(linearly) stable*** if

$$\text{re}(\sigma) \geq 0\,, \qquad \forall\ \sigma; \tag{8.86}$$

while it is ***(linearly) unstable*** if

$$\exists\sigma : \text{re}(\sigma) < 0. \tag{8.87}$$

Because the results of linear stability theory are of practical relevance, at least for knowing the onset of instability, the question of comparing the results of nonlinear energy stability to those of the linear stability method arises. In this respect one has to compare the spectral problem arising from (8.77) to (8.85). It is easily seen that the Euler-Lagrange equations of the variational problem (8.77) (in dimensional form) are {[59], pp. 43-44}

$$\begin{cases} \nu\Delta\mathbf{q} + \mathbf{D}\cdot\mathbf{q} = -\nabla\pi_0(\mathbf{x}), \\ \nabla\cdot\mathbf{q} = 0, \\ \mathbf{q} = 0\,. \end{cases} \tag{8.88}$$

At first glance, it does not appear to be easy to compare the two eigenvalue problems (8.85), (8.88). But the nonlinear energy stability of a steady basic flow may also be phrased in a different but equivalent way. In fact, setting

$$\mathcal{G}(\mathbf{u},\mathbf{v}) = \frac{\int_\Omega \mathbf{u}\cdot\mathbf{D}\cdot\mathbf{u}\,d\Omega + \nu\|\nabla\mathbf{u}\|^2}{\|\mathbf{u}\|^2}, \tag{8.89}$$

$\Omega$the following theorem immediately holds [194]

Theorem 8.7 - *Let the basic motion* $m_0 = (\mathbf{v},p)$ *be steady, and let*

$$\gamma = \min_{\mathbf{u}\in I} \mathcal{G} \tag{8.90}$$

exist. If γ *is positive, then the basic motion* m_0 *is unconditionally asymptotically stable according to*

$$\|\mathbf{u}(t)\|^2 \le \|\mathbf{u}_0\|^2 e^{-\gamma t}. \tag{8.91}$$

The existence of the minimum (8.90) can be proved following [190]. Further, as proved in [194] (for steady basic flows), $\gamma > 0$ is equivalent to $R < R_E$. Therefore instead of comparing (8.88) to (8.85), one may compare to (8.85) the Euler-Lagrange equations connected to the variational problem (8.90). It is an easy matter to establish that the spectral problem connected to (8.90) is

$$\begin{cases} \mu\mathbf{q} = \mathbf{q}\cdot\mathbf{D} - \nu\Delta\mathbf{q} + \nabla\pi_0 & \text{in } \Omega \\ \nabla\cdot\mathbf{q} = 0 & \text{in } \Omega \\ \mathbf{q} = 0 & \text{on } \partial\Omega, \end{cases} \tag{8.92}$$

to verify that the eigenvalues μ_i $(i = 1,2,\ldots)$ are positive, and that γ is given by the smallest of them (exercise 8.10).

Taking into account that

$$\nabla\mathbf{v} = \mathbf{D} + \underline{\Omega} \tag{8.93}$$

where $\Omega_{ij} = \frac{1}{2}(\partial_i v_j - \partial_j v_i)$ is the *vorticity tensor* of the basic flow m_0, let us consider the operators S and A such that $\forall\,\mathbf{a} \in L_2(\Omega)$

$$S\,\mathbf{a} = -\nu\Delta\mathbf{a} + \mathbf{a}\cdot\mathbf{D}, \tag{8.94}$$

$$A\,\mathbf{a} = \mathbf{a}\cdot\underline{\Omega} + \mathbf{v}\cdot\nabla\mathbf{a}. \tag{8.95}$$

It is an easy matter to verify that, with respect to the $L_2(\Omega)$-scalar product, S is symmetric while A is skew-symmetric (see [72] and exercise 8.11). Therefore (8.85) and (8.92) can be written respectively thus

$$\begin{cases} S(\mathbf{q}) + A(\mathbf{q}) + \nabla \pi_0 = \sigma \mathbf{q} & \text{in } \Omega, \\ \nabla \cdot \mathbf{q} = 0 & \text{in } \Omega, \\ \mathbf{q}(\mathbf{x}) = 0 & \text{on } \partial\Omega, \end{cases} \tag{8.96}$$

$$\begin{cases} S(\mathbf{q}) + \nabla \pi_0 = \mu \mathbf{q} & \text{in } \Omega, \\ \nabla \cdot \mathbf{q} = 0 & \text{in } \Omega, \\ \mathbf{q}(\mathbf{x}) = 0 & \text{on } \partial\Omega. \end{cases} \tag{8.97}$$

When Ω is bounded, or bounded in at least one direction, and only *normal mode* eigenfunctions (i.e. eigenfunction periodic along the directions in which Ω is unbounded) are considered, then in suitable function space, it can be proved that [72]

i) the eigenvalues of (8.96) can be ordered in a sequence $\{\sigma_n\}_{n\in \mathbb{N}}$ in such a way that $\text{re}(\sigma_i) \leq \text{re}(\sigma_j)$ for $i < j$;

ii) the eigenvalues of (8.97) can be ordered in a non-decreasing sequence

$$\mu_1 \leq \mu_2 \leq \ldots \leq \mu_n \leq . \tag{8.98}$$

Comparing (8.96) to (8.97) it follows that

a) the linear stability depends on the sign of the first eigenvalue of the operator $L = S + A$;

b) energy stability depends on the sign of the first eigenvalue of the symmetric part S of L;

c) when $A = 0$, there is coincidence between linear and nonlinear energy stability.

8.5.3 Nonlinear energy stability of unsteady flows

Let the basic motion $m_0 = (\mathbf{v}, p)$ be unsteady and let

$$\frac{1}{R_E} = \sup_{t\geq 0} \max_{\mathbf{u}\in I} \mathcal{F} \tag{8.99}$$

exist. Then, as is easily seen, the Serrin theorem 8.6 continues to hold with R_E given by (8.99). Concerning conditions which ensure the boundedness of $\mathcal{F}$ on $I \times \mathbb{R}_+$, it is an easy matter to show that the boundedness is guaranteed by one of the following conditions (see exercise 8.12):

$$\sup_{t\geq 0} \max_{\Omega} |\mathbf{D}(\mathbf{x},t)| \leq \infty, \qquad \sup_{t\geq 0} \max_{\Omega} |\mathbf{v}(\mathbf{x},t)| < \infty. \tag{8.100}$$

We remark that theorem 8.7 continues to hold also. In fact, when the basic flow is unsteady, the functional (8.89) depends explicitly on t and (8.90) gives

$$\gamma(t) = \min_{\mathbf{u}\in I} \mathcal{G}(\mathbf{u},\mathbf{v},t)\,. \tag{8.101}$$

From (8.72), (8.89) and (8.99) it turns out that

$$\frac{1}{2}\frac{d}{dt}\|\mathbf{u}\|^2 = -\gamma(t)\|\mathbf{u}\|^2$$

and hence

$$\|\mathbf{u}\|^2 = \|\mathbf{u}_0\| e^{-2\int_0^t \gamma(\tau)d\tau}\,. \tag{8.102}$$

Therefore, as pointed out in [194], *if*

$$\lim_{t\to\infty}\int_0^t \gamma(\tau)d\tau = \infty\,, \tag{8.103}$$

then the unsteady basic flow is unconditionally asymptotically stable [194].

8.6 STABILITY OF VISCOUS FLOWS IN UNBOUNDED REGIONS

8.6.1 Introduction

Let Ω be a domain unbounded in all directions (i.e. *not* contained between two parallel planes) and let us begin by specifying the difficulties that one encounters in generalizing the energy method to such domains. For the sake of simplicity, we consider the problem of uniqueness (which is essential for well posedness (Theor. 1.2)) in the case that Ω is the exterior of a bounded domain Ω_0. Denoting by S_r a sphere of radius r centered at a point $O \in \Omega_0$, let $\bar{r}(>0)$ be such that S_r contains within it Ω_0 for any $r > \bar{r}$. Set, for any $r > \bar{r}$,

$$\Sigma_r = \partial S_r\,,\ \Omega_r = S_r \cap \Omega,\ \partial\Omega_r = \partial\Omega_0 \cup \Sigma_r\,.$$

Considering the I.B.V.P. (8.70) – (8.71) on the unbounded domain $\Omega = \mathbb{R}_3/\Omega_0$, then, at least for smooth solutions, because $\partial\Omega = \partial\Omega_0$, it follows that

$$\frac{1}{2}\frac{d}{dt}\int_{\Omega_r} u^2 d\mathbf{x} = -\int_{\Omega_r}[\mathbf{u}\cdot\nabla(\mathbf{u}+\mathbf{v})\mathbf{u} + \mathbf{v}\cdot\nabla\mathbf{u}\cdot\mathbf{u} - \nu\mathbf{u}\cdot\Delta\mathbf{u}]d\mathbf{x} + \\ + \int_{\Sigma_r}\pi\mathbf{u}\cdot\mathbf{n}d\Sigma_r\,. \tag{8.104}$$

But it turns out that

$$\begin{cases} \displaystyle\int_{\Omega_r} (\mathbf{u}+\mathbf{v})\cdot\nabla\mathbf{u}\cdot u d\mathbf{x} = \frac{1}{2}\int_{\Sigma_r} u^2(\mathbf{u}+\mathbf{v})\cdot\mathbf{n}d\mathbf{x} \\ \displaystyle\int_{\Omega_r} \mathbf{u}\cdot\Delta\mathbf{u}d\mathbf{x} = \int_{\Sigma_r}\mathbf{u}\cdot\nabla\mathbf{u}\cdot\mathbf{n}d\mathbf{x} - \int_{\Omega_r}(\nabla\mathbf{u})^2 d\mathbf{x} \end{cases}$$

hence, in view of (8.104), we obtain

$$\frac{1}{2}\frac{d}{dt}\int_{\Omega_r} u^2 d\mathbf{x} = -\int_{\Omega_r}[\mathbf{u}\cdot\mathbf{D}\cdot\mathbf{u}+(\nabla\mathbf{u})^2]d\mathbf{x} + \\ + \int_{\Sigma_r}[\pi\mathbf{u} - \frac{1}{2}u^2(\mathbf{v}+\mathbf{u}) + \nu\mathbf{u}\cdot\nabla\mathbf{u}]\cdot\mathbf{n}d\mathbf{x}. \tag{8.105}$$

Then requiring, for instance,

$$\begin{cases} \mathbf{u}\in L^2(\Omega), \ \sup\limits_{\Omega\times\mathbb{R}_+}|\mathbf{D}| \le k = \text{positive constant} \\ \displaystyle\lim_{r\to\infty}\int_{\Sigma_r}[\pi\mathbf{u} - \frac{1}{2}u^2(\mathbf{v}+\mathbf{u}) + \nu\mathbf{u}\cdot\nabla\mathbf{u}]\cdot\mathbf{n}d\mathbf{x} = 0, \end{cases} \tag{8.106}$$

it follows that

$$\|\mathbf{u}\|_2^2 \le \|\mathbf{u}_0\|_2^2 e^{2kt} \tag{8.107}$$

and hence $\mathbf{u}_0 = 0 \Rightarrow \mathbf{u} = 0$ on Ω, $\forall t > 0$.[3] But (8.106) requires that the perturbation $(\mathbf{u},\pi)$ to the basic motion $(\mathbf{v},p)$ have suitable behaviour at large spatial distances.[4] In particular, the global energy perturbation $\|\mathbf{u}\|_2^2$ has to be finite for all time $t > 0$. But this is not consistent with the fact that the early perturbations to the basic flow at least, should have the same behaviour as the basic flow. In fact, there exist physically meaningful, basic flows which have an infinite (kinetic) energy.[5] This fact shows that the first problem that

---

[3]In passing, we notice that when Ω is bounded, one can require only $\sup\limits_{\Omega\times\mathbb{R}_+}|\mathbf{D}| < \infty$ for obtaining uniqueness.

[4]For instance, $(8.106)_3$ is satisfied by requiring that

$$|\pi\mathbf{u} - \frac{1}{2}u^2(\mathbf{v}+\mathbf{u}) + \nu\mathbf{u}\cdot\nabla\mathbf{u}| = O(r^{-2-\epsilon}).$$

[5]Let Ω be the exterior of the cylinder $\Omega = \{x^2+y^2 \le r_1^2, r_1 = \text{const.} > 0\}$ and assume that Ω_0 is rotating with angular velocity ω_1 about the z axis. If Ω is filled with a viscous incompressible fluid under the action of a conservative force, a steady flow having the velocity (in cylindrical coordinates)

$$v_r = v_z = 0, \qquad v_\theta = \frac{\omega_1 r_1^2}{r}$$

is admitted [191]. It is an easy matter to verify that v does not belong to $L_2(\Omega)$.

one has to solve when Ω is unbounded is to find "less restrictive" assumptions on the perturbations for which the energy relation (8.72) (or one similar), may hold. Passing now to the stability problem, one has to take into account that on unbounded domains:

i) the existence of the maximum (8.77), and also the boundedness of (8.75) are not guaranteed,

ii) the Poincaré inequality (8.80) (which ensures the asymptotic, in time, decay of the perturbation energy) does not hold (exercise 6.12).

We refer the readers to [72] for the stability problem in unbounded region Ω when $\partial\Omega$ is supposed to be rigid and moving in the prescribed way $(8.69)_2$.

In the subsequent sections, following [195] we consider the stability problem when more general boundary conditions are appended to (8.68) (subsect. 8.6.2), obtaining two energy relations holding for the exterior of starshaped domains (subsect. 8.6.3). Next we consider the case of a spatial sink and obtain a stability condition (subsect. 8.6.4).

8.6.2 A mixed boundary value problem: free boundary-like conditions

Let $\Omega \subset \mathbb{R}^3$ be unbounded and let $\partial\Omega = \Sigma \cup \Gamma$, $\Gamma \cap \Sigma = \emptyset$ be smooth with Σ rigid and Γ "free but invariable" (i.e. a system of coordinates can be chosen, on which it does not change). Instead of (8.69), we consider the mixed boundary conditions[6]

$$\begin{cases} \mathbf{v} = \mathbf{a}(\mathbf{x},t) & \text{on } \Sigma, \\ \mathbf{v}\cdot\mathbf{n} = \mathbf{a}_1(\mathbf{x},t), \quad \mathbf{D}(\mathbf{v})\mathbf{n} - (\mathbf{n}\cdot\mathbf{D}(\mathbf{v})\cdot\mathbf{n})\mathbf{n} = \mathbf{b} & \text{on } \Gamma \end{cases} \tag{8.108}$$

where $\mathbf{a}$, $\mathbf{a}_1$, $\mathbf{b}$ (with $\mathbf{b}\cdot\mathbf{n} = 0$) are prescribed vector fields, and $(8.108)_2$ – $(8.108)_3$ represent the slip condition. In particular, $(8.108)_3$ corresponds to prescribing the tangential component of the stress vector $\mathbf{t} = -p\mathbf{n} + 2\nu\mathbf{D}(\mathbf{v})\cdot\mathbf{n}$.

Passing to the I.B.V.P. (8.70) – (8.71), instead of (8.71), we have

$$\begin{cases} \mathbf{u}(\mathbf{x},0) = \mathbf{u}_0(\mathbf{x}) & \text{on } \Omega, \\ \mathbf{u} = 0 & \text{on } \Sigma, \\ \mathbf{u}\cdot\mathbf{n} = \mathbf{n}\cdot\mathbf{d}\times\mathbf{n} = 0 & \text{on } \Gamma\,. \end{cases} \tag{8.109}$$

In the sequel we shall be concerned with solutions of the above problem (8.70), (8.109) which satisfy the following energy equality

$$\begin{aligned} \frac{d}{dt}\int_\Omega f u^2 d\Omega =& 2\int_\Omega [f\mathbf{u}\cdot\nabla\mathbf{u}\cdot\mathbf{v} + (\nabla f\cdot\mathbf{u})\mathbf{u}\cdot\mathbf{v} + \pi\nabla f\cdot\mathbf{u} + \frac{\nu}{2}u^2\Delta f]d\Omega + \\ & - 2\int_\Omega f[(\mathbf{u}+\mathbf{v})\cdot\nabla\mathbf{u}\cdot\mathbf{u} + \nu\nabla\mathbf{u} : \nabla\mathbf{u}]d\Omega + \\ & + \nu\int_\Gamma [2f(\mathbf{u}\cdot\nabla\mathbf{n}\cdot\mathbf{u} - \mathbf{u}\cdot\nabla u_n) - u^2\frac{\partial f}{\partial\mathbf{n}}]d\Gamma \end{aligned} \tag{8.110}$$

[6]Either Σ or Γ may be empty. The boundary conditions on Γ and, especially that corresponding to $(8.109)_3$, on the perturbation $\mathbf{u}$, are called "stress-free boundary conditions".

where $f(\mathbf{x})$ is any weight function. Of course, to this class of solutions belong those solutions to (8.70), (8.109) whose behaviour at large spatial distances is suitably linked to that of f. Formally, the equality (8.110) can be obtained by integrating $(8.70)_1$ multiplied by $f\mathbf{u}$, and taking into account the following identities for divergence-free vectors $\mathbf{u}$ satisfying (8.109) :

$$\begin{aligned}
&\text{i) } f\mathbf{u}\cdot\Delta\mathbf{u} = \nabla\cdot(f\nabla\mathbf{u}\cdot\mathbf{u} - \frac{1}{2}\mathbf{u}^2\nabla f) + \frac{1}{2}\mathbf{u}^2\Delta f - f\nabla\mathbf{u} : \nabla\mathbf{u},\\
&\text{ii) } f\mathbf{u}\cdot\nabla\mathbf{v}\cdot\mathbf{u} = \nabla\cdot[f(\mathbf{u}\cdot\mathbf{v})\mathbf{u}] - f\mathbf{u}\cdot\nabla\mathbf{u}\cdot\mathbf{v} - (\nabla f\cdot\mathbf{u})\mathbf{u}\cdot\mathbf{v},\\
&\text{iii) } f\nabla\pi\cdot\mathbf{u} = \nabla\cdot(f\pi\mathbf{u}) - \pi\nabla f\cdot\mathbf{u},\\
&\text{iv) } \mathbf{n}\cdot\nabla\mathbf{u}\cdot\mathbf{u} = 2\mathbf{n}\cdot\mathbf{d}\cdot\mathbf{u} - \mathbf{u}\cdot\nabla\mathbf{u}\cdot\mathbf{n} = -\mathbf{u}\cdot\nabla\mathbf{u}\cdot\mathbf{n} =\\
&\qquad = -\mathbf{u}\cdot\nabla u_n + \mathbf{u}\cdot\nabla\mathbf{n}\cdot\mathbf{u}.
\end{aligned}$$

Remark 8.6 - *In the sequel we will consider only the case in which Γ is constituted by parts Γ_i of plane, spherical or cylindrical surfaces. Then it is easily shown that*

a) $\mathbf{u}\cdot\nabla u_n = 0, \mathbf{u}\cdot\nabla\mathbf{n}\cdot\mathbf{u} = 0$ *on a plane surface,*

b) $\mathbf{u}\cdot\nabla u_n = 0, \mathbf{u}\cdot\nabla\mathbf{n}\cdot\mathbf{u} = \pm\dfrac{u_\theta^2}{\rho_0}$ *on a cylindrical surface where ρ_0 is the radius (having the same sign),*

c) $\mathbf{u}\cdot\nabla u_n = 0, \mathbf{u}\cdot\nabla\mathbf{n}\cdot\mathbf{u} = \pm\dfrac{\mathbf{u}^2}{\rho_0}$ *on a spherical surface, where ρ_0 is the radius (having the same sign).*

In fact in the case a), let $z = 0$ be the equation of Γ_i. Then from $u_n = u_z \equiv 0$ on Γ_i it follows that $\dfrac{\partial u_z}{\partial x} = \dfrac{\partial u_z}{\partial y} \equiv 0$ and hence $\mathbf{u}\cdot\nabla u_n \equiv 0$ on Γ_i. Further, because $\mathbf{n} = \nabla z$ is constant, $\nabla\mathbf{n} = 0$. In the case b), let $\rho = \rho_0 = const. > 0$ be the equation of Γ_i in a cylindrical polar frame of reference (O, ρ, θ, z) and let $\mathbf{e}_\rho = \nabla\rho$. It follows that $\mathbf{n} = \pm\mathbf{e}_\rho$. From $u_n \equiv 0$ on Γ_i it turns out that $\dfrac{\partial u_n}{\partial\theta} = \dfrac{\partial u_n}{\partial z} \equiv 0$ on Γ_i. Hence $u_n = 0 \Rightarrow u_r = 0$ on Γ_i and hence $u_n \equiv 0 \rightarrow u_r \equiv 0$ on Γ_i and therefore $\mathbf{u}\cdot\nabla u_n = 0$. Further by a straightforward calculation it follows that $\mathbf{u}\cdot\nabla\mathbf{n}\cdot\mathbf{u} = \pm\mathbf{u}\cdot\nabla\mathbf{e}_\rho\cdot\mathbf{u} = \pm\dfrac{u_\theta^2(\rho_0, \theta, z)}{\rho_0}$. The results of c) above are obtained in a completely analogous manner [158].

8.6.3 Energy relation for the exterior of a starshaped bounded domain

Let C denote the class of starshaped convex domains whose boundary is constituted by the union of plane, cylindrical or spherical surfaces.[7] Let $\Omega_0 \subset C$

[7] Of course, C is not empty. In fact, C contains spheres, cylinders, cubes,...

and let Ω be the domain exterior to Ω_0. Our goal, in this section, is to obtain a suitable energy relation holding in Ω. For definiteness, we consider the case where Ω_0 is a starshaped bounded domain consisting of the union of a cylinder and two half spheres of the same radius r_0 (Fig. 8.1).

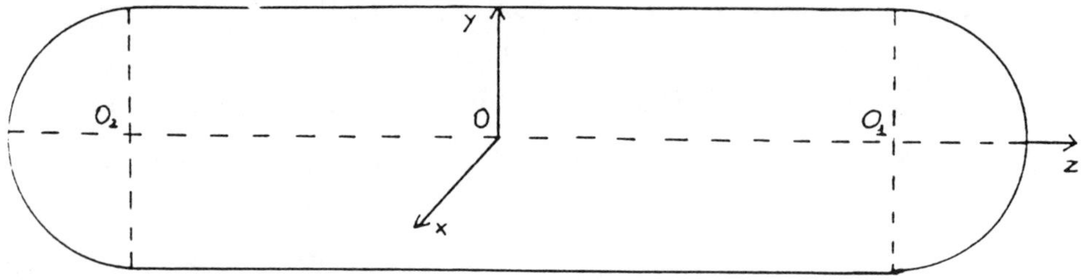

Figure 8.1

Denoting by O a point with respect to which Ω_0 is starshaped (Fig. 8.1), by C a "positive constant",[8] and setting $r = \|\mathbf{x}\|$, the following theorem holds

Theorem 8.8 - *Let* $(\mathbf{u}, \pi) \in \mathcal{P}$

$$\mathcal{P} = \{(\mathbf{u}, \pi) : |\mathbf{u}| \leq C, \, |\pi|(1+r)^\gamma \leq C, \, \gamma > \frac{1}{2}\} \qquad \text{in } \Omega \times [0, T] \tag{8.111}$$

and

$$|\mathbf{v}(\mathbf{x}, t)| \leq C\,. \tag{8.112}$$

Then, if

$$\exists\, t_0 \in [0, T) : \mathbf{u}(t_0) \in L^2(\Omega) \tag{8.113}$$

necessarily

$$\exists\, C(T) : \|\mathbf{u}(t)\|_2^2 + \int_0^T \|\nabla \mathbf{u}\|_2^2 d\tau \leq C\,, \tag{8.114}$$

and the energy relation holds

$$\|\mathbf{u}(t)\|_2^2 = \|\mathbf{u}(t_0)\|_2^2 - 2\int_{t_0}^t \left[\int_\Omega \mathbf{u} \cdot \mathbf{D} \cdot \mathbf{u} d\Omega + \nu \|\nabla \mathbf{u}\|_2^2\right] d\tau +$$
$$- \int_{t_0}^t \int_{\partial\Omega} u^2 \mathbf{v} \cdot \mathbf{n} d\sigma d\tau - \frac{1}{r_0} \int_{t_0}^t \left[\sum_{i=1}^2 \int_{\Gamma_i} u^2 d\sigma + \int_{\Gamma_3} (\mathbf{u} \cdot \mathbf{e}_\theta)^2 d\sigma\right] d\tau \tag{8.115}$$

where Γ_i $(i = 1, 2)$ *are the spherical parts of the boundary while* Γ_3 *represents the cylindrical part. Further* $\mathbf{e}_\theta$ *is the unit vector in the* θ *direction of the cylindrical polar frame of reference* (O, ρ, θ, z).

[8] The numerical values of C are inessential and its values may be different in a single calculation.

Proof. Taking into account the points b)-c) of the Remark 8.6, the proof can be obtained by choosing $f = \varphi e^{-\alpha r}, \alpha = \text{const.} > 0$, where φ is given by (6.59), and following the procedures used in proving theorem 6.5, or following [72], [158].[9]

Remark 8.7 - *Let* $\mathbf{v}_\infty$ *be a constant vector. On taking into account* $(8.109)_3$ *and*

$$\nabla \cdot \mathbf{u} = 0 \Rightarrow \mathbf{u} \cdot \mathbf{D} \cdot \mathbf{u} = \mathbf{u} \cdot \nabla \mathbf{v} \cdot \mathbf{u} = \mathbf{u} \cdot \nabla(\mathbf{v} - \mathbf{v}_\infty) \cdot \mathbf{u} = \\ = \nabla \cdot \{[(\mathbf{v} - \mathbf{v}_\infty) \cdot \mathbf{u}]\mathbf{u}\} - \mathbf{u} \cdot \nabla \mathbf{u} \cdot (\mathbf{v} - \mathbf{v}_\infty), \tag{8.116}$$

the energy relation (8.115) *is equivalent to*

$$\|\mathbf{u}(t)\|_2^2 = \|\mathbf{u}(t_0)\|_2^2 + 2\int_{t_0}^t \left[\int_\Omega \mathbf{u} \cdot \nabla \mathbf{u} \cdot (\mathbf{v} - \mathbf{v}_\infty) d\Omega - \nu \|\nabla \mathbf{u}\|_2^2\right] d\tau + \\ - \int_{t_0}^t \int_{\partial\Omega} \mathbf{u}^2 \mathbf{v} \cdot \mathbf{n} \, d\sigma d\tau \\ - \frac{1}{r_0} \int_{t_0}^t \left[\sum_{i=1}^2 \int_{\Gamma_i} u^2 d\sigma + \int_{\Gamma_3} (\mathbf{u} \cdot \mathbf{e}_\theta)^2 d\sigma\right] d\tau. \tag{8.117}$$

Remark 8.8 - *From the relation* (8.115), *on letting* $O_1O_2 \to 0$ *(see Figure* 8.1) *it follows that* Ω_0 *becomes the sphere of radius* r_0 *centered at* O *and* (8.115) *becomes*

$$\|\mathbf{u}(t)\|_2^2 = \|\mathbf{u}(t_0)\|_2^2 - 2\int_{t_0}^t \left[\int_\Omega \mathbf{u} \cdot \nabla \mathbf{v} \cdot \mathbf{u} d\Omega + \nu \|\nabla \mathbf{u}\|_2^2\right] d\tau + \\ - \int_{t_0}^t \left[\int_{\partial\Omega} (\mathbf{v} \cdot \mathbf{n} + \frac{1}{r_0}) \mathbf{u}^2 d\sigma\right] d\tau. \tag{8.118}$$

8.6.4 Stability criteria

Coupling the energy relations to the weighted inequalities (6.63), stability criteria for fluid motions occurring in the exterior of starshaped domains, under stress free boundary-like conditions can be obtained. As happens in bounded domains, the stability is linked to the supremum of the functional

$$\mathcal{F}(\mathbf{v}, \mathbf{u}, t) = \frac{-\int_\Omega \mathbf{u} \cdot \mathbf{D} \cdot \mathbf{u} d\Omega}{\int_\Omega \nabla \mathbf{u} : \nabla \mathbf{u} d\Omega}. \tag{8.119}$$

---

[9] It is an easy matter to verify that $\frac{\partial f}{\partial n} \geq 0$ on Γ_i.

In fact, from (8.115) it follows that

$$\begin{aligned}\|\mathbf{u}(t)\|_2^2 - \|\mathbf{u}(t_0)\|_2^2 = 2\int_{t_0}^t (\mathcal{F} - \nu)\|\nabla\mathbf{u}\|_2^2 d\tau + \\ - \int_{t_0}^t \int_{\partial\Omega} u^2 \mathbf{v}\cdot\mathbf{n}\, d\sigma d\tau + \\ - \frac{1}{r_0}\int_{t_0}^t [\sum_{i=1}^2 \int_{\Gamma_i} u^2 d\sigma + \int_{\Gamma_3} (\mathbf{u}\cdot\mathbf{e}_\theta)^2 d\sigma] d\tau\,,\end{aligned} \qquad (8.120)$$

and one is led to consider the basic motions $(\mathbf{v}, p)$ for which $\mathcal{F}$ is bounded. Let the assumptions of theorems 6.10 and 8.8 hold and let $r_0 = \inf_{P\in\partial\Omega} |OP| > 0$, where O is a point with respect to which Ω_0 is starshaped. There exist classes of basic motions, physically meaningful, for which (8.119) is bounded in $\mathcal{P}$. This is the case, for instance, for the class $C_1(k)$

$$C_1(k) = \begin{cases} (\mathbf{v}, p) : i)|\mathbf{v}| \le C,\ ii) m = \sup_{t\ge t_0}\{\sup_{\Omega} |\mathbf{D}| \left(\frac{r}{r_0}\right)^k\} < \infty, \\ k \in [2,3];\ iii) \left(\frac{4mr_0^k}{3-k}\mathbf{e}_r - \mathbf{v}\right)\cdot\mathbf{n} \le 0 \text{ a.e. on } \partial\Omega,\ \forall t \ge t_0. \end{cases}$$

Let $(\mathbf{v}, p) \in C_1(k)$. Then it follows that

$$(\mathcal{F} - \nu)\|\nabla\mathbf{u}\|_2^2 \le m r_0^k \int_\Omega \frac{u^2}{r^k} d\Omega - \nu \int_\Omega |\nabla\mathbf{u}|^2 d\Omega\,, \qquad (8.121)$$

hence, by (6.63)

$$(\mathcal{F} - \nu)\|\nabla\mathbf{u}\|_2^2 \le [m r_0^2 \left(\frac{2}{3-k}\right)^2 - \nu]\|\nabla\mathbf{u}\|_2^2 - \frac{2mr_0^k}{3-k}\int_{\partial\Omega} u^2 \nabla\mathbf{x}\cdot\mathbf{n}_0 d\sigma\,. \qquad (8.122)$$

Therefore, from (8.120) it turns out that

$$\begin{aligned}\|\mathbf{u}(t)\|_2^2 - \|\mathbf{u}(t_0)\|_2^2 \le 2[m r_0^2 \left(\frac{2}{3-k}\right)^2 - \nu]\int_{t_0}^t \|\nabla\mathbf{u}\|_2^2 d\tau + \\ + \int_{t_0}^t \int_{\partial\Omega} \left(\frac{4mr_0^k}{3-k}\mathbf{e}_r - \mathbf{v}\right)\cdot\mathbf{n} u^2 d\sigma d\tau + \\ - \frac{1}{r_0}\int_{t_0}^t [\sum_{i=1}^2 \int_{\Gamma_i} u^2 d\sigma + \int_{\Gamma_3} (\mathbf{u}\cdot\mathbf{e}_\theta)^2 d\sigma] d\tau\,.\end{aligned} \qquad (8.123)$$

Introducing the Reynolds number

$$Re = \frac{m r_0^2}{\nu}\,, \qquad (8.124)$$

it follows that *the condition*

$$Re < \left(\frac{3-k}{2}\right)^2 \qquad (8.125)$$

implies stability.

Remark 8.9 - *We notice that:*

i) a steady condition analogous to (8.15) *may be obtained by using the energy relation* (8.117)*;*

ii) the class of basic motions (is not empty and) contains at least the steady motions

$$(\mathbf{v}, p) \equiv \left(\frac{Q}{4\pi r^2}\nabla\mathbf{x}, -\frac{Q^2}{32\pi^2 r^4} + U + const.\right) \qquad (8.126)$$

where Q is a constant and U is the potential of the body force [191]*;*

iii) the case Ω_0 =sphere centered at O has been studied in [158].

8.6.5 A non-linear stability condition for a spatial sink

For definiteness we consider now the basic motion (8.126) in the case $Q < 0$ (i.e. O is a sink) and, for completeness, use (8.117).

Setting

$$V = 4\pi \max_{\Omega}[\mathbf{v}\left(\frac{r}{r_0}\right)^2], \qquad R_e = \frac{V r_0}{\nu} = \frac{|Q|}{\nu r_0}, \qquad (8.127)$$

it turns out that

$$V = \frac{Q}{r_0^2}, \quad |\mathbf{v} - \mathbf{v}_\infty| < \frac{V}{4\pi}\left(\frac{r_0}{r}\right)^2 = \frac{V r_0^2}{4\pi r^2} = \frac{\nu r_0 R_e}{4\pi r^2}, \qquad (8.128)$$

$$2\mathbf{u}\cdot\nabla\mathbf{u}\cdot|\mathbf{v} - \mathbf{v}_\infty| \leq \frac{u^2}{\nu}|\mathbf{v} - \mathbf{v}_\infty|^2 + \nu|\nabla\mathbf{u}|^2 \leq \frac{\nu r_0^2 R_e^2}{(4\pi)^2}\frac{u^2}{r^4} + \nu(\Delta\mathbf{u})^2. \qquad (8.129)$$

Taking into account that (6.63) implies

$$r_0^2 \int_\Omega \frac{u^2}{r^4} d\Omega \leq \int_\Omega \frac{u^2}{r^2} d\Omega \leq 4\int_\Omega (\nabla\mathbf{u})^2 d\Omega, \qquad (8.130)$$

from (8.117) it follows that

$$\|\mathbf{u}(t)\|_2^2 \leq \|\mathbf{u}(t_0)\|_2^2 + \nu\left(\frac{R_e^2}{4\pi^2} - 1\right)\int_{t_0}^t \|\nabla\mathbf{u}\|_2^2 d\tau. \qquad (8.131)$$

Therefore *the condition*

$$R_e < 2\pi \tag{8.132}$$

ensures the stability of the spatial sink in the $L^2(\Omega)$*-norm.*

Remark 8.10 - *Let us notice that* (8.132) *ensures simple stability only. It remains to consider the problem of asymptotic stability and the determination of the "best" conditions ensuring energy stability.*

8.7 THE ONSET OF NATURAL CONVECTION:BÈNARD PROBLEM

8.7.1 Introduction

Because of its relevance in many geographical and industrial applications, the onset of convection has attracted – in the past as nowadays – the attention of many authors. The essential feature of the phenomenon is as follows.

Let S be a horizontal layer of fluid in the rest state and heated from below in such a way that an adverse temperature gradient β is maintained in S. Because of thermal expansions, the fluid at the bottom expands as it becomes hotter. When β reaches a critical value β_c, the buoyancy overcomes gravity, the fluid rises and a pattern of cellular motion may be seen. This is the onset of natural convection. The phenomenon – although recognized earlier by Rumford (1797) and Thomson (1882) – is called Bènard convection because of his experiments (1900) on this phenomenon. Of course, at the onset of convection the rest state loses stability.

Concerning the critical value β_c of β, we point out that it is not the exact parameter for determining the onset of convection. In fact, the onset of convection depends also on the layer depth d, and the correct non-dimensional parameter for describing this *threshold phenomenon* is the Rayleigh number

$$R^2 = \frac{g\alpha\beta d^4}{k\nu}$$

where g denotes gravity and α, k and ν are the coefficients of volume expansion, thermal conductivity and kinematic viscosity respectively. Rayleigh showed that the onset of convection happens when R^2 surpasses a certain critical value R_c^2. We refer to [193] for the study of stability of the rest state by means of the method of linear stability. For non-linear energy stability we refer to [59] [71] and recall the result, first established by Joseph in 1965, of the coincidence of the Rayleigh critical number R_L^2 of linear stability and the Rayleigh critical number R_E^2 of non-linear stability in the L_2-norm.

For more general problems which include new effects, linear theories predict many interesting types of fluid behaviour which have been experimentally verified. In particular, the method of linear stability predicts stabilizing effects when the fluid is

i) rotating about a vertical axis,

ii) electrically conducting and embedded in a magnetic field (in the M.H.D. context).

In these cases, in order to recover the results of linear stability, many efforts have been made in non-linear energy stability theory, and "generalized" energies (i.e. Liapunov functions different from the usual L_2-norm) have been introduced by many authors [59] [190]. In the sequel, in order to give an idea of how these problems have been studied, we consider the stability of the rotating Bènard problem and review a work of Mulone & Rionero [196].

8.7.2 Perturbation equations to the nonconvecting stationary solution

Let us consider an infinite horizontal layer of homogeneous fluid, under the action of a vertical gravity field $\mathbf{g} = -g\mathbf{k}$, in which an adverse temperature gradient $\beta > 0$ is maintained. Moreover, let us suppose that the fluid is rotating about the vertical axis z with a constant angular velocity Ω. Let $d > 0, \Omega_d = \mathbb{R}^2 \times (0, d)$, and $Oxyz$ be a cartesian frame of reference, with unit vectors $\mathbf{i},\mathbf{j},\mathbf{k}$ respectively, rotating about the z axis with the same angular velocity. Let us assume that the layer is parallel to the xy plane.

Here we study the stability of a nonconvecting stationary solution (basic motion) $m_0 = (\mathbf{U}^*, T^* = -\beta z + T_0^*, p_1)$, where $\mathbf{U}^*, T^*$ and p_1 are the velocity, temperature and pressure (incorporating the centrifugal forces) fields respectively; T_0^* is the reference temperature (the temperature on $z = 0$). The fluid is confined between the planes $z = 0$ and $z = d$ with assigned temperatures $T^*(0) = T_0^*$ and $T^*(d) = -\beta d + T_0^*$.

The Oberbeck-Boussinesq equations which govern the evolution of a non-dimensional disturbance $(\mathbf{u}, \theta, p)$ to m_0 are [197]

$$\begin{aligned} &\mathbf{u}_t + \mathbf{u} \cdot \nabla \mathbf{u} = -\nabla p + R\theta \mathbf{k} + \Delta \mathbf{u} + T\mathbf{u} \times \mathbf{k} \\ &\nabla \cdot \mathbf{u} = 0 \\ &Pr(\theta_t + \mathbf{u} \cdot \nabla \theta) = Rw + \Delta \theta \,, \end{aligned} \tag{8.133}$$

in $\Omega_1 \times (0, \infty)$, where $\Omega_1 = \mathbb{R}^2 \times (0, 1)$, with initial conditions

$$\left.\begin{aligned} &\mathbf{u}(x, y, z, 0) = \mathbf{u}_0(x, y, z) \\ &\theta(x, y, z, 0 = \theta_0(x, y, z), \end{aligned}\right\} \quad (x, y, z) \in \Omega_1 \tag{8.134}$$

and stress-free boundary conditions

$$\begin{aligned} &w(x, y, z, t) = 0, \quad u_z(x, y, z, t) = v_z(x, y, z, t) = 0, \\ &\theta(x, y, z, t) = 0, \qquad \text{for} \quad z = 0, z = 1, t > 0. \end{aligned} \tag{8.135}$$

The subscripts z and t denote partial derivatives, $\mathbf{u}_0, \theta_0$ are assigned regular fields with $\nabla \cdot \mathbf{u}_0(\mathbf{x}) = 0, \mathbf{u} = (u, v, w)$, R^2 is the Rayleigh number and T^2, Pr are the Taylor and Prandtl numbers respectively:

$$T^2 = \frac{4\Omega^2 d^4}{\nu^2}, \quad Pr = \frac{\nu}{k}.$$

As is usual, we assume that the perturbations are periodic functions of x and y of periods $2\pi/a_1$ and $2\pi/a_2$, respectively, $(a_1 > 0, a_2 > 0)$ and denote by Ω_p the periodicity cell

$$\Omega_p = [0, \frac{2\pi}{a_1}] \times [0, \frac{2\pi}{a_2}] \times [0, 1]$$

and by $a = (a_1^2 + a_2^2)^{1/2}$ the wave number. Moreover, taking into account the fact that the stability of m_0 makes sense only in a class of solutions of (8.133) – (8.134) in which m_0 is unique, we exclude any other rigid solution requiring the "average velocity condition"

$$\int_{\Omega_p} u d\Omega_p = \int_{\Omega_p} v d\Omega_p = 0.$$

8.7.3 Linear Liapunov stability

Here we study the linear stability of the conduction solution (motionless state) by the Liapunov method.

Following [194], [198 – 199] we introduce the fields $\varsigma = (\nabla \times \mathbf{u}) \cdot \mathbf{k}, w, \theta$ and write the linear evolution equations of $\varsigma, \Delta w$, and θ as

$$\begin{aligned} \varsigma_t &= T w_z + \Delta\varsigma, \\ \Delta w_t &= R\Delta_1\theta - T\varsigma_z + \Delta\Delta w, \\ Pr\theta_t &= Rw + \Delta\theta, \end{aligned} \tag{8.136}$$

where

$$\Delta_1 = \frac{\partial^2}{\partial x^2} + \frac{\partial^2}{\partial y^2}.$$

Let u$ define the *natural field* ϕ (which follows from equation $(8.136)_2$:

$$\phi = R\Delta_1\theta - T\varsigma_z. \tag{8.137}$$

The system (8.136) then becomes

$$\begin{aligned} \varsigma_t &= T w_z + \Delta\varsigma, \\ \Delta w_t &= \phi + \Delta\Delta w, \\ Pr\theta_t &= Rw + \Delta\theta, \\ \phi &= R\Delta_1\theta - T\varsigma_z, \end{aligned} \tag{8.138}$$

and it implies the system

$$\begin{aligned}
\Delta w_t &= \phi + \Delta\Delta w, \\
\phi_t &= \frac{R^2}{Pr}\Delta_1 w - T^2 w_{zz} + \Delta\phi + R(\frac{1}{Pr} - 1)\Delta\Delta_1\theta, \\
Pr\theta_t &= Rw + \Delta\theta, \\
\phi &= R\Delta_1\theta - T\varsigma_z .
\end{aligned} \tag{8.139}$$

Equation $(8.139)_2$ is obtained in this way: multiply $(8.138)_1$ by $-T$ and take the derivative with respect to z, divide $(8.138)_3$ by Pr, take the Δ_1 and then subtract the equations so obtained. To the system (8.139) we add the further boundary conditions

$$\phi(x,y,z,t) = w_{zz}(x,y,z,t) = \theta_{zz}(x,y,z,t) = 0 \quad \text{on} \quad z = 0, z = 1,$$

and for which we refer to [193], [197].

When $Pr = 1$ the system (8.139) becomes

$$\begin{aligned}
\Delta w_t &= \phi + \Delta\Delta w, \\
\phi_t &= R^2\Delta_1 w - T^2 w_{zz} + \Delta\phi, \\
\theta_t &= Rw + \Delta\theta, \\
\phi &= R\Delta_1\theta - T\varsigma_z .
\end{aligned} \tag{8.140}$$

Since the first two equations do not contain the variable θ, then, in the first instance, we can consider separately the system

$$\begin{aligned}
\Delta w_t &= \phi + \Delta\Delta w, \\
\phi_t &= R^2\delta_1 w - T^2 w_{zz} + \Delta\phi .
\end{aligned} \tag{8.141}$$

Now we introduce the Liapunov function (in the fields ϕ and w)

$$E(t) = \frac{1}{2}\int_{\Omega_p} \{\phi^2 + \lambda|\nabla w|^2\} d\Omega_p \tag{8.142}$$

with $\lambda > 0$ an (energy) parameter that will be chosen later. We have

$$\dot{E} = R^2(\Delta_1 w, \phi) - T^2(w_{zz}, \phi) - \lambda(\phi, w) - [||\nabla\phi||^2 + \lambda||\Delta w||^2] \tag{8.143}$$

where

$$m = \max_{\mathcal{H}} \mathcal{F}(w, \phi), \tag{8.144}$$

$$\mathcal{F}(w,\phi) = \frac{R^2(\Delta_1 w, \phi) - T^2(w_{zz}, \phi) - \lambda(\phi, w)}{||\nabla\phi||^2 + \lambda||\Delta w||^2} \tag{8.145}$$

and
$\mathcal{H} = \{w, \phi$ regular in Ω_p, periodic in x, y of periods $2\pi/a_1$ in x and $2\pi/a_2$ in y with given a_1 and $a_2, w = \phi = 0$ on $z = 0, z = 1, w_{zz} = 0$ on $z = 0, z = 1, 0 < \|\nabla\phi\|^2 + \lambda\|\Delta w\|^2 < \infty\}$.

The existence of the maximum m can be proved using the theorem of Rionero [192]. It is easy to see that $m \geq 0$. For this, we observe that if $(w, 0) \in \mathcal{H}$ then $\mathcal{F}(w, 0) = 0$. Assuming

$$m < 1, \tag{8.146}$$

from (8.143) and the Poincaré inequality

$$2E(t)\pi^2 \leq \|\nabla\phi\|^2 + \lambda\|\Delta w\|^2,$$

we have

$$\dot{E}(t) \leq 2\pi^2(m-1)E,$$

and then integrating this last inequality we get

$$E(t) \leq E(0)\exp\{-2\pi^2(1-m)t\}. \tag{8.147}$$

The condition (8.146) is the *linear-energy stability condition.* It gives restrictions on the Rayleigh number R^2 which will ensure linear energy stability.

In order to obtain the subset $\mathcal{H}_1$ of $\mathcal{H}$ of the admissible functions (w, ϕ) for the maximum m, we consider the Euler-Lagrange equations of the variational problem (8.144) – (8.145). They are

$$\begin{aligned} &R^2\Delta_1 w - T^2 w_{zz} - \lambda w + 2m\Delta\phi = 0, \\ &R^2\Delta_1\phi - T^2\phi_{zz} - \lambda\phi + 2m\Delta\Delta w = 0. \end{aligned} \tag{8.148}$$

From these equations and the boundary conditions it follows that all the even derivatives of ϕ and w are zero on $z = 0, z = 1$, then, as a complete set of solutions of (8.148), we may choose the set $\mathcal{H}_1$ constituted by the *normal-mode* solutions in $\mathcal{H}$ of the following kind: $w_n = W_{0n}\sin n\pi z\exp\{i(a_1x + a_2y)\}$, $\phi_n = \phi_{0n}\sin n\pi z\exp\{i(a_1x + a_2y)\}$ with $W_{0n} \neq 0, \phi_{0n} \neq 0$ and the same a_1 and a_2 which define the periodicity cell Ω_p.

Along the solutions $(w_n, \phi_n) \in \mathcal{H}_1$, we have

$$\mathcal{F}(w_n, \phi_n) = \frac{\phi_{0n}W_{0n}(-R^2a^2 + n^2\pi^2T^2 - \lambda)}{\phi_{0n}^2(n^2\pi^2 + a^2) + \lambda W_{0n}^2(n^2\pi^2 + a^2)^2}. \tag{8.149}$$

From (8.148) it is easy to see that

$$\phi_{0n}^2 = \lambda W_{0n}^2(n^2\pi^2 + a^2);$$

then we get

$$\mathcal{F}(w_n,\phi_n) = \pm\frac{R^2a^2 - n^2\pi^2T^2 + \lambda}{2\sqrt{\lambda}(n^2\pi^2 + a^2)^{3/2}} \tag{8.150}$$

where + or − holds if $W_{0n}\phi_{0n} < 0$ or $W_{0n}\phi_{0n} > 0$ respectively. Now we define

$$\mathcal{H}_2 = \{(w_n,\phi_n \in \mathcal{H}_1 : W_{0n}\phi_{0n} < 0\}$$
$$\mathcal{H}_3 = \{(w_n,\phi_n \in \mathcal{H}_1 : W_{0n}\phi_{0n} > 0\}$$

and fix T^2, R^2, a^2 and λ such that

$$R^2a^2 - T^2\pi^2 + \lambda > 0\,. \tag{8.151}$$

For any $(w_n,\phi_n) \in \mathcal{H}_2$, we have

$$\mathcal{F}(w_n,\phi_n) = \frac{R^2a^2 - n^2\pi^2T^2 + \lambda}{2\sqrt{\lambda}(n^2\pi^2 + a^2)^{3/2}} \tag{8.152}$$

and then

$$m_2 = \max_{\mathcal{H}_2} \mathcal{F}(w_n,\phi_n) = \mathcal{F}(w_1,\phi_1) = \frac{R^2a^2 - \pi^2T^2 + \lambda}{2\sqrt{\lambda}(\pi^2 + a^2)^{3/2}}\,. \tag{8.153}$$

For any $(w_n,\phi_n) \in \mathcal{H}_3$, we have

$$\mathcal{F}(w_n,\phi_n) = \frac{n^2\pi^2T^2 - R^2a^2 - \lambda}{2\sqrt{\lambda}(n^2\pi^2 + a^2)^{3/2}}\,. \tag{8.154}$$

It is easy to see that the maximum m of $\mathcal{F}(w,\phi)$ cannot be attained in $\mathcal{H}_3$. In fact, for any $(w_n,\phi_n) \in \mathcal{H}_3$ we have

$$\mathcal{F}(w_n,\phi_n) < \frac{n^2\pi^2T^2}{2\sqrt{n}^3\pi^3} \le \frac{T^2}{2\pi\sqrt{\lambda}}\,.$$

If there existed $(w_{\overline{n}},\phi_{\overline{n}}) \in \mathcal{H}_3$ maximizing $\mathcal{F}$, i.e. $\mathcal{F}(w_{\overline{n}},\phi_{\overline{n}}) = m$, then, by choosing λ in such a way that

$$\frac{T^2}{2\pi\sqrt{\lambda}} < 1 \leftrightarrow \lambda > \frac{T^4}{4\pi^2}\,,$$

we should have stability for any Rayleigh number, but this is not true. To see this, we choose

$$R^2 > \frac{T^2\pi^2 + (\pi^2 + a^2)^3}{a^2} \tag{8.155}$$

and define

$$\alpha = \sqrt{\frac{R^2 a^2 - T^2 \pi^2}{(\pi^2 + a^2)}} - (\pi^2 + a^2), \quad \beta = (\pi^2 + a^2)[\alpha + (\pi^2 + a^2)],$$

$$\tilde{w} = C_0 e^{\alpha t} \sin \pi z e^{i(a_1 x + a_2 y)}, \quad \tilde{\phi} = -C_0 \beta \tilde{w}, \quad C_0 \neq 0,$$

to obtain

$$\tilde{E}(t) = E(\tilde{w}, \tilde{\phi}) = \frac{C_0^2 \pi^2}{a_1 a_2} [\beta^2 + \lambda (\pi^2 + a^2)^2] e^{2\alpha t},$$

where $E(\tilde{w}, \tilde{\phi})$ is the Liapunov function (8.142) evaluated on $(\tilde{w}, \tilde{\phi})$. Since, from (8.155), it follows that $\alpha > 0$, then $\tilde{E}(t) \to +\infty$ for $t \to +\infty$. Hence

$$m = m_2 = 1$$

for

$$R_0^2(a^2) = R_0^2(a^2, \lambda, T^2) = \frac{T^2 \pi^2 - \lambda + 2\sqrt{\lambda}(\pi^2 + a^2)^{3/2}}{a^2}. \tag{8.156}$$

The manner in which the solution for the *linear-energy critical stability parameter* R_{LE}^2 is to carried out is the following. For any fixed T^2 and given a^2 and λ, we must determine the maximum with respect to λ of the characteristic value of R_0^2. The minimum with respect to a^2 of the characteristic values obtained, is the critical Rayleigh number R_{LE}^2 of linear-energy stability

$$R_{LE}^2 = \min_{a^2} \max_{\lambda} R_0^2(a^2, \lambda, T^2). \tag{8.157}$$

Let us now define

$$f(\lambda) = R_0^2(a^2, \lambda, T^2) = \frac{T^2 \pi^2 - \lambda + 2\sqrt{\lambda}(\pi^2 + a^2)^{3/2}}{a^2}. \tag{8.158}$$

For a given $a^2, f(\lambda) > 0$ implies

$$\lambda < \lambda_1 = 2(\pi^2 + a^2)^{3/2} \sqrt{T^2 \pi^2 + (\pi^2 + a^2)^{3/2}} + T^2 \pi^2 + 2(\pi^2 + a^2)^3.$$

Now, in order to compute the maximum with respect to λ of $f(\lambda)$, we observe that

$$\lim_{\lambda \to 0} f(\lambda) = \frac{T^2 \pi^2}{a^2}, \qquad \lim_{\lambda \to \lambda_1} f(\lambda) = 0,$$
$$f'(\lambda) = 0 \quad \text{for} \quad \lambda = \overline{\lambda} = (\pi^2 + a^2)^3 \tag{8.159}$$

and then

$$\max_{\lambda} f(\lambda) = f(\bar{\lambda}) = \frac{T^2\pi^2 + (\pi^2 + a^2)^3}{a^2}. \tag{8.160}$$

For a given a^2 the last term of (8.160) coincides with the characteristic equation of classical linear stability (see Chandrasekhar [193], Chap. III, (128)). Then we find that

$$R_{LE}^2 = \min_{a^2} \frac{T^2\pi^2 + (\pi^2 + a^2)^3}{a^2} = \frac{T^2\pi^2 + (\pi^2 + a_c^2)^3}{a_c^2} = R_c^2 \tag{8.161}$$

where R_c^2 is the critical Rayleigh number for the onset of linear instability, and a^2 is the solution of the equation (see Chandrasekhar [193], \$27, (129) and (131))

$$2a^6 + 3a^4\pi^2 - \pi^4(\pi^2 + T^2) = 0.$$

8.7.4 Conditional nonlinear stability

Here we confine ourselves to presenting the results of nonlinear stability of the basic motionless state obtained in [196]. Introducing the Liapunov function

$$V(t) = E_0(t) + bE_1(t), \qquad b > 0, \tag{8.162}$$

where

$$E_0(t) = E(t) + \frac{\mu}{2}\|\Delta_1\theta\|^2, \qquad \mu = \text{const.} > 0 \tag{8.163}$$

$$E_1(t) = \frac{\|\nabla \mathbf{u}\|^2 + Pr\|\Delta\theta\|^2 + \|\nabla\varsigma\|^2 + \|\Delta w\|^2 + \|\nabla\Delta_1 \mathbf{u}\|^2}{2} \tag{8.164}$$

the following theorem holds

Theorem 8.9 - *There exists a computable constant A such that if*

$$0 < m < 1, \qquad V(0) < A^{-2}, \tag{8.165}$$

then

$$V(t) \leq V(0) exp\left\{\frac{-\pi^2(1-m)}{2}[1 - AV(0)^{1/2}]t\right\}. \tag{8.166}$$

Proof. The proof is too long to be given here. We refer interested readers to [196].

8.8 EXERCISES

Exercise 8.1 - Consider the I.B.V.P. (2.61) in the case of heat diffusion in the "cold ice" of glaciers (2.62). Show that the solutions are asymptotically the same in the L^2- norm, i.e. $\lim_{t\to\infty} ||\mathbf{u}||_2^2 = 0$.

Hint - Take into account (2.62) and that

$$u_t = u_{xx} + \frac{\partial^2}{\partial x^2}\int_0^u g(\xi)d\xi$$

where g is given by (2.62). Show that $1+g > 3/4$ and deduce that $||u||_2^2 \to 0$ (exponentially), when $t \to \infty$.

Exercise 8.2 - Under the assumptions of theorem 4.2, show that $||\nabla\mathbf{u}||_{L^2(\Omega)}$ $\leq ce^{-(\lambda d-m)t}$, where $c =$ const. > 0.

Hint - Set $\mathcal{E}(t) = \frac{1}{2}\int_\Omega(\nabla\mathbf{u})^2 d\mathbf{x} = \Sigma_{1=1}^n \mathcal{E}_i$ where $\mathcal{E}_i = \frac{1}{2}\int_\Omega(\nabla u_1)^2 d\mathbf{x}$. Then obtain

$$\frac{d\mathcal{E}_i}{dt} = \int_\Omega(\nabla u_i\cdot\nabla u_{it})d\mathbf{x} = \int_\Omega(\nabla u_i\cdot(D_i\nabla\Delta u_i + \nabla f_i)d\mathbf{x} =$$

$$D_i\int_{\partial\Omega}\frac{du_i}{d\mathbf{n}}\Delta u_i d\sigma - D_i\int_\Omega(\Delta u_i)^2 d\mathbf{x} + \Sigma_{ij}^{1,n}\Sigma_s^{1,m}\int_\Omega\frac{\partial u_i}{\partial x_s}\frac{\partial u_j}{\partial x_s}\frac{\partial f_i}{\partial u_j}d\Omega$$

Answer - From $\frac{du_i}{d\mathbf{n}} = 0$, $d = \inf(D_1,\ldots,D_m)$, $m = \max_{u_j}\sqrt{\Sigma_{ij}\left(\frac{\partial f_i}{\partial u_j}\right)}$ it follows that

$$\frac{d\mathcal{E}}{dt} \leq -d\int_\Omega(\Delta\mathbf{u})^2 d\mathbf{x} + m\int_\Omega(\nabla\mathbf{u})^2 d\mathbf{x}.$$

But (see Appendix) $\mathbf{u}\in W_2^2(\Omega)$, $\frac{du}{d\mathbf{n}} = 0$ on $\partial\Omega$ implies $||\Delta\mathbf{u}||_2^2 \geq \lambda||\nabla\mathbf{u}||_2^2$ where λ is the smallest positive eigenvalue of $-\Delta$ on Ω. Hence it turns out that

$$\frac{d\mathcal{E}}{dt} \leq 2(m-\lambda d)\mathcal{E}(t) \to ||\nabla\mathbf{u}||_2^2 \leq ||\nabla\mathbf{u}_0||_2^2 e^{-2(\lambda d-m)t}.$$

Exercise 8.3 - Under the assumptions of Exercise 8.2 show that

$$||\mathbf{u}-\mathbf{U}||_{L_\infty(\Omega)} \leq ce^{-(\lambda d-m)t}.$$

Hint - Take into account the result of Exercise 8.2 and that (see Appendix)

$$\lambda\|\mathbf{u}-\overline{\mathbf{u}}\|_2^2 \le \|\nabla\mathbf{u}\|_2^2, \qquad \forall t \ge 0.$$

Exercise 8.4 - The insect dispersal is generally modelled by reaction-diffusion equations, with constant diffusion coefficient. Reference [186] contains values for diffusion coefficients, measured from experiments on several insect species. However, in insect dispersal, when the diffusion increases because of the population pressure, the diffusion coefficient is no longer constant and it is an increasing function of the insect density. In such a case, taking into account (1.141) and assuming

$$\mathrm{j} = -D(\rho)\nabla\rho, \quad \sigma = 0, \quad D(\rho) > 0, \quad \frac{dD}{d\rho} > 0$$

there follows the dispersal equation

$$\rho_t = \nabla\cdot[D(\rho)\nabla\rho].$$

In a typical case,

$$D(\rho) = D_0\left(\frac{\rho}{\rho_0}\right)^n \qquad n \ge 0$$

wherein D_0 and ρ_0 are positive constants, the dispersal equation becomes, in one dimension

$$\rho_t = D_0\frac{\partial}{\partial x}\left[(\frac{\rho}{\rho_0})^n\nabla\rho\right], \qquad x \in I\!R\,.$$

Appending the initial conditions

$$\rho(x,0) = N\delta(x)$$

where $\delta(x)$ is the Dirac delta function and N the number of insects released at $x = 0$ at $t = 0$, verify that

i) for $n = 0$, a solution is given by

$$\rho(x,t) = \frac{Ne^{-x^2/(4D_0t)}}{2\sqrt{\pi D_0 t}}, \qquad t > 0;$$

ii) for $n > 0$, a solution is given by

$$\rho(x,t) = \begin{cases} \rho_0[\varphi(t)]^{-1}\left\{1-[\dfrac{x}{r_0\varphi}]^2\right\}^{1/n}, & |x| \le r_0\varphi(t) \\ 0, & |x| > r_0\varphi(t) \end{cases} \tag{8.167}$$

where

$$\varphi(t) = \left(\frac{2D_0(n+2)t}{r_0^2 n}\right)^{1/(2+n)},$$

$$r_0 = \frac{N\Gamma\left(\frac{1}{n}+\frac{3}{2}\right)}{\sqrt{\pi}n_0\gamma\left(\frac{1}{n}+1\right)},$$

Γ being the gamma function,

iii) the value of r_0 guarantees that the integral of n over all x is N.

Interpreting the solution (8.167) as a kind of wave, determine the wave front and its speed V of propagation.

Hint - The wave front is $x = r_0\varphi(t)$ and it propagates with the speed $r_0\dfrac{d\varphi}{dt}$.

Exercise 8.5 - Verify that the diffusion equation ($k =$ const.)

$$u_t = ku_{xx}$$

does not admit travelling wave solutions.

Hint - Look for solutions in the form $u(x,t) = u(\xi)$ with $\xi = x - ct$. It follows that $2k\dfrac{d^2u}{d\xi^2} + c\dfrac{du}{d\xi} = 0$. Solving, it follows...

Answer - $u(\xi) = a + be^{-\lambda\xi}$, where $\lambda = -c/k$ and a, b are integration constants. Requiring the boundedness of u $\forall\,\xi$ it follows $b = 0$ and hence $u(\xi)$ cannot be a travelling wave because it is constant.

Exercise 8.6 - Verify that $u = 0$ and $u = 1$ are respectively unstable and stable critical points of the kinetic equation

$$\frac{du}{dt} = u(1-u)$$

of the Fisher equation (8.56). Passing to the I.B.V.P.

$$\begin{cases} u_t = u_{xx} + u(1-u) & (x,t) \in I\!R \times I\!R_+ \\ u(x,0) = u_0(x), & x \in I\!R \end{cases}$$

do the points $u = 0$ and $u = 1$ remain, respectively, unstable and stable in the L_∞-norm?

Hint - For the O.D.E. it is enough to consider the linear stability. Linearizing about $u = 0$, it follows that

$$\frac{du}{dt} = u.$$

Linearizing about $u = 1$, it follows that $\dfrac{du}{dt} = -u$.

Answer - $\dfrac{du}{dt} = \pm 1 \rightarrow u = ke^{\pm t}$, k being a constant. Yes (see exercise 2.20.)

Exercise 8.7 - Verify that the equation

$$u_t = u(1 - u^q) + u_{xx}, \qquad q > 0$$

admits exact travelling wave solutions such that

$$u(x,t) = U(\xi), \quad \xi = x - ct, \quad U(-\infty) = 1, \quad U(\infty) = 0$$

by looking for solutions of the form $U = (1 + e^{b\xi})^{-s}$, where b and s are positive constants.

Hint - Find $U'' + cU' + U(1 - U^q) = 0$ and substitute...

Answer - $s = \dfrac{2}{q}$, $b = \dfrac{q}{\sqrt{2(2+q)}}$, $c = \dfrac{4+q}{\sqrt{2(2+q)}}$. See [47], pp. 287-288.

Exercise 8.8 - Denoting by a, d, α, β and γ positive constants, show that the equation

$$u_t = a(u - \alpha)(\beta - u)(u - \gamma) + Du_{xx} \quad (0 < \alpha < \beta < \gamma)$$

has exact travelling wave solutions such that

$$U(\xi) = u(x,t), \quad \xi = x - ct, \quad U(-\infty) = \gamma, \quad U(\infty) = \alpha.$$

Hint - Find $DU'' + cU' + a(U - \alpha)(\beta - U)(U - \gamma) = 0$. Solve the system

$$\begin{cases} DU'' + cU' + a(U - \alpha)(\beta - U)(U - \gamma) = 0 \\ U' = \lambda(U - \alpha)(U - \gamma) \end{cases}$$

choosing suitably the parameter λ.

Answer - $\lambda = \sqrt{\dfrac{a}{2D}}$, $c = (\alpha - 2\beta + \gamma)\sqrt{\dfrac{aD}{2}}$.

Exercise 8.9 - Let Ω be a bounded domain. Then, in the set (8.76), the following Poincaré inequality has been derived [190]:

$$\|\mathbf{u}\|_2^2 \leq \frac{d^2}{\alpha}\|\nabla \mathbf{u}\|_2^2, \qquad \forall \mathbf{u} \in I,$$

where d is the diameter of Ω and $\alpha \approx 80$. Show that: i) $R_E^* = \alpha \approx 80$ is a conservative estimate of (8.77), ii) the condition $R < R_E^*$ ensures exponential energy stability, whatever the geometry of Ω and whatever the form of the basic motion (universal stability), when the Reynolds number is defined as follows:

$$R = \frac{Md^2}{\nu}, \quad M = \lambda_m$$

λ_m being the maximum of the three eigenvalues of $\mathbf{D}$, maximazed over Ω.

Answer - i) $\mathcal{F} \leq \dfrac{m\|u\|_2^2}{\|\nabla u\|_2^2} \leq \dfrac{md^2}{\alpha}$ $\quad \forall \mathbf{u} \in \mathcal{F}$; ii) (8.78) implies $\dfrac{1}{2}\dfrac{d}{dt}\|\mathbf{u}\|_2^2 \leq$ $\leq \dfrac{\nu}{\alpha}(R-\alpha)\|\nabla\mathbf{u}\|^2$. Hence $R \leq \alpha$ implies $\dfrac{1}{2}\dfrac{d}{dt}\|\mathbf{u}\|u_2^2 \leq \dfrac{2\nu}{d^2}(R-\alpha)\|\mathbf{u}\|^2$, i.e. $\|\mathbf{u}\|_2^2 \leq \|\mathbf{u}_0\|^2 \exp[\dfrac{2\nu}{d^2}(R-\alpha)t]$.

Exercise 8.10 - Consider the eigenvalue problem (8.92); show that the eigenvalues are positive and verify that the smallest of them gives (8.90).

Hint - Following a standard procedure, multiply $(8.92)_1$ by the complex coniugate $\overline{\mathbf{q}}$ of $\mathbf{q}$ and integrate over Ω.

Answer - Let $\mathbf{q} = \mathbf{a} + i\mathbf{b}$, where $\nabla \cdot \mathbf{a} = \nabla \cdot \mathbf{b} = 0$ on Ω and $\mathbf{a} = \mathbf{b} = 0$ on $\partial\Omega$. On taking into account that $D_{ij} = D_{ji}$ it follows that

$$\mathbf{q}\cdot\mathbf{D}\cdot\overline{\mathbf{q}} = \Sigma_{rs}(a_r + ib_r)D_{rs}(a_s - ib_s) = \Sigma_{rs}(a_r D_{rs} a_s + b_r D_{rs} b_s) = \mathbf{a}\cdot\mathbf{D}\cdot\mathbf{a} + \mathbf{b}\cdot\mathbf{D}\cdot\mathbf{b}\,.$$

Further

$$\int_\Omega \overline{\mathbf{q}} \cdot \Delta\mathbf{q} d\Omega = \int_\Omega (\mathbf{a}\cdot\Delta\mathbf{a}\mathbf{b}\cdot\Delta\mathbf{b})d\Omega + i\int_\Omega (\mathbf{a}\cdot\Delta\mathbf{b} - \mathbf{b}\cdot\Delta\mathbf{a})d\Omega.$$

But, in view of the boundary conditions it follows that

$$\int_\Omega \mathbf{a}\cdot\Delta\mathbf{b} d\Omega = \int_\Omega \mathbf{b}\cdot\Delta\mathbf{b} d\Omega,$$

hence

$$\mu = \frac{\displaystyle\int_\Omega (\mathbf{a}\cdot\mathbf{D}\cdot\mathbf{a} + \mathbf{b}\cdot\mathbf{D}\cdot\mathbf{b} + \mathbf{a}\cdot\Delta\mathbf{a} + \mathbf{b}\cdot\Delta\mathbf{b})d\Omega}{\|\mathbf{q}\|_2^2} = \text{real number.}$$

Let $\mathbf{u}^* \in \mathcal{I}$ be the function minimizing $\mathcal{G}$ and let $\lambda = \mathcal{G}(\mathbf{u}^*, v)$. Because (8.92) are the Euler-Lagrange equations of the variational problem (8.90), then $\mathbf{u}^*$ is an eigensolution of (8.92) and λ is the corresponding eigenvalue. But if $\mathbf{q}$ is another solution to (8.92) and μ is the corresponding eigenvalue from (8.92) it follows that

$$\mu = \frac{\displaystyle\int_\Omega (\mathbf{q}\cdot\mathbf{D}\cdot\mathbf{q} - \nu\mathbf{q}\cdot\Delta\mathbf{q})d\Omega}{\|\mathbf{q}\|_2^2} = \mathcal{G}(\mathbf{u},\mathbf{v}) \leq \mathcal{G}(\mathbf{u}^*,\mathbf{v}) = \lambda\,.$$

Exercise 8.11 - Verify that, with respect to the $L_2(\Omega)$-scalar product, the operators

$$
\begin{aligned}
S\mathbf{a} &= -\nu\Delta\mathbf{a} + \mathbf{a}\cdot\mathbf{D} \\
A\mathbf{a} &= \mathbf{a}\cdot\underline{\Omega} + \mathbf{v}\cdot\nabla\mathbf{a}
\end{aligned}
$$

are respectively symmetric and skew-symmetric.

Hint - Take into account that $\mathbf{D}$ and $\underline{\Omega}$ are respectively symmetric and skew-symmetric and that $\Delta\mathbf{q}\cdot\mathbf{w} = \nabla\cdot[\nabla\mathbf{q}\cdot\mathbf{w}] - \nabla\mathbf{q} : \nabla\mathbf{w}$.

Answer - $\displaystyle\int_\Omega S\mathbf{a}\cdot\overline{\mathbf{q}}d\Omega = \int_\Omega \mathbf{a}\cdot S\overline{\mathbf{q}}d\Omega$; $\displaystyle\int_\Omega A\mathbf{a}\cdot\overline{\mathbf{q}}d\Omega = -\int_\Omega \mathbf{a}\cdot A\overline{\mathbf{q}}d\Omega$ where $\mathbf{a},\overline{\mathbf{q}} \in I \cup W_2^2$.

Exercise 8.12 - Let Ω be a bounded domain and let $(\mathbf{v}, p)$ be an unsteady basic flow. Show that each of the conditions (8.100) ensures the boundedness of (8.99).

Hint - Take into account the exercise 8.9, and that on the function set (8.76), one has $\displaystyle\int_\Omega \mathbf{u}\cdot\mathbf{D}\cdot\mathbf{u}d\Omega = -\int_\Omega \mathbf{u}\cdot\nabla\mathbf{u}\cdot\mathbf{u}d\Omega$.

Answer - $(8.100)_1$ implies that there exists a positive number k such that $\lambda_m(x,t) \le k$, on $\Omega\times I\!R_+$, when λ_m is the maximum of the three eigenvalues of $\mathbf{D}$. Hence, by the Poincaré inequality (8.80)

$$
\frac{\displaystyle\int_\Omega \mathbf{u}\cdot\mathbf{D}\cdot\mathbf{u}d\Omega}{||\nabla\mathbf{u}||_2^2} \le \frac{k||\mathbf{u}||_2^2}{||\nabla\mathbf{u}||_2^2} \le k\lambda\,.
$$

Analogously $(8.100)_2$ implies that there exists a positive number V such that $|v(\mathbf{x},t)| \le V$ on $\Omega\times I\!R_+$. Hence it follows that

$$
\begin{aligned}
&\frac{\displaystyle\int_\Omega \mathbf{u}\cdot\mathbf{D}\cdot\mathbf{u}d\Omega}{||\nabla\mathbf{u}||_2^2} = \frac{\displaystyle -\int_\Omega \mathbf{u}\cdot\nabla\mathbf{u}\cdot\mathbf{v}d\Omega}{||\nabla\mathbf{u}||_2^2} \le \frac{\displaystyle\int_\Omega |\mathbf{v}||\nabla\mathbf{u}||\mathbf{u}|d\Omega}{||\nabla\mathbf{u}||_2^2} \le \\
&\le \frac{V}{2}\frac{||\mathbf{u}||_2^2 + ||\nabla\mathbf{u}||_2^2}{||\nabla\mathbf{u}||_2^2} \le \frac{V}{2}(\lambda+1).
\end{aligned}
$$

Exercise 8.13 - In the absence of rotation, the Taylor number T is zero and the Oberbeck-Boussinesq equations (sect.8.7) become

$$
\begin{cases}
\mathbf{u}_t + \mathbf{u}\cdot\nabla\mathbf{u} = -\nabla p + R\theta\mathbf{k} \\
\nabla\cdot\mathbf{u} = 0 \\
Pr(\theta_t + \nabla\mathbf{u}\cdot\nabla\theta) = Rw + \Delta\theta\,.
\end{cases}
$$

When the rest planes $z = 0$ and $z = 1$ are assumed rigid, the boundary conditions become

$$\mathbf{u} = \theta = 0 \qquad \text{for} \quad z = 0, z = 1, t \geq 0.$$

Show that the eigenvalues of the linearized problem are real (principle of exchange of stabilities)

Hint - Set $\mathbf{u}(\mathbf{x},t) = e^{\sigma t}\mathbf{u}^*(\mathbf{x})$, $\theta(\mathbf{x},t) = e^{\sigma t}\theta^*(\mathbf{x})$, $p(\mathbf{x},t) = e^{\sigma t}p^*(\mathbf{x})$ where $\mathbf{u}^*, \theta^*, p^*$ are assumed periodic in the x and y directions. Then follow the hint of exercise $\mathcal{E}$.10.

Answer - Omitting the asterisks, it follows that

$$\begin{cases} \sigma\mathbf{u} = -\nabla p + R\theta\mathbf{k} + \Delta\mathbf{u} \\ \nabla \cdot \mathbf{u} = 0 \\ \sigma Pr\theta = Rw + \Delta\theta\,. \end{cases}$$

On multiplying the first and the last equation by the complex conjugate $\tilde{\mathbf{u}}$ and $\tilde{\theta}$ of $\mathbf{u}$ and θ respectively, taking into account the boundary conditions and integrating over a periodicity cell Ω_p, it follows that

$$\sigma \int_{\Omega_p} (\mathbf{u} \cdot \tilde{\mathbf{u}} + Pr\theta\tilde{\theta})d\mathbf{x} = R\int_{\Omega} (w\tilde{\theta} + \tilde{w}\theta)d\mathbf{x} - \int_{\Omega} (\nabla\theta \cdot \nabla\tilde{\theta} + \nabla\mathbf{u} : \nabla\tilde{\mathbf{u}})d\mathbf{x}.$$

Setting $f = f_1 + if_2$, $f_i (i = 1, 2)$ being real, it follows that

$$\sigma = \frac{2R\displaystyle\int_{\Omega} (w_1\theta_1 + w_2\theta_2)d\mathbf{x} - \Sigma_{i=1}^2(\|\mathbf{u}_1\|_2^2 + \|\theta_1\|_2^2)}{\Sigma_{i=1}^2(\|\mathbf{u}_i\|_2^2 + pR|\theta_i\|_2^2)}.$$

But the right hand side is real, hence σ is real.

Exercise 8.14 - Consider the Bénard problem in the absence of rotation (exercise 8.1.3) and show that

i) the critical Rayleigh number of linear stability R_L is the smallest eigenvalue R of the system

$$\begin{cases} R\theta\mathbf{k} + \Delta\mathbf{u} = \nabla p \\ \nabla \cdot \mathbf{u} = 0 \\ Rw + \Delta\theta = 0 \end{cases}$$

$$\mathbf{u} = \theta = 0 \qquad \text{for} \ z = 0, z = 1, t \geq 0\,,$$

ii) the critical Rayleigh number R_E of non-linear stability in the L_2-norm is equal to R_L and that $R < R_L$ ensures nonlinear, unconditional exponential stability in this norm.

Hint - Take into account exercise 8.12 and the procedure of subsection 8.5.1.

Answer - i) is an immediate consequence of exercise 8.13. In fact, because σ is real and the perturbations are assumed to be $e^{\sigma t}(\mathbf{u}, \theta, p)$, the threshold of instability is given by the smallest R for which there exists a steady perturbation ($\sigma = 0$) and the onset of convection manifests itself as stationary convection.

Denoting by Ω a periodicity cell and taking into account the homogeneous boundary conditions on $z = 0$ and $z = 1$, it follows that

$$\int_\Omega \mathbf{u} \cdot \nabla \mathbf{u} \cdot \mathbf{u} d\mathbf{x} = \frac{1}{2} \int_\Omega \mathbf{u} \cdot \nabla \theta^2 d\mathbf{x} = 0 .$$

On setting $E = \frac{1}{2}(\|\mathbf{u}\|_2^2 + Pr\|\theta\|_2^2)$, from the nonlinear equations (8.133) it turns out that

$$\frac{dE}{dt} = \left(R\frac{I}{D} - 1\right) D$$

where

$$I = 2 \int_\Omega w\theta d\mathbf{x}, \quad D = \|\nabla \mathbf{u}\|_2^2 + \|\nabla \theta\|_2^2 .$$

Therefore the critical Rayleigh number R_E of nonlinear stability in the L_2-norm is

$$\frac{1}{R_E} = \max_H \frac{I}{D}$$

where the functional space of the kinematically admissible perturbations is given by $H = \{\mathbf{u}, \theta$ regular in Ω, periodic in the x and y directions and such that $\nabla \cdot \mathbf{u} = 0$ in Ω and $\mathbf{u} = \theta = 0$ on $z = 0, z = 1\}$.

In fact, as is easily seen,

$$R < R_E \rightarrow E \leq E_0 \exp[-2\lambda_1 \left(\frac{R_E - R}{R_E}\right) t]$$

where λ_1 is the constant appearing in the Poincaré inequality.

The Euler-Lagrange equations for the maximum on $\frac{I}{D}$ on H are found to be

$$\begin{cases} R_E \theta \mathbf{k} + \Delta \mathbf{u} = \nabla p \\ \nabla \cdot \mathbf{u} = 0 \\ R_E w + \Delta \theta = 0 \end{cases}$$

together with homogeneous conditions on $z = 0$ and $z = 1$. This B.V.P. does not differ from the B.V.P. governing the steady perturbations of linear stability (sec. i)), hence ii) is completely proved.

Let us remark that the result of exercises 8.14-8.15 constitutes the theorem of Joseph mentioned in the subsection 8.7.1.

Chapter nine

HYPERBOLIC, RELATED EQUATIONS, AND OTHERS

9.1 INTRODUCTION

In this last chapter we consider some hyperbolic and related equations and others. The first section is devoted to the Monge-Ampère equation which is a celebrated example of an equation which can change character. Next (sect. 9.3) we introduce functionals which are good candidates for producing Liapunov functions for some hyperbolic P.D.E.s. The sections 9.4-9.5 are devoted respectively to an hyperbolic heat conduction equation and to Arnold's theorem concerning the nonlinear stability of steady plane flows of an ideal incompressible fluid. In section 9.6 we perform a complete qualitative analysis of a nonlinear integrodifferential equation of the particle transport theory. Finally in the last section we consider an I.B.V.P. for an *exterior*, elastic region where the elastic field is allowed to grow rapidly at infinity and derive therefor some estimates of the continuous dependence type.

9.2 MONGE-AMPÈRE EQUATIONS

9.2.1 Introduction

Whereas terminology varies somewhat, we shall follow the nomenclature of Gilbarg and Trudinger [201] in referring to the P.D.E. (in two independent variables x, y)

$$\psi_{xx}\psi_{yy} - \psi_{xy}^2 = f(x, y) \tag{9.1}$$

as the Monge-Ampère equation, it being understood that $f(x,y)$ is assigned. Equations which differ from this to the extent that f depends also upon ψ and/or its first derivatives are referred to by Gilbarg and Trudinger [201] as equations of Monge-Ampère type. We shall principally discuss aspects of the Monge-Ampère equations (in the above terminology) but shall also deal with some equations of Monge-Ampère type.

Let us note that the character of (9.1) – elliptic, hyperbolic etc. – is determined by the sign of the (assigned) function $f(x,y)$:

if $f < 0$ the equation is *hyperbolic*;
if $f = 0$ the equation is *parabolic*;
if $f > 0$ the equation is called *elliptic*

(although [e.g.[201]] conventional usage in the latter case requires, in addition, that ψ_{xx} and ψ_{yy} should both be positive; however, if they are not, it is clear that $(-\psi)_{xx}$ and $(-\psi)_{yy}$ are – an insubstantial difference).

We give one possible – and useful – interpretation of (9.1) which has applications in meteorology. Consider a continuum (e.g. fluid) in (steady or unsteady) two-dimensional flow: the x, y, rectangular cartesian components of velocity u, v, respectively, are both functions of x, y. If the velocity field is divergence-free – as happens, for example, when the continuum is of fixed density – u, v may be written in terms of a stream function $\psi(x,y)$:

$$u = \psi_y \qquad v = -\psi_x \tag{9.2}$$

and the stream function ψ satisfies the Monge-Ampère equation (9.1) where $f(x,y)$ is related to the rectangular cartesian components of the acceleration field $a_x(x,y), a_y(x,y)$ by

$$f = -\frac{1}{2}(a_{x,x} + a_{y,y}) \quad (= -\frac{1}{2}\nabla \cdot \mathbf{a})\,; \tag{9.3}$$

in the case of a perfect fluid of constant density ρ, in the absence of body forces, the acceleration field $\mathbf{a}$ is given in terms of the pressure $p(x,y)$ by

$$\mathbf{a} = -\nabla_1 p/\rho\,. \tag{9.4}$$

As the literature on the Monge-Ampère equation is extensive, we merely outline some of the most notable references.

Gilbarg and Trudinger [201] give an extensive discussion of equations of the Monge-Ampère type in the elliptic case, including existence questions, together with an extensive list of references; notable contributions cited therein include those of Alexandrov, Bakelman, Heinz, Lewy, P.L. Lions, Nirenberg and Pogorolev. Interesting estimates for the two-dimensional (elliptic) case have also been given by Talenti [202].

So far as equations of Monge-Ampère type in the hyperbolic case are concerned we may cite the treatment given by Courant and Hilbert [30]. Among the most notable contributions to this area – which have a particular relevance to the differential geometry of surfaces of negative curvature – are those of Efimov and his school e.g. [203]; for meteorological applications, see the work of Rozendorn e.g. [204].

The mixed case i.e. when f, occurring in (9.1), changes sign is problematic – even where local existence is concerned; e.g. see [205].

Theorems 9.1-9.5 are based largely upon Flavin [135] but the remaining results are hitherto unpublished.

We shall use the cross-sectional method to analyze both a Dirichlet type problem and a problem of the initial boundary value type, principally for (9.1) – assuming the existence of sufficiently smooth solutions. It will be evident on recalling Chapter four, that Lemma 4.3 – in particular, Lemmas 4.3a, 4.3b – provides a natural vehicle for conveying the cross-sectional estimates that we seek.

We shall hereunder recast slightly Lemmas 4.3a and 4.3b together with their geometrical contexts. It suits our purpose to refer to two slightly different contexts, corresponding to the Dirichlet type problem and the initial boundary value problem referred to above; the contexts are referred to (suggestively) as *Context* D and *Context* I respectively. Recording the slightly recast lemmas has the advantage of making the material in this section (largely) self-contained.

The geometrical context is as follows. We consider a plane region R characterized as follows (x, y denoting rectangular cartesian coordinates therein):

(a) It is bounded, ***inter alia***, by two curves C_+ (upper) and C_- (lower); these curves $y = y_+(x), y = y_-(x)$, respectively, have the properties:

(α) they are single valued functions defined in $0 \leq x \leq L$ (*Context* D) or in $0 \leq x$ (*Context* I),

(β) they are twice continuously differentiable with respect to x in $0 < x < L$ or in $0 < x$, as appropriate (i. *Contexts* D,I respectively),

(γ) $y_+(x) > y_-(x)$, except, possibly, at $x = 0$ and/or $x = L$ where the two values coincide (degenerate cases).

(b) It is also bounded by straight line segments (parallel to the y axis) Γ_0, Γ_L, with abscissae $x = 0, x = L$ respectively, either or both of which may, in degenerate cases, vanish (*Context* D)

or

It is bounded by a straight line segment Γ_0 with abscissa $x = 0$, and occupies $x \geq 0$ (no upper limit for x being envisaged). (*Context* I).

Let Γ_x be the straight line segment contained in R whose points have abscissa x.

A function $\psi(x, y)$ is considered in R which has the following properties:

(i) $\psi \in C^2[\overline{R}/(\Gamma_0 \cup \Gamma_L)]$ $\quad$ (*Context* D)

or

$\psi \in C^2[\overline{R}/\Gamma_0]$ $\quad$ (*Context* I),

and

$$\int_{\Gamma_x} \psi_y \psi_{yxx} dy \quad \text{exists and is a continuous function of } x$$

for $0 < x < L$ (*Context D*) *or* $0 < x$ (*Context I*).

(ii) It satisfies

$$\psi = \mathcal{F}_+(x) \quad \text{on } \mathcal{C}_+, \qquad \psi = \mathcal{F}_-(x) \quad \text{on } \mathcal{C}_- \tag{9.5}$$

– supposed twice continuously differentiable.

We now rephrase Lemma 4.3a in a manner appropriate to the (slightly) modified context.

Lemma 9.1 - *In the context specified above* $((a), (b), (i), (ii))$ *suppose that*

$$\mathcal{F}_+''(x) = \mathcal{F}_-''(x) = 0\,.$$

Defining

$$F(x) = \int \psi_y^2 dy\,, \tag{9.6}$$

one has – for $0 < x < L$ (*Context* D) *or* $0 < x$ (*Context* I) *– that*

$$F'(x) = 2\int \psi_y \psi_{yx} dy + \psi_y^2 y'\,]\,, \tag{9.7}$$

$$F''(x) = 2\int (\psi_{xx}\psi_{yy} - \psi_{xy}^2) dy - \psi_y^2 y''\,]\,. \tag{9.8}$$

It is understood above, and subsequently, that all integrals arising are over Γ_x – unless the contrary is stated; moreover, the symbol] is as explained in Chapters three and four. The necessary modifications in the proof given in Chapter four are minimal.

We now rephrase Lemma 4.3b in a manner appropriate to the slightly modified context; the necessary modifications are again slight.

Lemma 9.2 - *In the context outlined above* $((a),(b),(i),(ii))$ - *prior to Lemma* 9.1 - *suppose that*

$$(i) \qquad y_+'' < 0 \qquad y_-'' > 0$$

(i.e. the curves C_+, C_- *are* **strictly** *convex)*
or

$$(ii) \qquad y_+'' > 0 \qquad y_-'' < 0$$

(i.e. the curves C_+, C_- *are* **strictly** *concave).*
Then $F(x)$, *as defined in* (9.6), *satisfies – in addition to* (9.7) –

$$F''(x) \geq -2\int(\psi_{xx}\psi_{yy} - \psi_{xy}^2)dy + \frac{\mathcal{F}''^2}{y''}] \tag{9.9}$$

in the case of the boundary restrictions (*i*) *immediately above, and*

$$F''(x) \leq -2\int(\psi_{xx}\psi_{yy} - \psi_{xy}^2)dy + \frac{\mathcal{F}''^2}{y''}] \tag{9.10}$$

in the case of the boundary restrictions (*ii*) *immediately above.*

9.2.2 Problems of Dirichlet type

Suppose that $\psi(x,y)$ satisfies the Monge-Ampère equation (9.1) in R, R being the region defined above for Context D (i.e. $0 \leq x \leq L$);
that ψ has therein the smoothness properties envisaged for Case D, together with

$$\int_{\Gamma_x} \psi_y^2 dy \in C(0 \leq x \leq L)\,;$$

that on the lateral boundaries of R

$$\psi = \mathcal{F}_+(x) \quad \text{on } C_+\,, \qquad \psi = \mathcal{F}_-(x) \quad \text{on } C_-\,; \tag{9.11}$$

and that
either

$$\psi = \text{specified on} \quad \Gamma_0, \Gamma_L \quad \text{(when non-degenerate)} \tag{9.12$_1$}$$

(*pointwise specification*)[1]

---

[1] $\int_{\Gamma_x} \psi_y^2 dy$ can be taken to be zero in the degenerate cases at the relevant end point/s.

or

$$\int_{\Gamma_x} \psi_y^2 dy = \text{specified for} \quad \Gamma_0, \Gamma_L \quad \text{(when non-degenerate)} \qquad (9.12_2)$$

(*global specification.*)[2]

Remark 9.1 - *Let us note that specifying the global data* $\int \psi_y^2 dy$ (*on* Γ_0 *and/or* Γ_L) *is tantamount to specifying the global momentum flux* (*on the relevant segment/s*) *in the context of a continuum/fluid at constant density in steady flow.*

In the case of the standard Dirichlet problem (*pointwise specification of* ψ *on the boundary*) *for the elliptic equation it is known* [30] *that there are at most two solutions. When global specification of data – over* Γ_0, Γ_L *– obtains, one might expect a family of possible solutions. It is natural to speculate as to the relationship away from the ends of* (*suitable*) *pairs of such solutions – as suggested by Saint-Venant's principle.*

The following two estimates for Dirichlet problems follow readily from Lemmas 9.1 and 9.2 together with the property (4.2) of convex functions; they are embodied in Theorems 9.1 and 9.2.

Theorem 9.1 - *For the Dirichlet type problem for which*

$$y_+'' \leq 0, \qquad y_-'' \geq 0$$

(*convex lateral boundaries*) *and*

$$\mathcal{F}_+'' = 0, \qquad \mathcal{F}_+'' = 0,$$

one has:

$$F(x) = \int \psi_y^2 dy$$

satisfies

$$F''(x) \geq -\int 2f dy;$$

and consequently

$$F(x) \leq G(x) \qquad (9.13_1)$$

---

[2]Cfr. footnote 1

where

$$G''(x) = -2\int f dy \tag{9.13$_2$}$$

subject to

$$G(0) = F(0), \qquad G(L) = F(L), \tag{9.13$_3$}$$

both of which are available from data.

The equality signs hold above iff the lateral boundaries C_+, C_- are straight lines.

Theorem 9.2 - *For the Dirichlet type problem for which the lateral boundaries C_+, C_- are* **strictly** *convex ($y_+'' < 0$, $y_-'' > 0$) one has:*

$$F(x) = \int \psi_y^2 dy$$

satisfies

$$F(x) \leq G(x) \tag{9.14$_1$}$$

where $G(x)$ satisfies

$$G''(x) = -\int 2f\, dy + \frac{\mathcal{F}''^2}{y''}] \tag{9.14$_2$}$$

subject to

$$G(0) = F(0)\,, \qquad G(L) = F(L) \tag{9.14$_3$}$$

both of which are available from the data.

Remark 9.2 - *The existence theorems (e.g. [201]) proved for the Dirichlet problem in the elliptic case (the most natural case) all relate to a domain with convex boundary. It is of interest that convex boundaries arise naturally in the above two Theorems via Lemmas* 9.1, 9.2.

Let us recall ((4.111)) that a lower bound is also available for $F(x)$ – quite independently of the P.D.E. and the problem at hand –

$$F(x) \geq \frac{\{\mathcal{F}]\}^2}{l(x)}\,, \tag{9.15}$$

$l(x)$ being the width of the region at abscissa x. Let us note also that pointwise upper bounds for $|\psi|$ are, of course, available from the upper bounds for F.

Non-existence. We complete the treatment of Dirichlet type problems with a short discussion of non-existence. Similar considerations obtain in the case of initial boundary value problems subsequently discussed.

We commence with a simple result – of particular relevance to the case where zero conditions are prescribed on the lateral boundaries ($\mathcal{F}_+ = \mathcal{F}_- = 0$). If there exists a point $x_1 : 0 < x_1 < L$ for which $G(x_1) < 0$ where $G(x)$ is the solution of (9.13_2) subject to (9.13_3), then no solution of the problem exists. That this is so is obvious – for the existence of such a point would contradict the non-negativity of F. Moreover, it is easy to construct values of $f(x, y)$ for which $G < 0$. A particulary simple example arises when $F(0) = F(L) = 0$ – notably when Γ_0, Γ_L are degenerate (each vanishes): if

$$\int_R f\, dA < 0 \tag{9.16}$$

non-existence follows; this is readily verified.

A more general non-existence criterion can be obtained on using (9.15) together with upper estimates for $F(x)$. The following is evident therefrom.

Theorem 9.3 - *No solution exists of the problem of Dirichlet type, for which the lateral boundaries C_+, C_- are convex/strictly convex, if there exists a point $x : 0 < x < L$ for which*

$$\frac{\{\mathcal{F}]\}^2}{l(x)} > G(x) \tag{9.17}$$

where $G(x)$ is determined by (9.13_2)–(9.13_3) or (9.14_2)–(9.14_3) as appropriate.

9.2.3 Problems of Initial Boundary Value type

Suppose that $\psi(x, y)$ satisfies (9.1) in R, R being the region defined above for Context I ($x \geq 0$);

that ψ has therein the smoothness properties envisaged for Context I, together with the requirement

$$F(x) \in C^1(x : x \geq 0)$$

where $F(x)$ is defined by (9.6);

that on the lateral boundaries

$$\psi = \mathcal{F}_+(x) \quad \text{on } C_+ \,, \qquad \psi = \mathcal{F}_-(x) \quad \text{on } C_- \,; \tag{9.18}$$

and that

either

$$\psi, \psi_x = \text{specified on } \Gamma_0 \quad (\textit{pointwise specification}) \tag{9.19_1}$$

or

that on $x = 0$

$$F(0) = \text{specified}, \quad F'(0) = \text{specified } (\textit{global specification}). \tag{9.19_2}$$

In the case of pointwise specification, it is necessary to posit enhanced regularity up to Γ_0 for ψ in order that F' may be calculated from (9.7): for example, $\psi \in C^2(\overline{R})$ – although less stringent conditions would suffice.

The following two estimates for the problems of the ***initial boundary value type*** are consequences of Lemmas 9.1 and 9.2; they are embodied in Theorems 9.4 and 9.5.

Theorem 9.4 - *For the problem of the Initial Boundary Value type, for* 9.1, *for which*

$$y''_+ \geq 0\,, \qquad y''_- \leq 0$$

(concave lateral boundaries) and

$$\mathcal{F}''_+ = 0\,, \qquad \mathcal{F}''_- = 0$$

one has:

$$F(x) = \int \psi_y^2 \, dy$$

satisfies

$$F(x) \leq G(x) \tag{9.20$_1$}$$

where $G(x)$ satisfies

$$G''(x) = -\int 2f \, dy\,; \tag{9.20$_2$}$$

subject to

$$G(0) = F(0)\,, \qquad G'(0) = F'(0)\,, \tag{9.23$_3$}$$

both of which are calculable from data.

The equality signs hold above iff the lateral boundaries C_+, C_- are straight lines.

That this is so follows from

$$F'' \leq -\int 2f \, dy$$

(a consequence of Lemma 9.1), and the fact that $H(x) \in C^1(x : x \geq 0)$, $H''(x) \leq 0(x \geq 0)$, $H(0) = H'(0) = 0$ obviously imply that $H \leq 0\,(x \geq 0)$.

Theorem 9.5 - *For the problem of the Initial Boundary Value type, for* (9.1), *for which*

$$y''_+ > 0, \qquad y''_- < 0$$

(**strictly** *concave lateral boundaries), with general boundary conditions on the lateral boundaries, one has*

$$F(x) \leq G(x), \tag{9.21$_1$}$$

where $G(x)$ satisfies

$$G''(x) = -\int 2f\,dy + \frac{\mathcal{F}''^2}{y''}\Big] \qquad (x>0) \tag{9.21$_2$}$$

subject to

$$G(0) = F(0), \qquad G'(0) = F'(0) \tag{9.21$_3$}$$

both of which are available from the data.

This follows from Lemma 9.2 and the simple principle for $H(x)$ referred to in connection with Theorem 9.3.

Non-existence criteria. These follow using the same general approach adopted in connection with Dirichlet type problems.

We commence with a simple *explicit* non-existence criterion appropriate to boundary conditions on concave lateral boundaries ($y_+'' \geq 0$, $y_-'' \leq 0$) such that

$$\mathcal{F}_+'' = \mathcal{F}_-'' = 0\,.$$

Suppose that (min.=minimum)

$$\min_x \int f\,dy = -m \tag{9.22}$$

where m is a constant, then it is evidently a consequence of Lemma 9.1 – similar to Theorem 9.4 – that

$$F(x) \leq mx^2 + F'(0)x + F(0)\,. \tag{9.23}$$

If the data are such that $F(x) < 0$ at any point $x = x_1 (> 0)$ it is evident that the solution fails to exist thereat. Explicit criteria are easily obtained. We consider, in particular, hereunder, the case $m > 0$ – a plausible condition as initial boundary value problems are (most) suited to hyperbolic equations. Whereas the case $m < 0$ is less plausible – as elliptic equations do not provide a natural setting for initial boundary value problems (e.g. ill-posedness previously encountered) – nevertheless, it is of interest to consider it. Plainly the following explicit criteria are deducible from (9.12) and (9.23)

Theorem 9.6 - *The solution to problems of the initial boundary value type (under conditions specified immediately above) fails to exist for sufficiently large value of x (> 0) if either*
(i) $m > 0$ and the initial conditions are such that

$$F'(0) < 0\,, \qquad F'^2(0) - 4mF(0) > 0, \tag{9.24}$$

or

(ii) $m < 0$ *irrespective of the initial conditions.*

Remark 9.3 - *Theorem* 9.3 *continues to be valid, mutatis mutandis, in the context of problems of initial boundary value type.*

Other estimates for I.B.V.P. We now proceed to discuss other cross-sectional estimates for problems of the initial boundary value type (slightly) modified in respect of the smoothness required of the solution and in respect of the global characterization of initial data. We show how it is possible to use (the general form of) Lemma 4.3 (with some minor adjustments) to obtain $L^{2p}(p > 1)$ bounds for ψ_y and $_+\psi_y$ (the quantity $_+u$ being as defined in Chapter three, 3.2.3).

We shall be concerned again with the region $R(x \geq 0)$ envisaged for Context I. Let us first note that Lemma 4.3 holds in R for somewhat less restricted functions than originally envisaged: it continues to be valid in R for functions $\psi \in C^2(\overline{R}/\Gamma_0)$ and such that $\int \dot{\phi}\psi_{yxx}dy$ exists and is a continuous function of x $(x \geq 0)$.

We derive the required estimates through a sequence of Lemmas. Let us first note the following:

Let us *suppose*, henceforward, in relation to the r.h.s. of (9.1), that

$$\mu(x,y) = -f(x,y) \geq 0 \tag{9.25}$$

and *define*

$$F_p(x) = \int (\psi_y^2)^p dy, \tag{9.26}$$

$$_+F_p(x) = \int (_+\psi_y^2)^p dy, \tag{9.27}$$

where p is any integer such that $p > 1$. Lemma 4.3 (slightly modified as above) implies the following two related Lemmas:

Lemma 9.3a - *Suppose that* $\psi(x,y)$ *satisfies* (9.1) (*wherein* (9.25) *is understood*),

(1) *that* $$\psi(x,y) \in C^2(\overline{R}/\Gamma_0)$$

and

$$\int_{\Gamma_x} \psi_y^{2p-1}\psi_{yxx}dy$$

exists and is a continuous function of x for all $x > 0$;
(2) *that, in relation to the lateral boundaries*

$$(i)\quad y''_+ \geq 0, \qquad\qquad y''_- \leq 0,$$
$$(ii)\quad \mathcal{F}''_+ = \mathcal{F}''_- = 0.$$

Then $F_p(x,y)$ satisfies

$$F''_p(x) - 2p(2p-1)M_p(x)F_p^{1-1/p}(x) \leq 0 \qquad (0 < x) \tag{9.28}$$

where

$$M_p(x) = \{\int \mu^p(x,y)dy\}^{1/p} \tag{9.29}$$

μ being defined in (9.25).

Lemma 9.3b - *Suppose that*

(1) $$\psi(x,y) \in C^2(\overline{R}/\Gamma_0)$$

and

$$\int_{\Gamma_x} ({}_+\psi_y)^{2p-1}\psi_{yxx}dy$$

exists and is a continuous function of x for all $x > 0$;
(2) *that, in relation to the lateral boundaries,*

$$(i)\quad y''_+ \geq 0, \qquad\qquad y''_- \leq 0,$$
$$(ii)\quad \mathcal{F}''_+ \leq 0, \qquad\qquad \mathcal{F}''_- \geq 0.$$

Then ${}_+F_p(x,y)$ satisfies

$${}_+F''_p(x) - 2p(2p-1)M_p(x)\,{}_+F_p(x) \leq 0 \qquad (0 < x) \tag{9.30}$$

where $M_p(x)$ is defined in (9.29).

The proofs of these are similar. In the case of Lemma 9.3a, we have from Lemma 4.3,

$$F''_p \leq 2p(2p-1)\int \mu\psi_y^{2p-2}dy \leq 2p(2p-1)M_p(x)F_p^{1-1/p}$$

on using Hölder's inequality in the last step. Lemma 9.3b is similar.

We now give two results which show how upper bounds for F_p and ${}_+F_p$ follow from Lemmas 9.3a and 9.3b respectively. These are embodied in Theorems

9.7a and 9.7b which follow.

Theorem 9.7a - *Suppose that all conditions specified in Lemma* 9.3*a obtain, and in addition thereto that*

$F'_p(x)$ *is continuous in* $x \geq 0$, *and that the initial conditions are such that*

$$F_p(0) > 0\,, \qquad F'_p(0) \geq 0\,.$$

Then

$$F_p(x) \leq G(x) \tag{9.31$_1$}$$

where $G(x)(>0)$ *satisfies*

$$G'' - 2p(2p-1)M_p(x)G^{1-1/p} = 0 \tag{9.31$_2$}$$

$$G(0) = F_p(0)\,, \qquad G'(0) = F'_p(0). \tag{9.31$_3$}$$

Theorem 9.7b - *Suppose that all conditions specified in Lemma* 9.3*b hold, and in addition thereto that*

${}_+F'_p(x)$ *is continuous in* $x \geq 0$, *and that the initial conditions are such that*

$${}_+F_p(0) > 0\,, \qquad {}_+F'_p(0) \geq 0\,.$$

Then

$${}_+F_p(x) \leq G(x) \tag{9.32$_1$}$$

where $G(x)(>0)$ *satisfies*

$$G'' - 2p(2p-1)M_p(x)G^{1-1/p} = 0\,, \tag{9.32$_2$}$$

$$G(0) =_+ F_p(0)\,, \qquad G'(0) =_+ F'_p(0)\,. \tag{9.32$_3$}$$

The *proofs* of both of the foregoing are similar, and only the first need be dealt with. The proof can be made to depend on the following proposition (e.g. Chapter one of Protter and Weinberger [131]):
If $H(x)$, defined for $x \geq 0$, is such that $H(x) \in C^1(0 \leq x)$, and if $m(x)$ is uniformly bounded for $x \geq 0$, then

$$H'' - m^2(x)H \leq 0 \qquad \text{for } x > 0\,, \tag{9.33$_1$}$$

$$H(0) = H'(0) = 0, \tag{9.33$_2$}$$

implies that

$$H(x) \leq 0 \qquad \text{for } x \geq 0. \tag{9.33$_3$}$$

It can be made so dependent on writing

$$H(x) = F_p(x) - G(x)$$

giving

$$\begin{aligned} 0 \geq & H''(x) - M_p(x)\{F_p^{1-1/p}(x) - G^{1-1/p}(x)\} \\ = & H''(x) - M_p(x)(1 - 1/p)\{\theta F_p + (1-\theta)G\}^{-1/p} H \end{aligned}$$

(where $0 < \theta < 1$) using the Mean Value Theorem. It is clear that the coefficient of H in the last line is uniformly bounded in view of the properties of F_p, G. Thus the proof is complete.

In some circumstances it is possible to extract *explicit* upper bounds for $F_p(x),_+F_p(x)$ from Theorems 9.7a and 9.7b. The following are examples:

Theorem 9.8a - *Suppose that all the conditions specified in Theorem 9.7a obtain, and that in addition thereto*

$$M_p'(x) \geq 0, \tag{9.34$_1$}$$

$$F_p^{2-1/p}(0) - \{F_p'(0)\}^2/\{4p^2 M_p(0)\} \geq 0, \tag{9.34$_2$}$$

then

$$F_p^{1/(2p)}(x) \leq F_p^{1/(2p)}(0) + \int_0^x M_p^{1/2}(x)dx. \tag{9.35}$$

Proof. Multiplying (9.31$_2$) by $2G'$ yields

$$(G'^2)' - 4p^2 M_p(x)(G^{2-1/p})' = 0.$$

Successive integrations yield

$$\begin{aligned} G'^2(x) &= \{F_p'(0)\}^2 + 4p^2 \int_0^x M_p(x)(G^{2-1/p})'dx \\ &= 4p^2 M_p(x)G^{2-1/p} - 4p^2 M_p(0)[F_p^{2-1/p}(0) - \{F_p'(0)\}^2/\{4p^2 M_p(0)\}] \\ &\quad - 4p^2 \int_0^x M_p'(x)G^{2-1/p}dx. \end{aligned}$$

In view of $(9.34_1), (9.34_2)$ it follows that

$$G' - 2pM_p^{1/2}(x)G^{1-1/(2p)} \leq 0$$

whence the required result follows on integration.

The following arises in a similar manner.

Theorem 9.8b - *Suppose that all the conditions specified in Theorem* 9.7*b obtain, and that in addition thereto*

$$M_p'(x) \geq 0, \tag{9.36$_1$}$$

$$\{{}_+F_p(0)\}^{2-1/p} - \{{}_+F_p'(0)\}^2/\{4p^2M_p(0)\} \geq 0, \tag{9.36$_2$}$$

then

$$\{{}_+F_p(x)\}^{1/(2p)} \leq \{{}_+F_p(0)\}^{1/(2p)} + \int_0^x M_p^{1/2}(x)dx\,. \tag{9.37}$$

Remark 9.4 - *The equality sign holds in* (9.35), (9.37) *when* ψ *has the general form*

$$\psi = (1 + \nu y)^S(\alpha + \beta x)$$

where ν, s, α, β *are suitable constants; it also holds in a limiting case of the foregoing – when* ψ *has the form*

$$\psi = e^{\gamma y}(\alpha + \beta x)$$

where γ *is also a constant; these are readily verified.*

Finally we contemplate an interesting limiting case of Theorems 9.7a and 9.7b.

Theorem 9.9 - *Suppose that all the conditions posited in Theorems* 9.8*a and* 9.8*b hold for all sufficiently large allowable p, then one has respectively*

$$\sup_y |\psi_y(x,y)| \leq \sup_y |\psi_y(0,y)| + \int_0^x \sup_y \sqrt{\mu(x,y)}dy\,| \tag{9.38$_1$}$$

$$\sup_y {}_+\psi_y(x,y) \leq \sup_y {}_+\psi_y(0,y) + \int_0^x \sup_y \sqrt{\mu(x,y)}dy\,. \tag{9.38$_2$}$$

These follow on letting $p \to \infty$ in (9.35) and (9.37). The equality sign holds in both of these in the circumstances outlined in the previous remark.

9.2.4 Other Equations of Monge-Ampère type

Results similar to those obtained above for the Monge- Ampère equation can also be obtained for some similar equations (see Exercises). In particular, this is so for the class of equations of Monge-Ampère type

$$\psi_{xx}\psi_{yy} - \psi_{xy}^2 = \{\ddot{\phi}(\psi_y)\}^{-1} g(x,y) \tag{9.39}$$

where $g(x,y)$ is assigned, and $\phi(u)$ is any twice differentiable function of its argument which has the following properties:

$$\phi(u) > 0 \qquad \forall\, u, \text{ except that } \phi(0) = 0\,;$$

$$\ddot{\phi}(u) > 0\,;$$

in the above a superposed dot means differentiation with respect to the argument u. Lemma 4.3 is the obvious vehicle for such estimates for $\int \phi(\psi_y)dy$ – bearing in mind that for such functions ϕ

$$u\dot{\phi}(u) - \phi(u) \geq 0$$

– as seen before.

We finish up with an analysis of an initial value problem for a P.D.E. of Monge-Ampère type called the *equation of prescribed Gauss curvature*, and we obtain a *non-existence* result therefor. The equation is

$$\frac{\psi_{xx}\psi_{yy} - \psi_{xy}^2}{(1 + \psi_x^2 + \psi_y^2)^2} = -\mu(x,y) \tag{9.40}$$

and expresses the fact that a surface $z = \psi(x,y)$ has Gaussian curvature $-\mu$. We content ourselves with a simple result (which suffices to elucidate the basic principle) and do not give the strongest result possible. It is left as an exercise to explore stronger versions and generalizations.

Theorem 9.10 - *Suppose that ψ satisfies all conditions specified for problems of initial boundary value type* [(9.18) – (9.19$_2$) *and smoothness assumptions etc. preceding them*] *except that* (9.40), *wherein* $\mu > 0$, *replaces* (9.1), *and that in respect of the lateral boundaries*

$$y_+'' = y_-'' = 0\,, \qquad \mathcal{F}_+'' = \mathcal{F}_-'' = 0\,. \tag{9.41}$$

Defining

$$\nu(x) = \{\int \mu^{-1}(x,y)dy\}^{-1}, \tag{9.42}$$

suppose in addition that

$$\nu'(x) \le 0, \tag{9.43$_1$}$$

and that the initial conditions satisfy

$$F'(0) > \sqrt{\frac{2}{3}\nu(0)F^3(0)} > 0. \tag{9.43$_2$}$$

Then the solution fails to exist for values of x such that

$$\int_0^x \sqrt{\nu(\eta)}d\eta > \sqrt{6}F^{-1/2}(0). \tag{9.44}$$

Proof. It follows from (9.8), (9.40) that

$$F''(x) = 2\int \mu(1 + \psi_x^2 + \psi_y^2)^2 dy \tag{9.45}$$

whence, on using Schwarz's inequality and (9.6), that

$$F'' \ge 2\nu(x)F^2. \tag{9.46}$$

In view of this and $F'(0) > 0$ it follows that $F'(x) > 0$. Thus multiplying (9.46) by F' yields

$$(F'^2)' \ge \frac{2}{3}\nu(F^3)'$$

and integration yields

$$F'^2(x) \ge \frac{2}{3}\nu(x)F^3(x) - \frac{2}{3}\int_0^x \nu'(\eta)F^3(\eta)d\eta + \{F'^2(0) - \frac{2}{3}\nu(0)F^3(0)\}.$$

In view $(9.43_1), (9.43_2)$ and the positivity of F', this implies

$$F'(x) \ge \sqrt{\frac{2}{3}\nu(x)}\, F^{3/2}(x)$$

whence, on integration,

$$2F^{-1/2}(0) = \int_{F(0)}^{\infty} F^{-3/2}dF \ge \int_{F(0)}^{F(x)} F^{-3/2}dF \ge \sqrt{\frac{2}{3}}\int_0^x \sqrt{\nu(\eta)}d\eta.$$

Plainly this contradicts (9.44) and non-existence follows.

The general methodology of the foregoing is often referred to as the method of "blow-up"; it is reminiscent of Theorem 3.10, for example. The particular method used above is often referred to as the method of convergent integrals (see Straughan [35], for example).

Finally – with reference to Theorem 9.9 – we remark that Efimov and his school have proved many non-existence results for surfaces of negative Gaussian curvature e.g. how the region in the $x - y$ plane is bounded, on to which a C^2 surface of negative curvature may be projected.

9.3 LIAPUNOV FUNCTIONS FOR SOME HYPERBOLIC P.D.E.s

Let $\Omega \subset I\!R_3$ be an open bounded connected domain and let X be a function space constituted by the functions satisfying homogeneous conditions on $\partial\Omega$. We consider here some types of hyperbolic P.D.E.s which generate a dynamical system on X, equipped with a suitable norm, and introduce functionals which are good candidates for producing Liapunov functions. We begin by considering the I.B.V.P.

$$u_{tt} + 2\alpha A u_t + Bu = f(u) \qquad \mathbf{x} \in \Omega,\ t \geq 0, \tag{9.47}$$

$$u(\mathbf{x},0) = u_0(\mathbf{x}),\ \ u_t(\mathbf{x},0) = \overset{\bullet}{u}_0(\mathbf{x}) \qquad \mathbf{x} \in \overline{\Omega}, \tag{9.48}$$

$$u(\mathbf{x},t) = 0 \qquad \mathbf{x} \in \partial\Omega \tag{9.49}$$

where $f(0) = 0$, $\alpha =$ const. > 0 and where A, B are linear self-adjoint spatial operators, definite on a dense subset S of a real Hilbert space $\mathcal{H}$ constituted by sufficiently smooth functions of X. It is known that (9.47) – (9.49) generate a dynamical system and we refer the readers to [82]-[83] for the proof. We assume this and suppose that the solutions are smooth. Moreover – for the sake of simplicity and completeness – we let the inner product on $\mathcal{H}$ be defined by

$$(u_1, u_2) = \int_\Omega u_1 u_2 d\Omega \qquad (u_i \in \mathcal{H},\ i = 1,2).$$

It is easily seen that the functional

$$V = \frac{1}{2}\int_\Omega [u \cdot Bu + (u_t + \alpha Au)^2 + \alpha^2 (Au)^2] d\Omega \tag{9.50}$$

is a good candidate for producing Liapunov functions. In fact, along the solutions to the I.B.V.P. (9.47) – (9.49) it follows

$$\overset{\bullet}{V} = \frac{1}{2}\int_\Omega [u_t \cdot Bu + u \cdot Bu_t + 2(u_{tt} + \alpha Au_t)(u_t + \alpha Au)] d\Omega + \alpha^2 \int_\Omega Au \cdot Au_t d\Omega .$$

But

$$(u_t, Bu) = (u, Bu_t) \quad , \quad u_{tt} + \alpha A u_t = f - Bu - \alpha A u_t$$

hence it follows

$$\overset{\bullet}{V} = \int_\Omega [u_t \cdot Bu + (f - Bu - \alpha Au_t)(u_t + \alpha Au) + \alpha^2 Au \cdot Au_t] d\Omega ,$$

which gives

$$\overset{\bullet}{V} = -\alpha \int_\Omega (u_t \cdot Au_t + Au \cdot Bu) d\Omega + \int_\Omega (u_t + \alpha Au) f d\Omega . \tag{9.51}$$

We remark that if B is positive definite, i.e.

$$u \cdot Bu > 0 \qquad \forall u \in H\,, \quad u \neq 0\,, \tag{9.52}$$

then V is positive definite along the solutions to the I.B.V.P. (9.47) – (9.49) and that conditions ensuring $\overset{\bullet}{V} \leq 0$, immediately can be obtained from (9.51) in the homogeneous case $f = 0$. A relevant case is considered in section 9.4. We limit ourselves here in mentioning the simplest case $\{\Omega = (0,1),\ A = I,\ B = -\Delta\}$. For $f \equiv 0$, the equation (9.47) becomes then

$$u_{tt} + 2\alpha u_t = u_{xx}\,. \tag{9.53}$$

Then V and $\overset{\bullet}{V}$ are given by

$$V = \frac{1}{2}\int_0^1 [u_x^2 + (u_t + \alpha u)^2 + \alpha^2 u^2]dx \tag{9.54}$$

$$\overset{\bullet}{V} = -\alpha \int_0^1 (u_t^2 + u_x^2)dx\,. \tag{9.55}$$

Passing to the nonhomogeneous case

$$u_{tt} + 2\alpha u_t - u_{xx} = f(u)\,, \tag{9.56}$$

let us set

$$U(u) = \int_0^u f(\xi)d\xi \tag{9.57}$$

$$V_1 = -\int_0^1 U[u(\xi)]d\xi \tag{9.58}$$

$$W = V + V_1 \tag{9.59}$$

with V given by (9.54). From (9.51) it turns out that

$$\overset{\bullet}{W} = -\alpha \int_0^1 [u_t^2 + u_x^2 - uf(u)]dx\,. \tag{9.60}$$

Therefore,

$$\xi f(\xi) \leq 0 \qquad \forall\, \xi \in I\!R \tag{9.61}$$

ensures $\overset{\bullet}{W} \leq 0$, while $U \leq 0$ ensures $W \geq 0$. In particular, in the case $\alpha = 0$, the function

$$E = \frac{1}{2}\int_0^1 (u_x^2 + u_t^2)dx + V_1 \tag{9.62}$$

with V_1 given by (9.58), is constant along the solutions to (9.56) and represents the total energy integral.

For the sake of simplicity we remain in the case $\Omega = (0,1)$ and consider, instead of (9.47), the nonlinear equation

$$u_{tt} = [\varepsilon + \varphi(u_x)]u_{xx} + u_{xtx} \qquad x \in (0,1),\ t \geq 0 \tag{9.63}$$

with $\varepsilon = \text{const.} > 0$ and $\varphi \in L^1_{\text{loc}}(I\!R)$. A deep analysis of (9.63) and similar equations can be found in [206], [207], [208]. We limit ourselves here to noting that the functional

$$V = \frac{1}{2}\int_0^1 (u_t^2 + \varepsilon u_x^2)dx + \int_0^1 (\int_0^{u_x} \psi(s)ds)dx \tag{9.64}$$

with

$$\psi(u_x) = \int_0^{u_x} \varphi(\xi)d\xi \tag{9.65}$$

is a Liapunov function with respect to the smooth solutions.

In fact it follows

$$\frac{\varepsilon}{2}\frac{d}{dt}\int_0^1 u_x^2 dx = -\varepsilon\int_0^1 u_{xx}u_t dx = -\int_0^1 u_t(u_{tt} - \varphi(u_x)u_{xx} - u_{xtx})dx$$
$$= -\frac{1}{2}\frac{d}{dt}\int_0^1 u_t^2 dx + \int_0^1 u_t(\frac{\partial}{\partial x}\int_0^{u_x} \varphi(\xi)d\xi)dx - \int_0^1 u_{xt}^2 dx\,.$$

Hence taking into account that (9.49) implies

$$\int_0^1 u_t\left(\frac{\partial}{\partial x}\int_0^{u_x} \varphi(\xi)d\xi\right)dx = -\int_0^1 u_{xt}\left(\int_0^{u_x} \varphi(\xi)d\xi\right)dx =$$
$$= -\frac{d}{dt}\int_0^1 dx\int_0^{u_x} \psi(s)ds\,,$$

it follows

$$\dot{V} = -\int_0^1 u_{xt}^2 dx \tag{9.66}$$

i.e. V is a Liapunov function. Therefore $\int_0^1 dx\int_0^{u_x} \psi(s)ds$ positive definite implies that V is non-negative.

We end this section mentioning that – as already noted in the previous chapters – many instability results are also known for equation of (9.47) type, expecially backward in time [115], [209], [210], [211]. We limit ourselves here to recalling an instability theorem. Indicating by L_i $(i = 1, 2, 3)$, positive definite, symmetric linear operators, and considering, instead of (9.47), the equation

$$L_1 u_{tt} - L_2 u_t + L_3 u = 0 \qquad \mathbf{x} \in \Omega,\ t \geq 0 \tag{9.67}$$

with the I.B.C. (9.48) – (9.49), then the following instability theorem holds.

Theorem 9.11 - *Assume that exist*
i) two positive constants $\alpha > \beta > 0$ *such that*

$$\begin{cases} 2\alpha(L_1 w, w) - (L_2 w, w) = 0 \\ \\ (w, L_3 w) - \beta(w, L_2 w) \leq 0 \end{cases} \qquad \forall w \in S$$

ii) two functions $u_0, \overset{\bullet}{u}_0 \in S$ *such that*

$$L_0 = (\overset{\bullet}{u}_0 - \alpha u_0, L_1[\overset{\bullet}{u}_0 - \alpha u_0]) + \alpha^2(u_0, L_1 u_0) + (u_0, L_3 u_0) - \alpha(u_0, L_2 u_0) > 0\,.$$

Then it follows

$$2(u_t, L_1 u_t) + 3\alpha^2(u, L_1 u) \geq L_0 \exp(2\alpha t)\,.$$

The proof can be obtained by using the weighted transformation $u = W\exp(\alpha t)$. We refer to [210] for the details.

9.4 A HYPERBOLIC HEAT CONDUCTION EQUATION

We consider now the equation

$$\tau u_{tt} + (1 - k\Delta)u_t - k\Delta u = 0 \tag{9.68}$$

where τ and k are positive constants. This equation, when u represents the temperature, models heat conduction at low temperatures [210] and represents a generalized Maxwell-Cattaneo equation (see Remark 9.5). Because (9.68) is linear, the stability of a basic solution to same boundary data is equivalent to the stability of the zero solution to the I.B.V.P. (9.68), (9.48), (9.49). Comparing (9.48) to (3.1), and taking into account the homogeneous boundary conditions, it follows

$$\alpha = \frac{1}{2\tau}\,, \quad A = 1 - k\Delta\,, \quad B = -\frac{k}{\tau}\Delta\,. \tag{9.69}$$

It is an easy matter to verify that – because of (9.49) the operators $A = 1 - k\Delta$, $B = -\frac{k}{\tau}\Delta$ are linear and self-adjoint with respect to the scalar product introduced in (9.3). Now, for the sake of simplicity, and following [21], we consider the case $\Omega = (0,1)$. Then, from (9.50) – (9.51), it turns out that

$$V = \frac{1}{2}\int_0^1 \left\{ \frac{k}{\tau}u_x^2 + [u_t + \frac{1}{2\tau}(u - ku_{xx})]^2 + \frac{1}{4\tau^2}(u - ku_{xx})^2 \right\} dx\,, \tag{9.70}$$

$$\overset{\bullet}{V} = -\frac{1}{2\tau}\int_0^1 [u_t^2 + ku_{tx}^2 + \frac{k^2}{\tau}u_x^2 + \frac{k^2}{\tau}u_{xx}^2]dx\,. \tag{9.71}$$

By using the Poincaré inequality (2.26), it is an easy matter to show that

$$V \leq \frac{\lambda}{2}\int_0^1 [u_t^2 + ku_x^2 + ku_{tx}^2 + \frac{k^2}{\tau}u_{xx}^2]dx \tag{9.72}$$

with

$$\lambda = \max\left\{ \frac{k}{\tau} + \frac{3}{2\tau^2\pi^2}, \frac{1}{k\pi^2}, \frac{3}{2\tau} \right\}. \tag{9.73}$$

Hence it turns out that

$$V \leq V_0 e^{-\lambda t}\,. \tag{9.74}$$

On the other hand it follows that

$$\begin{aligned} V \geq & \frac{1}{2}\int_0^1 [\frac{k}{\tau}u_x^2 + \frac{1}{4\tau^2}(u - ku_{xx})]^2 dx = \\ = & \frac{1}{2}\int_0^1 \left[\frac{1}{4\tau^2}u^2 + \frac{k}{\tau}(1 + \frac{1}{2\tau})u_x^2 + \frac{k^2}{4\tau^2}u_{xx}^2 \right] dx\,. \end{aligned} \tag{9.75}$$

Therefore, from (9.74) – (9.75), the asymptotic exponential stability in the $W_2^2(0,1)$-norm follows, and this implies stability in the pointwise norm (Remark 2.7). In a completely different way and among several other results, the stability with respect to the W_2^1-norm has been obtained in [210].

It is an easy matter to verify that all the assumptions of the instability theorem 9.4 are satisfied by (9.68), backward in time. We refer to [210] for the details.

Remark 9.5 - *Let us consider the Cauchy problem for the one-dimensional parabolic heat conduction equation* (1.139) *in an infinite bar*

$$\begin{cases} T_t = kT_{xx} & x \in I\!R,\ t > 0 \\ T(x,0) = T_0(x) & x \in I\!R \end{cases} \tag{9.76}$$

where $k = const. > 0$, *is the thermal diffusivity. As is well known, the I.P.* (9.76) *admits the solution* {*see, for instance,* [213]}

$$T(x,t) = \frac{1}{\sqrt{4\pi\, kt}} \int_{-\infty}^{+\infty} T_0(\xi) \exp\left[-\frac{(x-\xi)^2}{4kt}\right] d\xi . \tag{9.77}$$

Assuming $T_0(x) \geq 0$, *but not identically zero, from* (9.77) *it follows that*

$$T(x,t) > 0 \qquad \forall\, x \in I\!R,\ \forall\, t > 0 . \tag{9.78}$$

In particular, even if $T_0(x)$ *is zero outside a bounded interval* $(0,l)$ ($l = const. > 0$), (9.78) *implies that at any instant* $t > 0$ (*no matter how close to zero*) *and at any point* x (*no matter how far from* $(0,l)$) *the temperature* T *is bigger than zero. This phenomenon may be interpreted in terms of the infinite diffusion velocity allowed by the parabolic heat equation* (9.76). *This is the so called* **paradox of infinite diffusion velocity**. *The paradox was overcome in* 1948, *by C.Cattaneo* [214] *who obtained a hyperbolic P.D.E. for the heat conduction containing, as a particular case, an equation already obtained, with different scope, by Maxwell in* 1867. *We recall that the diffusion velocity permitted by a hyperbolic equation is finite. For all these matters, we refer the reader to* [215].

9.5 NONLINEAR STABILITY OF NON-VISCOUS INCOMPRESSIBLE FLUID MOTIONS: ARNOLD'S THEOREMS.

We reconsider here the problem, introduced in 1.10, of the stability of a non-viscous incompressible fluid motion. Specifically, following the notations used in sect. 1.10, we reconsider the problem of the *nonlinear stability* of a steady planar motion $\mathbf{v}_* = \nabla^{\perp}\varphi$ with respect to planar perturbations $\mathbf{u} = \nabla^{\perp}\phi$, with $\varphi = \varphi(x_1, x_2)$ and $\phi = \phi(x_1, x_2, t)$ smooth enough. The domain S filled with the fluid, is assumed to be a simply connected domain of $I\!R^3$ and smooth enough to allow the use of the divergence theorem.

We begin by recalling two conservation laws and assume that the external force is conservative. Denoting by

$$E = \frac{1}{2}\int_S v^2\, dS \tag{9.79}$$

the (kinetic) energy of a regular motion $(\mathbf{v}, p)$, the following theorem holds (as expected).

Theorem 9.12 - *The energy E is conserved along the motions.*

Proof. In view of (1.72) – (1.73), it follows that

$$\frac{dE}{dt} = \int_S \mathbf{v}\cdot\mathbf{v}_t dS = -\int_S [\mathbf{v}\cdot\nabla\mathbf{v} + \nabla(p+U)]\cdot\mathbf{v}dS =$$
$$= -\int_S \nabla\left[\frac{v^2}{2}+p+U\right]\cdot\mathbf{v}dS = -\int_{\partial S}\left(\frac{v^2}{2}+p+U\right)\mathbf{v}\cdot\mathbf{n}d\Sigma = 0$$

and hence

$$\int_S v^2(\mathbf{x},t)dS = \int_S v_0^2(\mathbf{x})dS\,. \tag{9.80}$$

Theorem 9.13 - *The vorticity of planar flows is conserved along the particle paths.*

Proof. In view of $(1.79)_3$, (1.75) becomes

$$\Omega_t + \mathbf{v}\cdot\nabla\Omega = 0\,. \tag{9.81}$$

Therefore, the Lagrangian derivative of Ω is zero, i.e. along the motion of any particle

$$\Omega[\mathbf{x}(\mathbf{x}_0,t)] = \Omega(\mathbf{x}_0) \tag{9.82}$$

where $\mathbf{x}(\mathbf{x}_0,t)$ denotes the motion of the particle starting from $\mathbf{x}_0$ at time zero.

Remark 9.7 - *In view of* $(1.79)_2$ *and* (9.82), *it follows that*

$$\Delta\Psi[\mathbf{x}(\mathbf{x}_0,t)] = \Delta\Psi(\mathbf{x}_0) \tag{9.83}$$

and hence

$$\int_S f(\Delta\Psi)dS = const. \tag{9.84}$$

for any measurable function $f : \mathbb{R} \to \mathbb{R}$.

We are now in position to prove the following two theorems of Arnold [216-217].

Theorem 9.14 - *Let* $\mathbf{v}_* = \nabla^{\perp}\varphi$ *be a steady smooth basic motion, and let there exist two positive constants* α, β *such that*[3]

$$0 < \alpha \le \frac{\nabla\varphi}{\nabla\Delta\varphi} \le \beta < \infty\,. \tag{9.85}$$

[3]It is an easy matter to verify that $\nabla\varphi$ and $\nabla\Delta\varphi$ are collinear.

Then $\mathbf{v}_*$ *is nonlinearly stable, with respect to planar perturbations* $\mathbf{u} = \nabla^\perp \phi(x_1, x_2, t)$, *in the* $[\|\mathbf{u}\|_2 + \|\nabla \times \mathbf{u}\|_2]$ *norm.*

Proof. In view of (1.80), we have

$$\frac{\partial(\varphi, \Delta\varphi)}{\partial(x_1, x_2)} = 0,$$

and hence there exists a functional relation

$$\varphi = g(\Delta\varphi).$$

Setting

$$F(\xi) = \int_0^\xi g(s)ds,$$

with ξ belonging to the range of $\Delta\varphi$, it follows that

$$F'(\Delta\varphi) = g(\Delta\varphi) = \varphi,$$

and hence

$$\nabla\varphi = F''(\Delta\varphi) \cdot \nabla\Delta\varphi.$$

In view of (9.85), we have

$$\alpha \leq F'' = \frac{\nabla\varphi}{\nabla\Delta\varphi} \leq \beta. \tag{9.86}$$

We suppose that F, initially defined on the range of $\Delta\varphi$, is extended to all $\mathbb{R}$ in such a way as to be a smooth function satisfying (9.86) (exercise 9.12), and introduce the function

$$V = \frac{1}{2}\|\nabla\phi\|_2^2 + \int_S \{F[\Delta(\varphi + \phi)] - F(\Delta\varphi) - F'(\Delta\varphi)\Delta\phi\} dS. \tag{9.87}$$

Let us show that V is a Liapunov function. In fact, taking into account theorems 9.12 - 9.13, it follows that for any plane motion $\mathbf{v} = \nabla^\perp\Psi$

$$H(\Psi) = \frac{1}{2}\|\nabla\Psi\|_2^2 + \int_S F(\Delta\Psi)dS = \text{const.},$$

which implies

$$H(\varphi + \phi) - H(\varphi) = \int_S (\nabla\varphi \cdot \nabla\phi + \varphi\Delta\phi)dS + V = \text{const.} \tag{9.88}$$

Because φ is defined modulo a constant, we may assume $\varphi = 0$ on ∂S. Hence it turns out that

$$\int_S \nabla\varphi \cdot \nabla\phi dS = \int_S [\nabla \cdot (\varphi\nabla\phi) - \varphi\Delta\phi] dS = -\int_S \varphi\Delta\phi dS$$

and hence from (9.88) it follows that V is constant along the perturbations $\mathbf{u} = \nabla^\perp\phi$. We now show that assumption (9.85) implies that V is positive definite. In fact, in view of (9.86), from

$$\frac{1}{2}F''(\xi + \theta h)h^2 = F(\xi + h) - F(\xi) - F'(\xi)h \qquad 0 < \theta < 1 \tag{9.89}$$

with $\{h = \Delta\phi,\ \xi = \Delta\varphi\}$, it turns out that

$$\frac{1}{2}\alpha(\Delta\phi)^2 \le F[\Delta(\varphi + \phi)] - F(\Delta\varphi) - F'(\Delta\varphi)\Delta\phi \le \frac{1}{2}\beta(\Delta\phi)^2 \,. \tag{9.90}$$

Therefore we have

$$||\nabla\phi||_2^2 + \alpha||\Delta\phi||_2^2 \le 2V \le ||\nabla\phi||_2^2 + \beta||\Delta\phi||_2^2 \,. \tag{9.91}$$

In view of (9.91), it follows that V is positive definite. Further, because $V=$ const. along the perturbations $\mathbf{u} = \nabla^\perp\phi$, (9.91) implies

$$||\nabla\phi||_2^2 + \alpha||\Delta\phi||_2^2 \le 2V_0 \le ||\nabla\phi_0||_2^2 + \beta||\Delta\phi_0||_2^2 \,, \ \forall t > 0$$

and hence

$$||\nabla\phi||_2^2 + ||\Delta\phi||_2^2 \le \frac{\lambda}{\mu}\,[||\nabla\phi_0||_2^2 + ||\Delta\phi_0||_2^2] \,, \ \forall t > 0$$

with $\phi_0 = \phi(x_1, x_2, 0)$, $\lambda = \sup(1, \beta)$, $\mu = \inf(1, \alpha)$.

Theorem 9.15 - *Let $\mathbf{v}_* = \nabla^\perp\varphi$ be a steady smooth basic motion and let there exist two positive constants α, β such that*

$$0 < \alpha \le -\frac{\nabla\varphi}{\nabla\Delta\varphi} \le \beta < \infty \,. \tag{9.92}$$

Then $\mathbf{v}_*$ is nonlinearly stable in the $[||\mathbf{u}||_2 + ||\nabla \times \mathbf{u}||_2]$-norm at least with respect to planar perturbations $\mathbf{u} = \nabla^\perp\phi(x_1, x_2, t)$ such that $\forall t > 0$

$$||\nabla\phi||_2^2 \le k||\Delta\phi||_2^2 \tag{9.93}$$

where $k \in (0, \alpha)$ is a constant independent of ϕ.

Proof. In the case at hand, instead of (9.86), we have

$$\alpha \leq -F'' = -\frac{\nabla\varphi}{\nabla\Delta\varphi} \leq \beta .$$

Hence from (9.89), instead of (9.90), it follows that

$$-\frac{1}{2}\beta(\Delta\phi)^2 \leq F[\Delta(\varphi+\phi)] - F(\Delta\varphi) - F'(\Delta\varphi)\Delta\phi \leq -\frac{1}{2}\alpha(\Delta\phi)^2 ,$$

i.e.

$$\|\nabla\phi\|_2^2 - \beta\|\Delta\phi\|_2^2 \leq 2V \leq \|\nabla\phi\|_2^2 - \alpha\|\Delta\phi\|_2^2 .$$

In view of (9.93), we obtain

$$(\alpha - k)\|\Delta\phi\|_2^2 \leq -2V \leq \|\nabla\phi\|_2^2 + \beta\|\Delta\phi\|_2^2$$

and hence

$$\begin{cases} (\alpha - k)\|\Delta\phi\|_2^2 \leq -2V = -2V_0 \leq \|\nabla\phi_0\|_2^2 + \beta\|\Delta\phi_0\|_2^2 \\ (\alpha - k)\|\nabla\phi\|_2^2 \leq k(\alpha - k)\|\Delta\phi\|_2^2 \leq k[\|\nabla\phi_0\|_2^2 + \beta\|\Delta\phi_0\|_2^2] \end{cases}$$

which, for $\gamma = (k+1)/(\alpha - k)$, imply

$$\|\nabla\phi\|_2^2 + \|\Delta\phi\|_2^2 \leq \gamma[\|\nabla\phi_0\|_2^2 + \beta\|\Delta\phi_0\|_2^2] , \ \forall t > 0 .$$

Remark 9.7 - *The theorems* 9.14 – 9.15 *continue to hold for* $\alpha = 0$ *and for* S *multiply connected. We refer to* [218] *for the proof.*

9.6 A QUALITATIVE ANALYSIS OF A NONLINEAR INTEGRO-DIFFERENTIAL EQUATION OF THE PARTICLE TRANSPORT THEORY

9.6.1 Introduction

Apart from the behaviour of a viscoelastic body [18], a large class of phenomena can be modelled by integrodifferential equations. For instance, the behaviour of a dilute gas and the transport of particles (electrons in solids and plasmas, neutrons in nuclear reactors, phonons in superfluids, particles of a contaminant gas,...) are modelled by integrodifferential equations of the Boltzmann type [219]. We here consider the I.V.P.

$$f_t(\mathbf{z},t) + kQf(\mathbf{z},t) = k\int_{\mathbb{R}_3\times\mathbb{R}_3} \pi(\mathbf{x},\mathbf{y},\mathbf{z})f(\mathbf{x},t)f(\mathbf{y},t)d\mathbf{x}d\mathbf{y} \tag{9.94}$$

$$f(\mathbf{z},0) = f_0(\mathbf{z}) \tag{9.95}$$

where $(\mathbf{x},\mathbf{y},\mathbf{z}) \in I\!R_3 \times I\!R_3 \times I\!R_3$, $t \in [0,\infty)$ and k is a positive constant. The equation (9.94) models, in the scattering kernel theory and in the absence of removal and external forces [220-221], the evolution of the one particle distribution function f in a gas of particles, which diffuse through binary collisions in an infinite homogeneous medium. The velocities before collisions are denoted by $\mathbf{x},\mathbf{y}$ while $\mathbf{z}$ denotes the velocity after the collisions. The constant Q represents the number of particles injected into the medium

$$Q = \int_{I\!R_3} f_0(\mathbf{z})d\mathbf{z} = \int_{I\!R_3} f(\mathbf{z},t)d\mathbf{z}\,, \qquad \forall t \geq 0\,. \tag{9.96}$$

The function $\pi(\mathbf{x},\mathbf{y},\mathbf{z})$ is such that

$$\begin{cases} \pi(\mathbf{x},\mathbf{y},\mathbf{z}) \geq 0\,, & \text{on } I\!R_3^3 \\ \pi(\mathbf{x},\mathbf{y},\cdot) \in L_1(I\!R_3)\,, & \forall\,(\mathbf{x},\mathbf{y}) \in I\!R_3^2 \\ \displaystyle\int_{I\!R_3} \pi(\mathbf{x},\mathbf{y},\mathbf{z})d\mathbf{z} = 1\,, & \forall\,(\mathbf{x},\mathbf{y}) \in I\!R_3^2 \\ \pi(\mathbf{x},\mathbf{y},\mathbf{z}) = \pi(\mathbf{y},\mathbf{x},\mathbf{z})\,, & \forall\,(\mathbf{x},\mathbf{y},\mathbf{z}) \in I\!R_3^3 \end{cases} \tag{9.97}$$

and is called the ***scattering probability distribution***. Our aim is to perform a qualitative analysis of (9.94). Specifically, in 9.6.2 we obtain a global existence and uniqueness theorem, while in 9.6.3 we consider the problem of stability of the solutions obtaining a general theorem ensuring conditional stability. Finally, in 9.6.4, we give conditions ensuring unconditional stability.

9.6.2 Global existence and uniqueness

The global in time existence and uniqueness theorem has been given in [222]. We here follow [223] and begin by noting that the Cauchy problem (9.94)–(9.95) is equivalent to

$$f = Af \tag{9.98}$$

where A is the nonlinear operator

$$\begin{aligned} A :\varphi \in E \to A\varphi = e^{-kQt} f_0 + \\ e^{-kQt} k \int_0^t e^{kQ\tau} d\tau \int_{I\!R_3 \times I\!R_3} \pi(\mathbf{x},\mathbf{y},\mathbf{z})\varphi(\mathbf{x},\tau)\varphi(\mathbf{y},\tau)d\mathbf{x}d\mathbf{y}\,. \end{aligned} \tag{9.99}$$

Let T be any positive constant and let E denote the Banach space

$$\begin{aligned} E = \{\varphi(\mathbf{x},t) :\ I\!R_3 \times [0,T] \to I\!R;\ \varphi(\cdot,t) \in L_1(I\!R_3), \\ \forall t \in [0,T]; \varphi(\mathbf{x},\cdot) \in C[0,T],\ \forall \mathbf{x} \in I\!R_3\} \end{aligned} \tag{9.100}$$

equipped with the norm

$$|||\varphi||| = \max_{[0,T]} ||\varphi||_t \tag{9.101}$$

where $||\cdot||_t$ is the L_1-norm at the time t. Furthermore, we denote by B the closed set

$$B = \{\varphi \in E : |||\varphi||| \leq Q\} \tag{9.102}$$

and introduce the *weighted norm*

$$|||\varphi|||_W = |||e^{-kQt}\varphi||| \tag{9.103}$$

which is equivalent to (9.101).[4] When E is equipped with the norm (9.103), the set B will be denoted by B_*.

Theorem 9.16 - *For any positive T, in the space E equipped with the norm (9.103), A is a contractive operator such that $AB_* \subset B_*$. Hence the global existence and uniqueness of solution to the I.V.P.* (9.94) – (9.95) *follows.*

Proof. Let $\varphi \in B$. Then, $\forall\, t \in [0,T]$, it follows that

$$||A\varphi||_t \leq Qe^{-kQt} + ke^{-kQt}\int_0^t e^{kQ\tau}||\varphi||_\tau^2 d\tau \leq Qe^{-kQt}[1 + \int_0^t d(e^{kQ\tau})] = Q\,,$$

i.e. $|||\varphi||| \in B \rightarrow |||A\varphi||| \in B$. Because $|||\cdot|||_w \leq |||\cdot|||$, $AB \subset B \rightarrow AB_* \subset B_*$.
Let $\varphi, \psi \in B$. Then it follows that

$$A\varphi - A\psi = ke^{-kQt}\int_0^t e^{kQ\tau}d\tau \int_{\mathbb{R}_3^2} \pi[\varphi(\mathbf{x},\tau) + \psi(\mathbf{x},\tau)][\varphi(\mathbf{y},\tau) - \psi(\mathbf{y},\tau)]d\mathbf{x}d\mathbf{y}$$

which implies

$$\begin{aligned}
||e^{-kQt}(A\varphi - A\psi)||_t \leq\, & ke^{-2kQt}\int_0^t e^{kQ\tau}||\varphi + \psi||_\tau \cdot ||\varphi - \psi||_\tau d\tau \leq \\
\leq\, & 2kQe^{-2kQt}\int_0^t e^{2kQ\tau} \cdot e^{-kQ\tau}||\varphi - \psi||_\tau d\tau \leq \\
\leq\, & e^{-2kQt}|||\varphi - \psi|||_w (e^{2kQt} - 1) = \\
=\, & (1 - e^{-2kQt})|||\varphi - \psi|||_w \,.
\end{aligned}$$

Hence it turns out that

$$|||A\varphi - A\psi|||_w \leq (1 - e^{-2kQT})|||\varphi - \psi|||_w\,,$$

[4]In fact one has

$$e^{-kQT}|||\varphi||| \leq |||\varphi|||_W \leq |||\varphi|||\,.$$

i.e. A is a contractive mapping in the $|||\cdot|||_w$- norm.

Taking into account that B closed implies that B^* is also closed, the proof of theorem 9.16 is completed by means of the Caccioppoli-Banach fixed point theorem [224].

9.6.3 Stability in the L_1-norm

Here, following [225], we study the stability of a "basic" solution $f(\mathbf{z},t)$ to the I.V.P. (9.94) – (9.95). Let $g(\mathbf{z},t) = f(\mathbf{z},t) + u(\mathbf{z},t)$ be the "perturbed" solution corresponding to the initial data

$$g(\mathbf{z},0) = f_0(\mathbf{z}) + u_0(\mathbf{z})\,. \tag{9.104}$$

In view of (9.96), for any $t \geq 0$, it follows that

$$\int_{\mathbb{R}_3} g(\mathbf{z},t)d\mathbf{z} = \int_{\mathbb{R}_3} f(\mathbf{z},t)d\mathbf{z} = Q\,, \tag{9.105}$$

$$\int_{\mathbb{R}_3} u(\mathbf{z},t)d\mathbf{z} = 0\,. \tag{9.106}$$

Further, we easily obtain the following I.V.P. for the "perturbation" u

$$u_t + kQu = k\int_{\mathbb{R}_3^2} \pi(\mathbf{x},\mathbf{y},\mathbf{z})[f(\mathbf{x},t) + g(\mathbf{x},t)]u(\mathbf{y},t)d\mathbf{x}d\mathbf{y} \tag{9.107}$$

$$u(\mathbf{z},0) = u_0(\mathbf{z})\,. \tag{9.108}$$

Setting

$$\Omega_1(t) = \{\mathbf{z} \in \mathbb{R}_3 : u(\mathbf{z},t) \geq 0\},\ \Omega_2(t) = \mathbb{R}_3 - \Omega_1(t)\,, \tag{9.109}$$

$$E = 2\|u\| = \int_{\mathbb{R}_3} |u|d\mathbf{z}\,, \tag{9.110}$$

from (9.106) it turns out that

$$\int_{\Omega_1} |u|d\mathbf{z} = \int_{\Omega_2} |u|d\mathbf{z} = \|u\|\,, \tag{9.111}$$

and the following theorems hold.

Theorem 9.17 - *Along the solutions to* (9.107) – (9.108) *there holds*

$$\frac{dE}{dt} + kQE = k\int_{\mathbb{R}_3^2} [f(\mathbf{x},t) + g(\mathbf{x},t)]u(\mathbf{y},t)d\mathbf{x}d\mathbf{y}\cdot$$

$$\cdot\int_{\mathbb{R}_3} [\text{sign } u(\mathbf{z},t)]\cdot \pi(\mathbf{x},\mathbf{y},\mathbf{z})d\mathbf{z}\,. \tag{9.112}$$

Proof. From (9.107), taking into account that $|u|_t = (\text{sign}\, u)u_t$, it turns out that

$$\int_{\Omega_1(t)} |u|_t dz + kQ\|u\| = k\int_{\mathbb{R}_3^2}[f(\mathbf{x},t)+g(\mathbf{x},t)]u(\mathbf{y},t)d\mathbf{x}d\mathbf{y}\cdot \cdot\int_{\Omega_1(t)}\pi(\mathbf{x},\mathbf{y},\mathbf{z})dz, \tag{9.113}$$

$$\int_{\Omega_2(t)} |u|_t dz + kQ\|u\| = -k\int_{\mathbb{R}_3^2}[f(\mathbf{x},t)+g(\mathbf{x},t)]u(\mathbf{y},t)d\mathbf{x}d\mathbf{y}\cdot \cdot\int_{\Omega_2(t)}\pi(\mathbf{x},\mathbf{y},\mathbf{z})dz\,. \tag{9.114}$$

But

$$\int_{\Omega_1(t)}|u|_t dz + \int_{\Omega_2(t)}|u|_t dz = \int_{\mathbb{R}_3}|u|_t dz = \frac{d}{dt}\int_{\mathbb{R}_3}|u|dz = \frac{dE}{dt}, \tag{9.115}$$

hence from (9.113) – (9.114), (9.112) immediately follows.

Theorem 9.18 - *Denoting by I the set*

$$I = \{\varphi(\mathbf{z}) : |\varphi| \in L_1(\mathbb{R}_3), \int_{\mathbb{R}_3}\varphi dz = 0, \int_{\mathbb{R}_3}|\varphi|dz \le 2Q\}, \tag{9.116}$$

and by F and G the functionals

$$F : (\varphi, t) \in I \times \mathbb{R}_+ \to 2k\frac{\int_{\mathbb{R}_3^2} f(\mathbf{x},t)\varphi(\mathbf{y})d\mathbf{x}d\mathbf{y}\int_{\mathbb{R}_3}[\text{sign}\,\varphi(\mathbf{z})]\pi(\mathbf{x},\mathbf{y},\mathbf{z})dz}{\int_{\mathbb{R}_3}|\varphi(\mathbf{z})|dz}, \tag{9.117}$$

$$G : \varphi \in I \to k\frac{\int_{\mathbb{R}_3^2}\varphi(\mathbf{x})\varphi(\mathbf{y})d\mathbf{x}d\mathbf{y}\int_{\mathbb{R}_3}[\text{sign}\,\varphi(\mathbf{z})]\pi(\mathbf{x},\mathbf{y},\mathbf{z})dz}{\left(\int_{\mathbb{R}_3}|\varphi(\mathbf{z})|dz\right)^2}, \tag{9.118}$$

along the solutions to (9.107) – (9.108) *one has*

$$\frac{dE}{dt} = [F(u,t) - kQ]E + G(u)E^2\,. \tag{9.119}$$

Proof. - From (9.105) – (9.106) it follows that $u \in I$. Then, on taking into account (9.117) – (9.118), (9.119) immediately follows from (9.112).

Theorem 9.19 - *Defining*

$$\alpha = \frac{1}{kQ} \sup_{I \times I\!R_+} F(u,t), \ \beta = \frac{1}{k} \sup_I G(u), \ \lambda = (1-\alpha)Q - \beta E_0 \,, \tag{9.120}$$

if one has

$$\alpha < 1, \ E_0 < \frac{1-\alpha}{\beta} Q \tag{9.121}$$

then

$$E(t) \leq E_0 e^{-k\lambda t} \,. \tag{9.122}$$

Proof. First of all, let us notice that α and β are bounded. In fact from (9.117) – (9.118) immediately one obtains

$$F(u,t) \leq 2k \int_{I\!R_3} f(\mathbf{x},t) d\mathbf{x} \int_{I\!R_3} \pi(\mathbf{x},\mathbf{y},\mathbf{z}) dz$$

$$G(u,t) \leq k \int_{I\!R_3} \pi(\mathbf{x},\mathbf{y},\mathbf{z}) dz \,.$$

Therefore, taking into account (9.97), immediately it follows that $\alpha \leq 2$, $\beta \leq 1$. Further, from (9.119) – (9.120) it turns out that

$$\frac{dE}{dt} \leq kQ(\alpha - 1)E + k\beta E^2 \,. \tag{9.123}$$

On taking into account the procedure of exercise 2.2, one easily obtains from (9.122) from (9.121) and (9.123).

9.6.4 Unconditional stability

For some particular functions π, it is possible to avoid the condition (9.121) on the initial data.

Theorem 9.20 - *Let* $\exists h : I\!R_3 \to I\!R$, $h \in L_1(I\!R_3)$ *such that*

$$\delta = \operatorname*{ess\ sup}_{(\mathbf{x},\mathbf{y}) \in I\!R_3^2} \int_{I\!R_3} |\pi_1(\mathbf{x},\mathbf{y},\mathbf{z})| dz \tag{9.124}$$

where

$$\pi_1 = \pi(\mathbf{x},\mathbf{y},\mathbf{z}) - h(\mathbf{z})\,. \tag{9.125}$$

Then $\delta = 1/2$ implies unconditional stability in the L_1-norm, while $\delta < 1/2$ implies unconditional asymptotic (exponential) stability.

Proof. In view of (9.106), it follows that

$$\int_{\mathbb{R}_3^2} [f(\mathbf{x},t) + g(\mathbf{x},t)]u(\mathbf{y},t)d\mathbf{x}d\mathbf{y} \int_{\mathbb{R}_3} [\mathrm{sign}u(z,t)]\, h(\mathbf{z})d\mathbf{z} = 0\,.$$

Therefore (9.112) implies

$$\frac{dE}{dt} + kQE \le k\int_{\mathbb{R}_3} [f(\mathbf{x},t) + g(\mathbf{x},t)]|u(\mathbf{y},t)|d\mathbf{x}d\mathbf{y} \int_{\mathbb{R}_3} |\pi_1(\mathbf{x},\mathbf{y},\mathbf{z})|d\mathbf{z}\,. \tag{9.126}$$

But in view of (9.105) and (9.110), the R.H.S. of (9.126) does not exceed $2\delta QE$, hence it follows that

$$\frac{dE}{dt} \le (2\delta - 1)kQE\,, \tag{9.127}$$

which proves the theorem.

Let us define now

$$\pi_0(\mathbf{z}) = \operatorname*{ess\,inf}_{(\mathbf{x},\mathbf{y})\in\mathbb{R}_3^2} \pi(\mathbf{x},\mathbf{y},\mathbf{z})\,. \tag{9.128}$$

The following theorem holds.

Theorem 9.21 - *If there exists a measurable set $I \subset \mathbb{R}_3$ such that*

$$\pi_0 \in L_1(I),\ \varepsilon = \int_I \pi_0(\mathbf{z})d\mathbf{z} \ge \frac{1}{2}\,, \tag{9.129}$$

then $\varepsilon = 1/2$ implies unconditional stability in the L_1-norm, while $\varepsilon > 1/2$ implies unconditional (exponential) stability.

Proof. The assumptions of theorem 9.20 are satisfied on choosing

$$h(\mathbf{z}) = \begin{cases} \pi_0(\mathbf{z}), & \mathbf{z} \in I \\ 0, & \mathbf{z} \in \mathbb{R}_3 - I\,. \end{cases} \tag{9.130}$$

In fact, it follows that

$$\int_{\mathbb{R}_3} |\pi_1(\mathbf{x},\mathbf{y},\mathbf{z})|d\mathbf{z} = \int_{\mathbb{R}_3} \pi(\mathbf{x},\mathbf{y},\mathbf{z})d\mathbf{z} - \int_I \pi_0(\mathbf{z})d\mathbf{z} = 1 - \varepsilon\,.$$

Remark 9.8 - *A distribution function f is said to be Maxwellian if*

$$f : \mathbf{z} \in I\!R_3 \to M(\mathbf{z}) = \left(\frac{\beta}{\pi}\right)^{3/2} e^{-\beta z^2}.$$

Let π be very close to a Maxwellian $M(\mathbf{z})$, i.e.

$$\begin{cases} \pi(\mathbf{x},\mathbf{y},\mathbf{z}) = M(\mathbf{z}) + \varepsilon\pi^*(\mathbf{x},\mathbf{y},\mathbf{z}), & \varepsilon = const. \in I\!R, \\ \lim_{\varepsilon \to 0} \int_{I\!R_3} |\pi - M(\mathbf{z})| d\mathbf{z} = 0 . \end{cases}$$

Then it is an easy matter to verify that, choosing $h(\mathbf{z}) = M(\mathbf{z})$, the assumptions of theorem 9.20 *are satisfied for ε sufficiently small. We refer the interested readers to* [225] *where other stability criteria are given. (See exercises* 9.16 – 9.18.)

9.7 I.B.V.P. IN EXTERIOR REGIONS

Estimates for problems in exterior/infinite regions, involving both elliptic and parabolic equations, have already been derived in this book. The interesting feature of these problems is that the solution is allowed to grow rapidly at infinity. A common feature of the derivation of these estimates is the use of weighted integral methods. It is natural, therefore, to consider analogous issues for problems of a type normally (though not necessarily) associated with hyperbolic equations i.e. initial boundary value problems.

We commence with an I.B.V.P. for an exterior, infinite region, occupied by elastic material, in the presence of conditions of exponential growth at infinity. The class of materials considered includes certain nonlinear elastic materials as well as linear elastic materials; their characterization implies that the strain energy is (essentially) positive definite. We derive a simple estimate of the continuous dependence type which does not appear to be well known. We remark that there exist estimates of a similar type for nonlinear elastic materials whose strain-energy functions are not positive definite everywhere. The derivation of these is much more elaborate and entail additional hypotheses. Among these are [227-229]. Also see [230] and [72] for other estimates of this general type.

We complete the section by showing that a weighted Lagrange identity method can be used to obtain a continuous dependence type estimate in connection with an I.B.V.P. for an exterior region containing linear elastic material whose strain energy is not assumed to be positive definite.

Let Ω be a three-dimensional, unbounded region with smooth boundary $\partial\Omega$. It is occupied by elastic material of a type described subsequently. Rectangular cartesian coordinates are denoted by x_i, the usual indicial notation and summation convention are used; suffices following a comma denote partial differentiation-numerical suffices with respect to the relevant spatial variables, a suffix t with respect to the time variable.

Suppose that the rectangular cartesian components of displacement $u_i(\mathbf{x},t)$ satisfy

$$\rho u_{i,tt} = \left(\frac{\partial W}{\partial u_{i,j}}\right)_{,j} \qquad \text{in } \Omega\times(0,T] \tag{9.131}$$

where T denotes the relevant time span, $\rho(\mathbf{x})$ the density (in the undeformed medium) and $W(u_{r,s})$ the strain energy whose properties are defined hereunder. We consider classical solutions of (9.131) subject to boundary conditions

$$u_i(\mathbf{x},t) = 0 \qquad \text{on } \partial\Omega\times(0,T) \tag{9.132}$$

and subject to the initial conditions

$$u_i(\mathbf{x},0) = f_i(\mathbf{x}), \quad u_{i,t}(\mathbf{x},0) = g_i(boldx), \qquad \mathbf{x}\in\Omega \tag{9.133}$$

Moreover, we prescribe the asymptotic conditions

$$u_i = O(e^{\alpha r}), \quad u_{i,t} = O(e^{\alpha r}), \quad \frac{\partial W}{\partial u_{i,j}} = O(e^{\alpha r}) \tag{9.134}$$

as $r\to\infty$, where α is a prescribed constant (r being defined by $r=(x_ix_i)^{1/2}$).

We now present a simple estimate for a class of elastic materials (which includes linear elastic material) which does not appear to be well known.

For this simple estimate we prescribe the following material behaviour:

(i) There exists a positive constant k such that

$$\left(\frac{\partial W}{\partial u_{i,j}}\right)\left(\frac{\partial W}{\partial u_{i,j}}\right) \le kW(u_{r,s}). \tag{9.135$_1$}$$

(This implies that $W \ge 0$, and, without loss of generality, that $W = 0$ iff $u_{i,j}\equiv 0$.)

(ii) There exists a (positive) constant ρ_0 such that

$$\rho(\mathbf{x}) \ge \rho_0 > 0. \tag{9.135$_2$}$$

Defining the weighted energy

$$E(t) = \int_\Omega g[\frac{1}{2}\rho u_{i,t}u_{i,t} + W(u_{r,s})]dV \tag{9.136}$$

where

$$g(r) = e^{-\lambda r} \tag{9.137$_1$}$$

and λ is a constant such that

$$\lambda > 2\alpha \tag{9.137$_2$}$$

ensuring, in view of (9.134), the existence of (9.136) and all other integrals arising in the analysis. We proceed to obtain a first order differential inequality for $E(t)$ whence we obtain the required estimate.

Differentiation of (9.136) and use of (9.131) gives

$$\begin{aligned}\frac{dE}{dt} &= \int_\Omega g[u_{i,t}\left(\frac{\partial W}{\partial u_{i,j}}\right)_{,j} + \left(\frac{\partial W}{\partial u_{i,j}}\right)u_{i,t}]dV \\ &= \int_{\partial\Omega\cap\partial\Omega(\infty)} g u_{i,t}\frac{\partial W}{\partial u_{i,j}}\nu_j ds \\ &\quad - \int_\Omega g'(r)\frac{x_j}{r}\frac{\partial W}{\partial u_{i,j}}u_{i,t}dV\end{aligned} \tag{9.138}$$

where integration by parts has been used in the last step, ν_i denoting the components of the unit outward normal. The boundary terms in (9.138) vanish in view of (9.132), (9.134), (9.137). Use of the Cauchy inequality in the remaining term yields

$$\frac{dE}{dt} \leq \lambda\gamma\int_\Omega g[\frac{1}{2}\rho u_{i,t}u_{i,t} + \left(\frac{1}{2\rho\gamma^2}\right)\left(\frac{\partial W}{\partial u_{i,j}}\right)\left(\frac{\partial W}{\partial u_{i,j}}\right)]dV \tag{9.139}$$

where γ is any positive constant. Choosing γ suitably, with (9.135) and (9.136) in mind, we obtain

$$\frac{dE}{dt} \leq \lambda\gamma E \tag{9.140$_1$}$$

where

$$\gamma = \sqrt{\frac{k}{2\rho_0}}\,. \tag{9.140$_2$}$$

Straightforward integration yields the estimate embodied in the following theorem.

Theorem 9.22 - *The weighted energy* $E(t)$, *defined by* (9.136), (9.137), *for classical solutions of* (9.131) – (9.134) *in the context of the constitutive assumptions* (9.135), *satisfies the estimate*

$$E(t) \leq E(0)\exp(\lambda\gamma t) \tag{9.141}$$

where λ *is a constant defined by* (9.137_2) *and* γ *is the constant defined by* (9.140_2).

Remark 9.9 - *The estimate* (9.141) *is a continuous dependence type estimate for the null solution. It will be noted that the constitutive assumption* (9.135_1) *includes linear elastic materials. Consequently, the above theorem implies uniqueness and continuous dependence for linear elastic material under the stated condions.*

The proof continues to be valid for suitably defined *weak* solutions. Moreover, the theorem is also valid if zero traction conditions replace (9.132) i.e. $\dfrac{\partial W}{\partial u_{i,j}}\nu_j = 0$. We now consider a linear elastic material whose energy function is not positive definite. We consider an initial boundary value problem for the infinite region Ω (previously defined) in the presence of exponential growth at infinity. We show how a weighted version of the Lagrange identity approach (Chapter five) may be used to derive a continuous dependence type estimate – along the lines of [231]. We use the same general notation as above. The equations of motion are

$$(c_{ijkl}u_{k,l})_{,j} = \rho(\mathbf{x})u_{i,tt} \qquad (\mathbf{x},t) \in \Omega \times (0,2T], \tag{9.142$_1$}$$

where the non-homogeneous elasticities (assumed continuously differentiable) have the symmetry

$$c_{ijkl} = c_{jikl}\,. \tag{9.142$_2$}$$

The displacement components u_i satisfy the initial conditions

$$u_i(\mathbf{x},0) = f_i(\mathbf{x}), \quad u_{i,t}(\mathbf{x},0) = h_i(\mathbf{x}) \qquad \mathbf{x} \in \Omega \tag{9.143$_1$}$$

wherein f_i, h_i are assumed to satisfy

$$\int_\Omega (f_i f_i + h_i h_i)d\mathbf{x} \le M^2 \tag{9.143$_2$}$$

for some positive constant M, together with the boundary conditions

$$u_i = 0 \qquad \text{on} \quad \partial\Omega \times [0,T]\,. \tag{9.143$_3$}$$

It is supposed that the following asymptotic conditions are satisfied at large spatial distances:

$$(i) \qquad |u_i|, \quad |c_{ijkl}u_{k,l}| = o(e^{\alpha r}) \quad \text{as} \quad r \to \infty \tag{9.144}$$

where α is a positive constant,

$$(ii) \qquad \lim_{\alpha \to 0} \alpha^2 S(T) = 0 \tag{9.145$_1$}$$

wherein

$$S(t) = \int_0^{2t} \int_\Omega (e^{-\alpha r}/\rho) c_{ijkl} u_{k,l} c_{ijpq} u_{p,q} d\mathbf{x} d\eta\,. \tag{9.145$_2$}$$

Conditions guaranteeing (9.145$_1$) have been investigated and include, for example,

$$|c_{ijkl} u_{k,l} c_{ijpq} u_{p,q}| = O(r^{-1/2-\varepsilon}), \quad \varepsilon > 0, \text{ as } r \to \infty. \tag{9.146}$$

Additionally, the displacements are supposed to be restricted to the class

$$\int_0^T \int_\Omega \rho u_i(\mathbf{x},\eta) u_i(\mathbf{x},\eta) d\mathbf{x} d\eta \le M_1^2 \tag{9.147}$$

where M_1 is a positive constant.

We proceed to prove the following theorem, of the continuous dependence type, using a weighted Lagrange identity method.

Theorem 9.23 - *Classical solutions of* (9.142), (9.143), *satisfying the asymptotic conditions* (9.144), (9.145), *and further restricted by* (9.147), *satisfy the estimate*

$$\int_0^{2t} \int_\Omega \rho u_i(\mathbf{x},\eta) u_i(\mathbf{x},\eta) d\mathbf{x} d\eta \le e^{2T} T\Big[\int_\Omega \rho f_i f_i d\mathbf{x} +$$
$$T^{-1/2} M_1 \{ (\int_\Omega \rho f_i f_i d\mathbf{x})^{1/2} + T^{1/2} (2\int_\Omega \rho h_i h_i d\mathbf{x})^{1/2} \}\Big], \ \forall t \in (0,T]. \tag{9.148}$$

Proof. Let us introduce the weight function

$$g(\mathbf{x}) = e^{-\alpha r}, \qquad r = |\mathbf{x}|, \tag{9.149}$$

where α is a positive constant, and let $w_i(\mathbf{x},t), v_i(\mathbf{x},t)$ be solutions to (9.142) etc. We then have the following identity

$$\int_0^t \int_\Omega \rho(\mathbf{x}) g(\mathbf{x}) [w_{i,\eta}(\mathbf{x},\eta) v_{i,\eta}(\mathbf{x},\eta) - w_{i,\eta}(\mathbf{x},\eta) v_{i,\eta}(\mathbf{x},\eta)] d\mathbf{x} d\eta = 0 \tag{9.150}$$

which after an integration with respect to time, gives the basic weighted Lagrange identity:

$$\int_\Omega \rho(\mathbf{x}) g(\mathbf{x}) [w_i(\mathbf{x},t) v_{i,t}(\mathbf{x},t) - w_{i,t}(\mathbf{x},t) v_i(\mathbf{x},t)] d\mathbf{x}$$
$$= \int_0^t \int_\Omega \rho(\mathbf{x}) g(\mathbf{x}) [w_i(\mathbf{x},\eta) v_{i,\eta\eta}(\mathbf{x},\eta) - w_{i,\eta\eta}(\mathbf{x},\eta) v_i(\mathbf{x},\eta)] d\mathbf{x} d\eta$$
$$+ \int_\Omega \rho(\mathbf{x}) g(\mathbf{x}) [w_i(\mathbf{x},0) v_{i,t}(\mathbf{x},0) - w_{i,t}(\mathbf{x},0) v_i(\mathbf{x},0)] d\mathbf{x}\,. \tag{9.151}$$

We set

$$w_i(\mathbf{x},\tau) = u_i(\mathbf{x},\tau)\,, \quad v_i(\mathbf{x},\tau) = u_i(\mathbf{x}, 2t-\tau)] \qquad 0 \le \tau < 2t\,, \tag{9.152}$$

so that (9.151) becomes

$$2\int_\Omega \rho g u_i(t)u_{i,t}(t)d\mathbf{x} = \int_0^t \int_\Omega \rho g[u_{i,\eta\eta}(\eta)u_i(2t-\eta) - u_i(\eta)u_{i,\eta\eta}(2t-\eta)]d\mathbf{x}d\eta$$
$$+ \int_\Omega \rho g[f_i u_{i,t}(2t) - h_i u_i(2t)]d\mathbf{x} \tag{9.153}$$

where we have omitted dependence of functions on their spatial argument, and also employed the initial conditions (9.143_1). We next eliminate the inertial terms on the right of (9.153) by means of (9.142_1) and then integrate spatially by parts the resulting expression to obtain

$$2\int_\Omega \rho g u_i(t)u_{i,t}(t)d\mathbf{x} =$$
$$\int_0^t \int_\Omega g_{,j}[u_i(\eta)c_{ijkl}u_{k,l}(2t-\eta) - u_i(2t-\eta)c_{ijkl}u_{k,l}(\eta)]d\mathbf{x}d\eta$$
$$+ \int_\Omega \rho g[f_i u_{i,t}(2t) - h_i u_i(2t)]d\mathbf{x} \tag{9.154}$$

where we have used (9.144). We next apply the arithmetic-geometric mean inequality to the first term on the right of (9.154) to get

$$\int_0^t \int_\Omega g_{,j}u_i(\eta)c_{ijkl}u_{k,l}(2t-\eta)d\mathbf{x}d\eta$$
$$\le \int_0^t \int_\Omega |g_{,j}u_i(\eta)c_{ijkl}u_{k,l}(2t-\eta)|d\mathbf{x}d\eta$$
$$\le \frac{1}{2}\alpha^2 \int_0^t \int_\Omega (g/\rho)c_{ijkl}u_{k,l}(2t-\eta)c_{ijpq}u_{p,q}(2t-\eta)d\mathbf{x}d\eta$$
$$+ \frac{1}{2}\int_0^t \int_\Omega \rho g u_i(\eta)u_i(\eta)d\mathbf{x}d\eta$$
$$= \frac{1}{2}\alpha^2 \int_t^{2t} \int_\Omega (g/\rho)c_{ijkl}u_{k,l}(\eta)c_{ijpq}u_{p,q}(\eta)d\mathbf{x}d\eta$$
$$+ \frac{1}{2}\int_0^t \int_\Omega \rho g u_i(\eta)u_i(\eta)d\mathbf{x}d\eta\,. \tag{9.155}$$

By similarly treating the second term on the right of (9.154), and combining the resulting estimate with (9.155), we may deduce from (9.154)

$$2\int_\Omega \rho g u_i(t)u_{i,t}(t)d\mathbf{x} \le \frac{1}{2}\alpha^2 S(t)$$
$$+ \frac{1}{2}\int_0^{2t} \int_\Omega \rho g u_i(\eta)u_i(\eta)d\mathbf{x}d\eta + \int_\Omega \rho g[f_i u_{i,t}(2t) - h_i u_i(2t)]d\mathbf{x}\,, \tag{9.156}$$

where $S(t)$ is given by (9.145_2).

Now we introduce the functions $F(t)$ and $G(t)$ respectively defined by

$$F(t) = \int_\Omega \rho g u_i(t) u_i(t) d\mathbf{x} \tag{9.157}$$

and

$$G(t) = \int_0^{2t} F(\eta) d\eta\,. \tag{9.158}$$

It follows immediately that

$$\frac{d}{dt} G(t) = 2F(2t)\,. \tag{9.159}$$

We next integrate (9.156) over $(0, 2\tau)$, where $t < \tau < T/4$. This yields

$$\begin{aligned}
\frac{1}{2}\frac{d}{d\tau} G(\tau) &\le \frac{1}{2}\alpha^2 \int_0^{2\tau} S(\eta) d\eta + \frac{1}{2}\int_0^{2\tau} \left(\int_0^{2t} \int_\Omega \rho g u_i(\eta) u_i(\eta) d\mathbf{x} d\eta \right) dt \\
&+ \frac{1}{2}\int_\Omega \rho g f_i u_i(4\tau) d\mathbf{x} + \frac{1}{2}\int_\Omega \rho g f_i f_i d\mathbf{x} - \int_0^{2\tau}\int_\Omega \rho g h_i u_i(2\eta) d\mathbf{x} d\eta
\end{aligned} \tag{9.160}$$

$$\begin{aligned}
&\le \alpha^2 \tau S(2\tau) + \tau G(\tau) + \frac{1}{2}\int_\Omega \rho g f_i f_i d\mathbf{x} + \frac{1}{2}\int_\Omega \rho g f_i u_i(4\tau) d\mathbf{x} \\
&+ \left(\frac{1}{2} T \int_\Omega \rho g h_i h_i d\mathbf{x} \right)^{1/2} \left(\int_0^{2\tau} \int_\Omega \rho g u_i(2\eta) u_i(2\eta) d\mathbf{x} d\eta \right)^{1/2}.
\end{aligned} \tag{9.161}$$

Now we appeal to condition (9.147), and then rewrite (9.161) as

$$\frac{d}{d\tau}\left(e^{-2\tau} G(\tau)\right) \le 2\alpha^2 e^{-2\tau} \tau S(\tau) + e^{-2\tau} \int_\Omega \rho g f_i u_i(4\tau) d\mathbf{x} + c e^{-2\tau} \tag{9.162}$$

where

$$c = \int_\Omega \rho g f_i f_i d\mathbf{x} + M_1 \left(\int_\Omega \rho g h_i h_i d\mathbf{x} \right)^{1/2}. \tag{9.163}$$

An integration of (9.162) over $(0, s)$, where $\tau < s < T/4$, then gives the inequality

$$\begin{aligned}
G(s) &\le \alpha^2 S(T) T^2 e^{2T} + e^{2T} \int_0^s \int_\Omega \rho g f_i u_i(4\tau) d\mathbf{x} d\tau + c T e^{2T} \\
&\le \alpha^2 S(T) T^2 e^{2T} + e^{2T} \left(T \int_\Omega \rho g f_i f_i d\mathbf{x} \right)^{1/2} M_1 + c T e^{2T}\,.
\end{aligned}$$

Finally, we let $\alpha \to 0$ and use (9.145_1) to complete the proof.

Remark 9.10 - *Uniqueness of solution follows at once from* (9.148). *However, it will be evident that condition* (9.147) *is superfluous in this connection.*

9.8 EXERCISES

Exercise 9.1
(i) Prove the non-existence criterion (9.16) under the stated conditions – using the approach indicated in the text.
(ii) Consider a (simply connected) domain D with smooth, convex boundary ∂D, and *any* suitably smooth function $\psi(x, y)$ defined therein which takes a constant value on the boundary ∂D. Prove that

$$\int_D (\psi_{xx}\psi_{yy} - \psi_{xy}^2)dA = \int_{\partial D} (\nabla_1 \psi)^2 \rho^{-1}\, ds$$

where ρ is the radius of curvature. Show that the criterion (9.16) also follows from this.

Exercise 9.2 - Let R denote the circular domain $0 \leq x^2 + y^2 \leq a^2$ where $a(> 0)$ is constant. Consider the Dirichlet problem for the Monge-Ampère equation

$$\psi_{xx}\psi_{yy} - \psi_{xy}^2 = f(x, y) \qquad \text{in } R\,,$$

subject to

$$\psi = 0 \qquad \text{on } \partial R\,,$$

where

$$f(x, y) = \phi(r)$$

($r = \sqrt{x^2 + y^2}$) – $\phi(r)$ being continuous. Prove, by direct integration, that a necessary condition for the existence of *radially symmetric* C^2 solutions is

$$\int_0^r r\phi(r)dr \geq 0$$

or

$$\int_{R_r} f\, dA \geq 0 \qquad \forall r\,,$$

where R_r is a disc of radius r concentric with R.
Is this consistent with the (trivial) criterion for non-existence mentioned in the preceding question

$$\int_R f\, dA < 0\ ?$$

Considering the particular case

$$\phi(r) = \frac{\sin r}{r} \qquad r > 0\,,$$
$$\phi(0) = 0,$$

find the complete solution by integration.

Answer $\psi = \pm 4\{\cos(a/2) - \cos(r/2)\}$.

Exercise 9.3 - In the context of the P.D.E. of Monge-Ampère type

$$\psi_{xx}\psi_{yy} - \psi_{xy}^2 = \{\ddot{\phi}(\psi_y)\}^{-1} g(x,y)$$

where ϕ and g are as specified in the first paragraph of section 9.2.4, derive estimates for $\int_{\Gamma_x} \phi(\psi_y)dy$ analogous to those obtained in the text – both for a Dirichlet type problem and for a problem of the initial boundary value type. Also obtain non-existence criteria analogous to those obtained in the text.

Exercise 9.4

(i) Obtain analogues of Lemmas 9.1 and 9.2 when $F(x)$ is defined by

$$F(x) = \int_{\Gamma_x} y^{-1}\psi_y^2 dy$$

where the lower curve C_- is supposed to be in $y > 0$ (strictly).

(ii) Show how it may be used to obtain cross-sectional estimates – both for Dirichlet and initial boundary value problems – for the P.D.E. of Monge-Ampère type:

$$(y^{-1}\psi_y)_y\psi_{xx} - y^{-1}\psi_{xy}^2 + \frac{1}{2}[y^{-2}\psi_x^2]_y = \pi(x,y).$$

> (An equation of this nature governs the stream function $\psi(x,y)$ for *axisymmetric* flow of a continuum of constant density (or, more generally, one whose velocity field is divergence-free), where (y,x) correspond respectively to the radial and axial coordinates of cylindrical polar coordinates (normally denoted by r, z respectively). Moreover, π is a quantity related to pressure in the case of perfect fluid flow.)

(iii) Show how pointwise bounds are obtained from the relevant estimates.

(iv) Show how non-existence criteria are obtained from the relevant estimates.

Exercise 9.5 - Obtain a strengthening of Theorem 9.10 by retaining more terms in (9.46).

Exercise 9.6 - Show how generalizations of Theorem 9.10 may be obtained if the lateral boundaries are assumed to be such that

$$y_+'' < 0, \qquad y_-'' > 0$$

(i.e. lateral boundaries are strictly convex) and if $\mathcal{F}_+$, $\mathcal{F}_-$ are unrestricted, apart from smoothness requirements.

Hint - p.24 of Straughan [35] may be of assistance.

Exercise 9.7 - Consider the I.B.V.P.

$$\begin{aligned} &u_{tt} - u_{xxt} = [f(u_x)]_x && x \in (0,1], \ t > 0 \\ &u(x,0) = u_0(x), \ \ u_t(x,0) = \overset{\bullet}{u}_0(x) && x \in [0,1] \\ &u(0,t) = u(1,t) = 0 && t \geq 0 \end{aligned}$$

with $f \in L^1_{\text{loc}}(I\!R)$. Show that, at least for smooth solutions,

$$V = \frac{1}{2}\int_0^1 u_t^2 dx + \int_0^1 (\int_0^{u_x} f(\xi)d\xi)dx$$

is a Liapunov function. Consider, in particular, the case (9.63), and verify that (9.64) immediately follows.

Answer - $\overset{\bullet}{V} = -\int_0^1 (u_{xt})^2 dx$. In the case (9.63), $f = \varepsilon u_x + \int_0^{u_x} \varphi(\xi)d\xi$.

Exercise 9.8 - Consider the I.B.V.P. of exercise 9.7 with

$$f = \frac{u_x}{\sqrt{1+u_x^2}} \, .$$

Study the stability of the rest state.

Hint - The I.B.V.P. of exercise 9.7 arises in the study of purely longitudinal motions of a homogeneous bar of uniform cross-section. We refer the reader to [82,83] and [208].

Exercise 9.9 - Denoting by $\|\cdot\|$ the L^2-norm, and by α, β, k positive constants, consider the I.B.V.P.

$$\begin{aligned} &u_{tt} + g(u_t) + \alpha u_{xxxx} = (\beta + 2k\|u_x\|^2)u_{xx} && x \in (0,1), \ t > 0 \\ &u(x,0) = u_0(x), \ \ u_t(x,0) = \overset{\bullet}{u}_0(x) && x \in [0,1] \\ &u(0,t) = u(1,t) = u_x(0,t) = u_x(1,t) = 0 && t \geq 0, \end{aligned}$$

where $g \in C(I\!R)$ and is such that $\xi g(\xi) > 0$. Determine a Liapunov function.

Hint - Consider the function

$$V = \frac{1}{2}[\|u_t\|^2 + \beta\|u_x\|^2 + \alpha\|u_{xx}\|^2 + k\|u_x\|^4] \, .$$

Answer - At least along strong solutions, it follows that

$$\overset{\bullet}{V} = -\int_0^t u_t\, g(u_t)dt \le 0\,.$$

We refer the reader to [39], pp. 148-151, for a deeper analysis.

Exercise 9.10 - Consider the I.B.V.P.

$$\begin{aligned} &u_{tt} + u_t - u_{xx} = f(u) - g(x), && x \in (0,1) \\ &u(x,0) = u_0(x), \quad u_t(x,0) = \overset{\bullet}{u}_0(x) && x \in [0,1] \\ &u(0,t) = u(1,t) = 0 && t \ge 0 \end{aligned}$$

and verify that

$$V = \frac{1}{2}\int_0^1 [u_x^2 + u_t^2 - \int_0^u f(\xi)d\xi + g(x)u]dx$$

is a non-increasing function, at least along smooth solutions.

Answer - $\overset{\bullet}{V} = -\displaystyle\int_0^1 u_t^2 dx.$

Exercise 9.11 - Consider the I.B.V.P.

$$\begin{cases} u_{tt} + 2bu_t + cu + \gamma(x)u^k = au_{xx}\,, \delta & x \in (0,1),\, t \in I\!R_+ \\ u(x,0) = u_0(x), \quad u_t(x,0) = \overset{\bullet}{u}_0(x) & x \in [0,1] \\ u(0,t) = u(1,t) = 0 & t \ge 0 \end{cases}$$

where a, b, c, k are positive constants and $\gamma \in L^{2p}[0,1], p =$ const. $\ge 1/2$, is a nonnegative prescribed function. Show that, at least with respect to smooth perturbations, the rest state is asymptotically and exponentially stable in the W_2^1-norm.

Hint - Show that

$$V = \int_0^1 [(c-b)^2u^2 + au_x^2 + (bu + u_t)^2]dx\,,$$

is a Liapunov function.

Answer - See [226].

Exercise 9.12 - Let $f \in C^2[x_1, x_2]$ where $x_1 x_2$ are constants and let

$$\alpha \le f''(x) \le \beta\,, \qquad x \in [x_1, x_2]$$

$\alpha \leq \beta$ being positive constants.

Show that $\exists F \in C^2(\mathbb{R})$ such that

$$F(x) = f(x) \qquad \forall\, x \in [x_1, x_2]$$
$$\alpha \leq F''(x) \leq \beta \qquad \forall\, x \in \mathbb{R}\,.$$

Answer - It is easily verified that f can be extended as follows:

$$F = \begin{cases} \frac{1}{2} f''(x_1)(x^2 - x_1^2) + [f'(x_1) - f''(x_1)x_1](x - x_1) + f(x_1), & x \leq x_1 \\ f(x) & x \in [x_1, x_2] \\ \frac{1}{2} f''(x_2)(x^2 - x_2^2) + [f'(x_2) - f''(x_2)x_2](x - x_2) + f(x_2), & x \geq x_2\,. \end{cases}$$

Exercise 9.13 - Consider a parallel shear basic flow $\mathbf{v}^* = v(y)\mathbf{i}$ of an inviscid fluid between the two horizontal flat plates $y = \pm d$, $\mathbf{i}$ being the unit vector of the horizontal axis x. It is well known that a *necessary* condition for the linear instability of $\mathbf{v}^*$ (with respect to normal mode perturbations) is that v'' changes sign on $[-d, d]$ (***Rayleigh's inflection point theorem***). Show that when v'' does not change sign on $[-d, d]$ then $\mathbf{v}^*$ is nonlinearly stable in the norm $\|\mathbf{u}\|_2^2 + \|\nabla \times \mathbf{u}\|_2^2$, at least with respect to planar perturbations $\mathbf{u}(x, y, t) = [u_1(x, y, t), u_2(x, y, t), 0]$ periodic in the x-direction.

Hint - Take into account theorem 9.14 and Remark 9.7.

Answer - From $\mathbf{v}^* = \nabla^{\perp}\varphi$, it follows that $\varphi = \varphi(y), v = \nabla\varphi, \nabla\Delta\varphi = v''$. By a translation of the frame of reference in the x-direction, $\mathbf{v}$ and hence $\nabla\varphi$ are defined modulo a constant. Therefore, when $v''(y)$ does not change sign, we may assume $\dfrac{v}{v''} \geq 0$, $\forall\, y \in [-d, d]$. Therefore when $\dfrac{v}{v''} \in C[-d, d]$, (9.86) is implied.

Exercise 9.14 - Let $M(\mathbf{z})$ be a Maxwellian distribution function (Remark 9.8). Verify that, for any $\beta > 0$, $\displaystyle\int_{\mathbb{R}_3} M(\mathbf{z}) d\mathbf{z} = 1$.

Hint - Recall that the value of the Gauss integral is $\sqrt{\pi}/2$ and work in spherical polar coordinates.

Exercise 9.15 - Assume $\pi(\mathbf{x}, \mathbf{y}, \mathbf{z}) = \pi(\mathbf{z})$ and solve the I.V.P. (9.94)–(9.95). In particular find the solution when $\pi(\mathbf{z}) = M(\mathbf{z})$ is a Maxwellian distribution function.

Answer - $f = (f_0 - Q)e^{-kQt} + Q\pi(\mathbf{z})$; $f = (f_0 - Q)e^{-kQt} + Q\left(\dfrac{\beta}{\pi}\right)^{3/2} e^{-\beta z^2}$.

Exercise 9.16 - Suppose that $\overline{\pi}_0 = \operatorname*{ess\,inf}_{(x,y)\in \mathbb{R}_3^2} \pi(\mathbf{x},\mathbf{y},\mathbf{z})$. Show that if there exists a measurable domain I such that

$$\overline{\pi}_0 \in L_1(I), \qquad \varepsilon = \int_I \overline{\pi}_0(\mathbf{z})d\mathbf{z} \geq \frac{1}{2}$$

then $\varepsilon > 1/2$ implies asymptotic exponential L^1-stability of the solution to the I.V.P. (9.94) – (9.95).

Hint - Choose

$$h(z) = \begin{cases} \overline{\pi}_0(z) & z \in I, \\ 0 & 0 \notin I \end{cases}$$

and use theorem 9.21.

Exercise 9.17 - Suppose that $\overline{\pi}_0 = \operatorname*{ess\,sup}_{(x,y)\in \mathbb{R}_3^2} \pi(\mathbf{x},\mathbf{y},\mathbf{z}), \pi_0(z) \in L_1(\mathbb{R}_3), \varepsilon = \int_{\mathbb{R}_3} \pi_0(\mathbf{z})d\mathbf{z} < 3/2$. Show that the solution to the I.V.P. (9.94) – (9.95) is asymptotically exponentially stable in the $L^1(\mathbb{R}_3)$-norm.

Hint Choose $h(z) = \pi_0$ and use theorem 9.21.

Exercise 9.18 - Let $\exists p \geq 1$ and a measurable set I such that

$$\pi_0(z) \in L^1(I); \quad \overline{\pi}_0(z) \geq \frac{1}{p}\pi_0(z), \ \forall \mathbf{z}; \quad \frac{1}{p}\int_I \pi_0(z)dz > \frac{1}{2}$$

where π_0 and $\overline{\pi}_0$ are defined in exercises (9.16) – (9.17). Show that the solution to the I.V.P. (9.94) – (9.95) is asymptotically exponentially stable in the $L^1(\mathbb{R}_3)$-norm.

Hint - Choose

$$h(z) = \begin{cases} \frac{1}{p}\pi_0(z), & z \in I \\ 0, & z \notin I \end{cases}$$

and use theorem 9.21.

APPENDIX: FUNDAMENTAL INEQUALITIES

A.1 Cauchy inequality or arithmetic-geometric inequality

Let $a, b \in I\!R$. Then $(|a| - |b|)^2 \geq 0$ implies the *Cauchy inequality* or the *arithmetic-geometric inequality*:

$$ab \leq |ab| \leq \frac{1}{2}(a^2 + b^2).$$

Hence, $\forall\, \varepsilon > 0$, there follows (the *generalized or weighted Cauchy inequality*)

$$ab = \frac{a}{\sqrt{\varepsilon}} \cdot \sqrt{\varepsilon} b \leq \frac{1}{2}\left(\frac{a^2}{\varepsilon} + \varepsilon b^2\right).$$

A.2 Hölder integral inequality

Let $\Omega \subset I\!R_3$ be a measurable region and let f and g be integrable functions on Ω. Then the following Hölder inequality holds [232]

$$\int_\Omega |fg| d\Omega \leq \left(\int_\Omega |f|^p d\Omega\right)^{1/p} \left(\int_\Omega |g|^q d\Omega\right)^{1/q},$$

where p and q are positive numbers such that

$$\frac{1}{p} + \frac{1}{q} = 1.$$

Choosing $p = q = 2$, the Schwarz inequality

$$\int_\Omega |fg| d\Omega \leq \|f\|_2 \, \|g\|_2$$

immediately follows.

A.3 Sobolev inequality

Let Ω be a bounded domain in $\mathbb{R}_3$ with smooth boundary $\partial\Omega$, then for integrable (as needed) functions u with $u = 0$ on $\partial\Omega$ the following inequality holds

$$\int_\Omega \mathbf{u}^4 d\mathbf{x} \leq c\|\mathbf{u}\|_2 \, \|\nabla \mathbf{u}\|_2^3 \tag{1}$$

where c is a constant independent of Ω.

Inequality (1) follows immediately from

$$\left(\int_\Omega \mathbf{u}^2 d\mathbf{x}\right)^{1/3} \leq k \int_\Omega (\nabla \mathbf{u})^2 d\mathbf{x} \tag{2}$$

where $k = [4/(\pi^2\sqrt{3})]^{1/3}$, which is a particular case of a general Sobolev embedding inequality (see [233], pp. 148-157). In fact by the Schwarz inequality it turns out that

$$\int_\Omega \mathbf{u}^4 d\mathbf{x} \leq \|u\|_2 (\int_\Omega \mathbf{u}^6 d\mathbf{x})^{1/2} \tag{3}$$

and hence (2) implies (1), with $c \geq k^{3/2}$.

When $\Omega = [0, 1]$ and $u \in C_0^1([0, 1])$, the following Sobolev inequality holds

$$\int_0^1 \mathbf{u}^4 d\mathbf{x} \leq \frac{1}{\pi^2} \left[\int_0^1 (u')^2 dx\right]^2 . \tag{4}$$

In fact, $\forall x \in [0, 1]$, it follows that

$$\frac{1}{2} u^2(x) = \int_0^x u u_x dx \leq \int_0^x |u| \, |u_x| dx$$

$$\frac{1}{2} u^2(x) = -\int_x^1 u u_x dx \leq \int_x^1 |u| \, |u_x| dx .$$

Hence, taking into account (2.26), it turns out that

$$u^2(x) \leq \int_0^1 |u| \, |u_x| dx \leq \|u\|_2 \, \|\nabla u\|_2 \leq \frac{1}{\pi} \|\nabla u\|_2^2 .$$

On squaring and integrating on $[0, 1]$, (4) immediately follows.

A.4 Poincaré inequality

Let Ω be a bounded domain in $\mathbb{R}_3$ with smooth boundary $\partial\Omega$. Then in $C_0^2(\Omega)$ the following inequality holds:

$$\|u\|_2 \leq c(\Omega)\|\nabla u\|_2 \tag{5}$$

$c(\Omega)$ being a constant depending on the geometry and size of the domain Ω.

The proof of (5) can be accomplished as follows. First one proves the existence of the $\min\limits_{u\in H_0^1(\Omega)} \dfrac{\|\nabla u\|_2}{\|u\|_2}$ by using the direct method of variational calculus {See [45] vol.III, [233], and exercise 2.5.} Then, in order to obtain the value of the constant c, one considers the Euler-Lagrange equation of the problem [235], i.e.

$$\Delta u + \lambda u = 0 \qquad \text{in } \Omega \tag{6}$$

$$u = 0 \qquad \text{on } \partial\Omega. \tag{7}$$

It follows that the lowest $\{c(\Omega)\}^{-1}$ for which (5) holds is the lowest eigenvalue of the eigenvalue problem (6) – (7) (i.e. the principal eigenvalue of $-\Delta$).

Upper bounds for $c(\Omega)$ are well-known. For instance if l is the smallest distance between two parallel planes containing Ω then $c(\Omega) \leq l^2/2$ {[71], I, pp.13-14}. In the case of vector valued functions, divergence free in Ω, an upper bound for $c(\Omega)$ is given by d^2/α, where $\alpha \approx 80$ and d is the diameter of Ω {exercise 8.9}.

Let Ω be bounded in at least one direction and assume that it is contained between the planes $z = \pm l/2$. Let I denote the set of functions $u \in C_0^2(\Omega)$ periodic in the x and y directions of periods a and b respectively. Then inequality (5) continues to hold on any "cell of periodicity" i.e. any domain V like $[0,a]\times[0,b]\times[z_1,z_2]$, $z_i = z_i(x,y), i = 1,2$, being the intersections $\partial V \cap \partial\Omega$. In this case $l^2/2$ continues to be un upper bound of $C(V)$. If Ω is the layer $z \in [0,1]$ then $V = [0,a] \times [0,b] \times [0,1]$ is a cell of periodicity and one has $c(V) \leq 1/\pi^2$. In fact from (exercise 2.4)

$$\int_0^1 u^2(x,y,z)dz \leq \frac{1}{\pi^2}\int_0^1 \left(\frac{\partial u}{\partial z}\right)^2 dz \qquad \forall\,(x,y) \in [0,a]\times[0,b]$$

integrating on $[0,a]\times[0,b]$, there immediately follows $c(V) \leq 1/\pi^2$.

The Poincaré inequality (5) continues to hold with different boundary conditions. In the next section we will consider some of them in the case in which Ω is an interval or a two-dimensional domain.

Denoting by λ_1 the smallest positive eigenvalue of (6) with the boundary condition $\dfrac{du}{dn} = 0$, we end this section by recalling the following Poincaré inequalities {[5], pp.112-113 [62], [235]}:

i) Suppose $u \in W_2^1(\Omega)$, and $\dfrac{du}{d\mathbf{n}} = 0$ on $\partial\Omega$. Then

$$\|u - \overline{u}\|_2^2 \leq \frac{1}{\lambda_1}\|\nabla u\|_2^2 \tag{8}$$

where $\overline{u} = \dfrac{1}{\text{meas}\,\Omega}\displaystyle\int_\Omega u dx$.

ii) Suppose $u \in W_2^2(\Omega)$, and $\dfrac{du}{d\mathbf{n}} = 0$ on $\partial\Omega$. Then

$$\|\nabla u\|_2^2 \leq \frac{1}{\lambda_1}\|\Delta u\|_2^2\,. \tag{9}$$

A.5 Inequalities in one and two dimensions

A.5.1 Any function $\phi(x)$ defined on $0 \leq x \leq l$ such that $\phi(x)$ is continuous in $[0, l]$ and $\phi'(x) \in L^2[0, l]$ satisfies the inequalities given hereunder under the stated conditions

$$\int_0^l \phi'^2 dx \geq \pi^2 l^{-2} \int_0^l \phi^2 dx \tag{10}$$

either when

$$\phi(0) = \phi(l) = 0 \tag{a}$$

(exercise 2.4) or

$$\int_0^l \phi dx = 0\,. \tag{b}$$

The equality sign occurs in case (a) iff $\phi = K\sin(\pi x/l)$ where K is an arbitrary constant here (and elsewhere in this section); the equality occurs in case (b) iff $\phi = K\cos(\pi x/l)$.

One also has

$$\int_0^l \phi'^2 dx \geq \frac{1}{4}\pi^2 l^{-2} \int_0^l \phi^2 dx \tag{11}$$

either when

$$\phi(0) = 0 \qquad \text{or} \qquad \phi(l) = 0\,;$$

the equality sign occurs when $\phi = K\sin(\pi x/2l)$ in the former case.

A suitable reference for all of the above and those of A.5.2-A.5.5 is Mitrinović et al [62], Chapter two.

A.5.2 Any function $\phi(x)$ defined on $0 \le x \le l$ such that $\phi \in C^2[0,l]$ satisfies the inequalities given hereunder under the stated conditions

$$\int_0^l \phi''^2 dx \ge 4\pi^2 l^{-2} \int_0^l \phi'^2 dx \tag{12}$$

when

$$\phi(0) = \phi'(0) = \phi(l) = \phi'(l) = 0.$$

The inequality sign holds iff ϕ is the eigenfunction corresponding to the lowest eigenvalue $(4\pi^2 l^{-2})$ of the eigenvalue problem

$$\psi'''' + \lambda\psi'' = 0$$

where

$$\psi(0) = \psi'(0) = \psi(l) = \psi'(l) = 0\,.$$

One also has

$$\int_0^l \phi''^2 dx \ge \pi^2 l^{-2} \int_0^l \phi'^2 dx \tag{13}$$

where

$$\phi(0) = \phi(l) = 0\,.$$

It will be noted that this is a consequence of case (b) (see, also, exercise 2.6).

A.5.3 We now give a result which includes some of those of A.5.1 as special cases: Let both $p(x)(>0)$, $q(x)(\ge 0) \in C[a,b]$, $r(x)(>0) \in C^1[a,b]$. Suppose $\alpha_a, \alpha_b, \beta_a, \beta_b$ are non-negative constants such that $\beta_a^2 + \alpha_a^2 > 0$, $\beta_b^2 + \alpha_b^2 > 0$. Let λ_1 be the lowest (positive) eigenvalue of the Sturm-Liouville problem

$$(r\psi')' + (\lambda p - q)\psi = 0\,,$$

$$\beta_a\psi(a) - \alpha_a r(a)\psi'(a) = 0\,, \quad \beta_b\psi(b) + \alpha_b r(b)\psi'(b) = 0\,.$$

Any function $\phi(x)$ such that $\phi \in C^1[a,b]$ whose end values are unrestricted except that

$$\phi(a) = 0 \quad \text{if} \quad \alpha_a = 0\,, \qquad \phi(b) = 0 \quad \text{if} \quad \alpha_b = 0\,,$$

satisfies

$$\int_a^b (r\phi'^2 + q\phi^2)dx + A(\phi) + B(\phi) \ge \lambda_1 \int_a^b p\phi^2 dx \tag{14}$$

where

$$A(\phi) = \begin{cases} 0 & \text{if } \alpha_a = 0\,, \\ \dfrac{\beta_a}{\alpha_a}\phi^2(a) & \text{if } \alpha_a \neq 0\,, \end{cases}$$

$$B(\phi) = \begin{cases} 0 & \text{if } \alpha_b = 0\,, \\ \dfrac{\beta_b}{\alpha_b}\phi^2(b) & \text{if } \alpha_b \neq 0\,. \end{cases}$$

The equality sign holds in the foregoing inequality iff ϕ is the eigenfunction corresponding to the lowest (positive) eigenvalue of the Sturm-Liouville problem given above.

A.5.4 In what follows, $\mathcal{D}$ shall denote a bounded plane domain with sufficiently smooth boundary $\partial\mathcal{D}$.

A function ϕ defined in $\mathcal{D}$, such that $\phi \in C^1(\overline{\mathcal{D}})$, satisfies the following inequalities under the stated conditions:

$$\int_{\mathcal{D}} (\nabla_1\phi)^2 d\mathbf{x} \geq \overline{\lambda} \int_{\mathcal{D}} \phi^2 d\mathbf{x} \tag{15}$$

where $\overline{\lambda}$ is a suitable positive constant,
either when

$$\phi = 0 \qquad \text{on the boundary} \quad \partial\mathcal{D}, \tag{a}$$

in which case $\overline{\lambda} = \lambda_1$ – the lowest (positive) eigenvalue of

$$\nabla_1^2\Phi + \lambda\Phi = 0 \qquad \text{in } \mathcal{D}\,,$$

$$\Phi = 0 \qquad \text{on } \partial\mathcal{D}\,,$$

or

$$\int_{\mathcal{D}} \phi d\mathbf{x} = 0\,, \tag{b}$$

in which case $\overline{\lambda} = \nu_1$ – the lowest *positive* eigenvalue of

$$\nabla_1^2\Phi + \nu\Phi = 0 \qquad \text{in } \mathcal{D}\,,$$

$$\frac{\partial\Phi}{\partial n} = 0 \qquad \text{on } \partial\mathcal{D}\,,$$

$\partial/\partial n$ denoting the normal derivative.

The equality sign holds in cases (*a*) and (*b*) iff ϕ coincides with the eigenfunction corresponding to the relevant eigenvalue.
(Note: The eigenvalues λ_1, ν_1, are often called the "fixed membrane" eigenvalue and the "free membrane" eigenvalue respectively.)

We now record an analogue, in the current context, of (13):

$$\int (\nabla^2 \phi)^2 dA \geq \lambda_1 \int (\nabla \phi)^2 dA \tag{16}$$

for all functions ϕ defined in the bounded region $\mathcal{D}$ such that $\phi \in C^2(\overline{\mathcal{D}})$ and which vanish on the boundary $\partial \mathcal{D}$.

A.5.5 The following inequality arises frequently in the text:
For any continuous function $\psi(y)$ defined in $0 \leq y \leq l$, and such that $\psi(0) = \psi(l) = 0$ such that $\psi' \in L^2[0, l]$

$$\psi^2(\xi) \leq \xi(1 - \xi/l) \int_0^l \psi'^2 dy \tag{17}$$

for any ξ, $0 < \xi < l$.
The equality sign holds iff

$$\begin{aligned} \psi &= y/\xi, & 0 \leq y \leq \xi, \\ \psi &= (1 - y/l)(1 - \xi/l)^{-1}, & \xi \leq y \leq l \end{aligned}$$

– apart from an arbitrary multiplicative constant. (The inequality is essentially a Schwarz inequality.) See Straughan [35], for example.

An inequality of a similar nature arises (from Schwarz's inequality) if one of the zero end conditions is dropped.

A.6 A point-wise estimate for functions $\varphi \in C^1(\mathbb{R})$

Let c be a positive constant and let I_c be the set of functions

$$\varphi : \mathbb{R} \to \mathbb{R} \quad , \quad \varphi \in C^1(\mathbb{R}) \quad , \quad |\varphi'| \leq c .$$

Then $\varphi \in I_c$ *and* h=const.>0, imply the estimate

$$|\varphi(x)| < k \left[\int_x^{x+h} \varphi^2 \, dt + \left(\int_x^{x+h} \varphi^2 \, dt \right)^{1/2} \right]^{1/2}, \qquad \forall x \in \mathbb{R} \tag{18}$$

where $k^2 = \max(1/h, 2ch^{1/2})$.

In fact, let $\xi \in [x, x+h]$. By the Schwarz inequality it follows that

$$\varphi^2(x) = \varphi^2(\xi) - \int_x^\xi \frac{d}{dt}\varphi^2 \, dt \le \varphi^2(\xi) + 2ch^{1/2}\left(\int_x^{x+h} \varphi^2 \, dt\right)^{1/2}$$

Integrating with respect to ξ over $[x, x+h]$, it turns out that

$$\varphi^2(x) = h^{-1}\int_x^{x+h} \varphi^2 \, dt + 2ch^{1/2}\left(\int_x^{x+h} \varphi^2 \, dt\right)^{1/2}$$

and hence (16) immediately follows.

If $\varphi \in L^2(\mathbb{R})$, then (18) implies

$$\sup_{\mathbb{R}} |\varphi| < k\left[\|\varphi\|^2 + \|\varphi\|\right]^{1/2} . \tag{19}$$

A.7 A comparison theorem

There exist many comparison theorems involving *partial* differential inequalities which are analogous to that contained in Theorem 4.4. The following is one such result which is relevant to parabolic operators:

Let $H(x,t)$ be any function which satisfies

(*i*) $H(x,t)$ is continuous in $0 \le x \le L$, $t \ge 0$,

(*ii*) $H_t - H_{xx} \le 0$ in $0 < x < L$, $t > 0$,

(*iii*) $H(0,t) = H(L,t) = 0$ for $t \ge 0$.

Then

$$H(x,t) \le 0 .$$

See Protter and Weinberger [131], for example.

REFERENCES

1. Fichera G., I difficili rapporti fra l'analisi funzionale e la fisica matematica, *Rend. Acc. Naz. Lincei, Matematica e Applicazioni*, s.IX, vol. I, fasc. 2, 161, 1990.
2. Hadamard J., *Lectures on Cauchy's Problem in linear P.D.E.*, New York, Dover, 1952.
3. Hormander, L.,*Linear Partial Differential Operators*, Academic Press, Springer, N. Y., 1963.
4. Lewy H., An example of a smooth linear partial differential equation without solution, *Ann. Math.*, 66, 155-158, 1957.
5. Smoller J., *Shock waves and Reaction - Diffusion Equations*, Springer-Verlag, 1983, n.258 of "A Series of Comprehensive Studies in Mathematics".
6. Dafermos C. M., Asymptotic behaviour of solution of hyperbolic balance laws, in *Bifurcation Phenomena in Mathematical Physics (C. Bardos and D. Bessis, Eds.), D. Reidel, Dordrecht*, 1979, 521.
7. Knops R.J., Wilkes E.W., *Theory of Elastic Stability*, Handbuch der Physik, vol. VI a/3 (1973), Springer-Verlag.
8. Walker J.A., *Dynamical Systems and Evolution Equations. Theory and applications*, Plenum Press, New York, 1980.
9. Belleni Morante A., *Applied Semigroups and Evolution Equations*, Oxford Math. Monographs, Clarendon Press, 1979; and, *A concise guide to semigroups and evolution equations*, Series on Advances in Math. Appl. Sc., vol. 19, World Scientific.
10. Mc Bride A.C., *Semigroups of Linear Operators: an Introduction*, Pitman Research Notes in Math. Series 156.
11. Chernoff P., Marsden J., On continuity and smoothness of group action, *Bull. Am. Math. Soc.*, 76, 1044, 1970.

12. Muratori P., Teoremi di unicitá per un problema relativo alle equazioni di Navier- Stokes, *Boll. U.M.I.* (4), 4, 1971.
13. Galdi G.P., Rionero S., On the best conditions on the gradient of pressure for uniqueness of viscous flows in the whole space, *Pacific J. of Math.*, 104, 1, 77, 1983.
14. Fichera G., Avere una memoria tenace crea gravi problemi, *Arch. Rat. Mech. Anal.*, 70, 245, 1979.
15. Fichera G., Sul principio della memoria evanescente, *Rend. Sem. Mat. Un. Padova*, 68, 245, 1982.
16. Coleman B., Noll W., Foundations of linear viscoelasticity, *Rev. Modern Physics*, 33, 2, 239, 1961.
17. Fichera G., *Problemi analitici nuovi nella Fisica Matematica classica*, Quaderno C.N.R. - G.N.F.M.; Lessons given at the IX Summer School of Math. Physics - Ravello, Villa Rufolo 10-28 Sept. 1984.
18. Fabrizio M., Morro A., *Mathematical Problems in Linear Viscoelasticity*, SIAM Studies in Appl. Math. vol. 12, 1992.
19. Graffi D., On the fading memory, *Appl. Anal.*, 15, 295, 1983.
20. Morro A., Fabrizio M., On uniqueness in linear viscoelasticity: a family of counterexamples, *Quart. Appl. Math. XLV*, 2, 321, 1987.
21. Giorgi C., Lazzari B., Uniqueness and stability in linear viscoelasticity: some counterexamples, *Proc. V Int. Meeting "Waves and stability in continuous media", Rionero S. editor, World Sc. Series on Adv. Math. Appl. Sc.*, 4, 146.
22. Knops R.J., Payne L.E., Uniqueness theorems in linear elasticity, *Springer tract in Nat. Phil.*, 19, 1971.
23. Carbonaro B., Russo R., Sharp uniqueness classes in linear elastodynamics, *J. of Elasticity*, 27, 37, 1992.
24. Carbonaro B., Russo R., Singularity problems in linear elastodynamics, *Ricerche di Matematica.*
25. Russo R., On the traction problem in linear elastostatics, *J. of Elasticity*, 27, 57, 1992.
26. Tykhonov A., Theorems d'unicité pour l'equation de la chaleur, *Mat. Sbornik*, 42, 199, 1935.
27. Friedman A., *Partial Differential Equations of Parabolic Type*, Prentice-Hall Inc., London.
28. Flavin J.N., Almost uniqueness for a non-linear Dirichlet problem, *Proc. R. Ir. Acad.*, 93 A, No. 2, 203, 1993.
29. Ames W.F., *Nonlinear Partial Differential Equations in Engineering*, Math. in Sc. and Eng. v. 18, Academic Press, 1965.
30. Courant, R., Hilbert, D.: *Methods of Mathematical Physics*, vol. II, p. 324, New York Interscience, 1962.

31. Kaplan S., On the growth of solutions of quasi-linear parabolic equations, *Comm. Pure Appl. Math.*, VXVI, 305, 1963.
32. Lavrentev M.M., Romanov V.G., Sisatskij S.P., *Problemi mal posti in Fisica Matematica e Analisi*, Pubb. n.12 Ist. Anal. Globale a Appl. Firenze, 1983.
33. Tikhonov A., Arsénine V., *Méthodes de résolution de problemes mal posés*, Edition Mir., 1976.
34. Payne L.E., *Improperly posed Problems in Partial Differential Equations*, SIAM regional conference series in Applied Math., vol 22, 1975.
35. Straughan B., *Instability, Nonexistence and Weighted Energy Methods in Fluid Dynamics and related theories*, Pitman Research notes in Mathematics, 74, 1982.
36. Ames K.A., Straughan B., *Contemporary Non-standard and Improperly posed Problems for Partial Differential Equations*, Academic Press, Series in Appl. Math. and Eng. (to appear).
37. Movchan A.A., Stability of processes with respect to two metrics, *J. Appl. Math. Mech.*, 24, 1960.
38. Liusternik L.A., Sobolev V.J., *Elements of Functional Analysis*, Frederick Ungar Publ. Comp., New York, II printing 1965, 69.
39. Hale J.K.: *Asymptotic behaviour of dissipative systems*, Math. Surveys and Monographs n.25, Amer. Math. Soc. Providence, Rhode Island, 1988.
40. Temam R., *Infinite-Dimensional Dynamical Systems in Mechanics and Physics*, Appl. Math Sc. 68, Springer-Verlag, 26, 1988 (Sec 1.4).
41. Payne L.E., On geometric and modeling perturbations in P.D.E.s, in *Nonclassical Continuum Mechanics, R.J. Knops & A.A. Lacey editors, London Math. Soc.* Lecture Note Series, 122, 1987.
42. Payne L.E., On stabilizing ill-posed problems against errors in geometry and modeling, Proc. Conf. on Inverse and Ill-posed Problems, Strobl. H. Engel and C.W. Grsetsch (eds), Academic Press, 399.
43. Payne L.E., Continuous dependence on geometry with applications to Continuum Mechanics, *Proc. Canadian Appl. Math. Soc. Meeting*, Vancouver, June, 1988.
44. Truesdell C. ed, *Encyclopedia of Physics*, Springer-Verlag, Berlin, 1973.
45. Zeidler E., *Nonlinear functional analysis and its application*, Springer-Verlag, 1986.
46. Reed M., Simon B., *Methods of Modern Mathematical Physics*, Academic Press, 1979, vols I-IV.
47. Murray J.D., *Mathematical Biology*, Biomathematics Texte, 19, Springer-Verlag, 1989.
48. Segel L.A., *Modelling Dynamic Phenomena in Molecular and Cellular Biology*, Cambridge: Cambridge University Press, 1984.
49. Fisher R.A., The wave of advance of advantageous genes, *Ann. Engenics* 7, 353, 1937.

50. Fife P.C., *Mathematical Aspects of Reacting and Diffusing Systems*, Lecture Notes in Biomathematics, 28, Springer-Verlag, Berlin, 1979.
51. Gurtin M.E., Mac Camy R.C., On the diffusion of biological populations, *Mathematical Biosciences*, 33, 35, 1977.
52. Grusa K.U., *Mathematical Analysis of Nonlinear Dynamics Processes*, Pitman Research Notes in Mathematics Series 176, Longman Scientific & Technical, 1988.
53. Leung A.W., *Systems of nonlinear P.D.E.s. Applications to Biology and Engineering*, Math. and its Applications, Kluwer Ac. Publ., 1989.
54. Capasso V., *Mathematical structures of epidemic system*, Lecture Notes in Biomathematics, Springer-Verlag, 1993.
55. Rothe F., *Global solutions of Reaction-diffusion Systems*, Lecture Notes in Mathematics 1072, Springer-Verlag, 1984.
56. Krasnov M.L., Kiselev A.I., Makarenko G.I., *Equazioni Integrali*, Ed. Mir., 1983.
57. Cosserat, E., and F., Sur la deformation infinitement petite d'une ellipsoide élastique, *C.R. Acad. Sc. Paris*, 127, 315, 1898.
58. Yudovich V.I., *The Linearization Method in Hydrodynamical Stability Theory. Translation of Mathematical Monographs*, Amer. Math. Soc., 74, Providence, Rhode Island, 1989.
59. Straughan, B., *The Energy Method, Stability and Nonlinear Convection*, Appl. Math. Sciences 91, Springer-Verlag, 1992.
60. Payne L.E., Straughan B., Error estimates for the temperature of a piece of cold ice, given data on only part of the boundary, *Nonlinear Analysis, Theory, Methods and Applications*,Vol.14, N.5, 443, 1990.
61. Foias C., Prodi G., Sur le comportement global des solutions non stationnaries des équations de Navier-Stokes en dimension 2, *Rend. Sem. Mat. Univ. Padova*, 39, 1, 1967.
62. Mitrinović D.S., Pecărić J.E., Fink A.M., *Inequalities involving functions and their integrals and derivatives*, Kluwer Academic Publishers, Dordrecht, 1991.
63. Liapunov A.M., *Probléme general de la stabilité du mouvement.* The paper, published in 1893 in Comm. Math. Soc. of Kharkow, was translated into French by Davaux E. and published in the Ann. Fac. Sci. Toulose, 9, 203-474 (1907) and reprinted as Vol. 17 in Ann. Math. Studies, Princeton, 1949.
64. La Salle S.P., Lefschetz S., *Stability by Liapunov's Direct Method with Applications*, Academic Press, New York, 1961.
65. Yoshizawa T., *Stability theory by Liapunov's second method*, Math. Soc. of Japan - Tokyo, 1966.
66. Hahn W., *Theory and applications of Liapunov's direct method*, Prentice-Hall Int. Series in Appl. Math. 1963.
67. Cesari L., *Asymptotic behaviour and stability problems in O.D.E.*, Springer-Verlag, Berlin (1959).

68. Zubov V.I., *Methods of A.M. Lyapunov and their applications*, Noordhoff L.T.D., Groningen, 1964.
69. Movchan A.A., *The direct method of Liapunov in stability problems of elastic systems*, J. Appl. Math. Mech. **23** (1959).
70. Kelvin, Lord, *On the stability of steady and of periodic fluid motion*, Phil. Mag. (5) **23**, 459-464; 529-539 (1887).
71. Joseph D.D., *Stability of fluid motions*, Vol. I-II Springer-Verlag (1976).
72. Galdi G.P., Rionero S., *Weighted Energy Methods in Fluid Dynamics and Elasticity*, Lect. Notes in Math. 1134, Springer-Verlag, 1985.
73. La Salle J.P., *The stability of Dynamical Systems*, C.B.M.S. Regional Conference Series in Appl. Math., S.I.A.M., Philadelphia, 1976.
74. Hale J.K., Dynamical systems and stability, *J. of Math. Anal. and Appl.*, 26, 39, 1969.
75. Dafermos,C.M., An invariance principle for compact processes, *J. Diff. Eqs.* 9, 239, 1971.
76. Dafermos, C.M., Uniform processes and semicontinuous Liapunov functionals, *J. Diff. Eqs.* 11, 401, 1972.
77. Dafermos, C.M., Slemrod, M., Asymptotic behaviour of nonlinear contraction semigroups, *J. Funct. Anal.* 13, 97, 1973.
78. Haraux, A., Comportment a l'infini pour certains systems dissipatifv nonlineaires, *Proc. Royal Soc. Edinburgh* 84 A, 213, 1979.
79. Baillon, J.B., Brezis, H., Une remarque sur le comportment asymptotique des semigroups non lineaires, *Houston J. Math.* 2, 5, 1976.
80. Ball, J.M., On the asymptotic behaviour of generalized processes with applications to nonlinear evolution equations, *J. Diff. Eqs.* 27, 224, 1978.
81. Slemrod, M., Weak asymptotic decay via "relaxed invariance principle" for a wave equation with nonlinear, non-monotone damping, *Proc. Royal Soc. Edinburgh,* 113A, 87, 1898.
82. Marcati, P., Abstract stability theory and applications to hyperbolic equations with time dependent dissipative force fields, *Comp. & Math. with Appl.* 12A, 541, 1986.
83. Marcati, P., Stability for second order abstract evolution equations, *Nonlinear Anal. Theory, Math. Appl.* 8, n.3, 237, 1984.
84. Salvadori, L., *Famiglie ad un parametro di funzioni di Liapunov nello studio della stabilitá*, Symp. Math. 4, Academic Press, 1971.
85. Salvadori, L., Uso di due indici nel problema della stabilitá, *Conf. Sem. Mat. Univ. Bari* 146, 1977.
86. Fielder, B., Mallet-Paret J., A Poincaré-Bendixon theorem for scalar reaction-diffusion equations, *Arch. Rat. Mech. Anal.* 107, 325, 1989.
87. Ladyzenskaja O.A., Solonnikov V.A., *Linear and quasilinear equations of parabolic type*, Vol. 23, Transl. Math. Monogr. A.M.S., Providence, Rhode Island (1968).

88. Vaghi C., *Soluzioni limitate o quasi periodiche di tipo parabolico non lineare*, Bollettino U.M.I., Serie IV, Anno I,n.4-5, 559 (1968).
89. Amann H., Quasilinear parabolic systems under nonlinear boundary conditions, *A.R.M.A.*, 92, 186; Dynamic theory of quasilinear parabolic equations II. Reaction-diffusion systems, *Diff. & Integr. Equ.*, vol. 3, 1, 1990; Dynamic theory of quasilinear parabolic systems III. Global existence, *Math. Z.* 202, 1989.
90. Maiellaro M., Rionero S., On the stability of Couette-Poiseuille flows in the anisotropic M.H.D. via the Liapunov Direct method, *Rend. Acc. Sc. Fisiche Matem. Napoli*, 1995 (to appear).
91. Okubo, A., *Diffusion and Ecological problems: Mathematical Models*, Biomathematics vol.10, Springer-Verlag, 1980.
92. Bear, J., *Dynamics of fluids in porous media*, Elsevier, 1972.
93. Anvarbek M. Meirmanov, *The Stefan Problem*, De Gruyter Expositions in Mathematics 3, Walter de Gruyter, Berlin, New York, 1992.
94. Aronson, D., Crandall, M.G., Peletier, L.A., Stabilization of solutions of a degenerate nonlinear diffusion problem, *Nonlinear Analysis, Theory, Methods and Applications*, vol.6, n.10, 1001, 1982.
95. Benilan P., Crandall M.G., The Continuous dependence of solutions of $u_t - \Delta(\varphi(u)) = 0$, *Indiana Un. Math.* vol. 30, n.2, 1, 1981.
96. Benilan P., Crandall M.G., Pierre M., Solutions of the porous medium equation in $I\!R^N$ optimal conditions on initial data, *Indiana Un. Math.* vol. 33, n.1, 51, 1984.
97. Di Benedetto, E., *Degenerate Parabolic Equations*, Springer-Verlag, Universitext, 1993.
98. Sobolev, S.L., *Applications of functional analysis in Mathematical Physics*, American Math. Soc., Providence, Rhode Island, 1983. Transl. of Math. Monographs, vol. 7.
99. Knowles, J.K., On Saint-Venant's principle in the two-dimensional linear theory of elasticity, *Arch. Ration. Mech. Anal.,* 21, 1, 1965.
100. Toupin, R.A., Saint-Venant's principle, *Arch. Ration. Mech. Anal.*, 18, 83, 1965.
101. Horgan, C.O. and Knowles, J.K., Recent developments concerning Saint-Venant's principle, *Adv. in Appl. Mech.,* 23, 179, 1983.
102. Horgan, C.O., Recent developments concerning Saint-Venant's principle: an update, *Appl. Mech. Rev.,,* 42, 295, 1989.
103. Flavin, J.N., Knops, R.J., and Payne, L.E., Decay estimates for the constrained elastic cylinder of variable cross-section, *Quart. Appl. Math.*, 47, 325, 1989.
104. Horgan, C.O. and Payne, L.E. Decay estimates for a class of nonlinear boundary value problems in two dimensions, *SIAM J. Math. Anal.,* 20, 782, 1989.
105. Flavin, J.N., Knops, R.J., and Payne, L.E., Asymptotic behaviour of solutions to semi-linear elliptic equations on the half-cylinder, Z. Angew. Math. Phys., 43, 405, 1992.

106. Knowles, J.K., On the spatial decay of solutions of the heat equation, Z. Angew. Math. Phys., 22, 1050, 1971.
107. Bellman, R.A., A property of summation kernels, *Duke Math. J.*, 15, 1013, 1948.
108. Beckenbach, E.F. and Bellman, R.A., *Inequalities*, Springer-Verlag, Berlin - Heidelberger - New York, 1965, Chapter 4.
109. Kalashnikov, A.S., Some problems of the qualitative theory of non-linear degenerate second-order parabolic equations, *Russian Math. Surveys*, 42:2, 169, 1987.
110. Peletier, L.A., The porous medium equation, in *Applications of non-linear analysis in the physical sciences,* Pitman, London, 1981, 229.
111. Levine, H.E. and Payne, L.E., Non existence theorems for the heat equation with nonlinear boundary conditions and for the porous medium equation backward in time, *Jour. Diff. Eq.*, 16, 319, 1974.
112. Degtyarev, L.M., Zakharov, V.E., and Rudakov, L.I., Two examples of Langmuir wave collapse, *Soviet Phys. JEPT*, 41, 57, 1975.
113. Glassey, R.T., On the blowing up of solutions to the Cauchy problem for the nonlinear Schrödinger equation, *J. Math. Phys.*, 18, 1794, 1977.
114. Knops, R.J., Levine, H.A., and Payne, L.E., Nonexistence, instability and growth theorems for solutions of a class of abstract nonlinear equations with applications to non-linear elastodynamics, *Arch. Rational Mech. Anal.*, 55, 52, 1974.
115. Levine, H.A., Instability and nonexistence of global solutions to nonlinear wave equations of the form $Pu_{tt} = -Au + F(u)$, *Trans. Amer. Math. Soc.*, 192, 1, 1974.
116. Knops, R.J. and Straughan, B., Non-existence of global solutions to nonlinear Cauchy problems arising in mechanics, in *Trends in Applications of Pure Mathematics to Mechanics*, Pitman, London, 1976, 187.
117. Levine, H.A., The role of critical exponents in blow up theorems, *SIAM Review*, 32, 262, 1990.
118. Agmon, S:, Uricité et convexité dans les problémes differentiels, *Sem. Math. Sup.*, University of Montreal Press, 1966.
119. Agmon, S. and Nirenberg, L., Lower bounds and uniqueness theorems of differential equations in a Hilbert space, *Comm. Pure Appl. Math.*, 20, 207, 1967.
120. Levine, H., Logarithmic convexity and the Cauchy problem for abstract second order differential inequalities, *J. Diff. Eqns.*, 8, 34, 1969.
121. Knops, R.J. and Payne, L.E., On the stability of solutions to the Navier-Stokes equations backward in time, *Arch. Rational Mech. Anal.*, 29, 331, 1968.
122. Knops, R.J. and Payne, L.E., Continuous data dependence for the equations of classical elastodynamics, *Proc. Cambridge Philos. Soc.*, 66, 481, 1969.
123. Knops, R.J. and Payne, L.E., Growth estimates for solutions of evolution-

ary equations in Hilbert space with applications to elastodynamics, *Arch. Rational Mech. Anal.*, 41, 363, 1971.

124. Levine, H., Continuous data dependence, regularization and three lines theorem for the heat equation with data in a space like direction, *Ann. de Mathematiques Pures et App.*, 134, 267, 1983.
125. Galdi, G.P. and Rionero, S., Continuous data dependence in linear elastodynamics on unbounded domains without definiteness conditions on the elasticities, *Proc. Royal Soc. Edinburgh*, 93A, 299, 1983.
126. Payne, L.E. and Straughan, B., Comparison of viscous flow backward in time with small data, *Int. J. Nonlinear Mech.*, 24, 209, 1989.
127. Payne, L.E. and Straughan, B., Order of convergence estimates on the interaction term for a micropolar fluid, *Int. J. Engng. Sci.*, 27, 837, 1989.
128. Payne, L.E. and Straughan, B., Effects of errors in the initial-time geometry on the solution of the heat equation in an exterior domain, *Q. J. Mech. Appl. Math.*, 43, 75, 1990.
129. Payne, L.E. and Straughan, B., Improperly posed and non-standard problems for parabolic differential equations in *Elasticity, Mathematical Methods and Application: The Ian Sneddon 70th Birthday Volume* (eds. G.Eason and R.W.Odgen), Ellis Horwood, 1990.
130. Straughan, B., Continuous dependence on the heat source and non-linear stability in penetrative convection, *Int. J. Non-Linear Mechanics*, 26, 221, 1991.
131. Protter, M.H. and Weinberger, H., *Maximum Principles in Differential Equations*, Prentice-Hall, New York, 1967.
132. Horgan, C.O., Payne, L.E., and Wheeler, L.S., Spatial decay estimates in transient heat conduction, *Q. Appl. Math.*, 42, 119, 1984.
133. Flavin, J.N., Knops, R.J., Some spatial decay estimates in continuum dynamics, *J. Elasticity*, 17, 249, 1987.
134. Flavin, J.N., Knops, R.J., Some convexity considerations for a two-dimensional traction problem, *Z. Angew. Math. Phys.*, 39, 166, 1987.
135. Flavin, J.N., Upper estimates for a class of non-linear partial differential equations, *J. Math. Anal. Appl.*, 144, 128, 1989.
136. Flavin, J.N., Rionero, S., Decay and other estimates for an elastic cylinder, *Q. J. Mech. Appl. Math.*, 46, 299, 1993.
137. Carleman, T., Sur un inégalité différentielle dans le théorie des fonctions analytiques, *C.R. Acad. des Sci. de Paris*, 196, 995, 1933.
138. Ericksen, J.L., Problems for infinite elastic prisms; Saint-Venant problem for elastic prisms, in *System of Nonlinear Partial Differential equations*, Ball, J.M. ed., University Press, Oxford, 1982, 81.
139. Knowles, J.K., Remarks on a question of Ericksen concerning elastostatic fields of Saint-Venant type, *Arch. Rational Mech. Analysis*, 90, 249, 1985.
140. Payne, L.E., On the stability and growth of solutions for classes of initial and boundary value problems, in *Mathematical methods and Models in Mechan-*

ics, Banach Publication No.15, 1985, 465.

141. Brun, L., Sur l'unicité en thermoélasticité dynamique et diverses expressions analogues a la formule de Clapeyron, *C.R. Acad. Sci. Paris*, 2584, 1965.
142. Brun,L., Méthodes énergétiques dans les systémes évolutifs linéaries, Premier partie: Séparation des énergies. Deuxiéme partie: Theoremes d'unicité, *J. Mecanique*, 8, 125, 1969.
143. Knops, R.J., Logarithmic convexity and other techniques applied to problems in continuum mechanics, in *Symp. on Non-well-posed Problems and Logarithmic Convexity*, Knops, R.J., Ed., Springer Lecture Notes 316, 31, 1973.
144. Knops, R.J. and Payne, L.E., Growth estimates for solutions of evolutionary equations in Hilbert space with applications to elastodynamics, *Arch. Rational Mech. Anal.*, 41, 363, 1971.
145. Payne, L.E., Some general remarks on non-well-posed problems for partial differential equations, in *Symp. on Non-well-posed Problems and Logarithmic Convexity*, Knops, R.J., Ed., Springer Lecture Notes 316, 1, 1973.
146. Maxwell, J.C., *Theory of Heat*, IVTA Edition, 1891, 264.
147. Levine, H.A., An equipartition of energy theorem for weak solutions of evolutionary equations in Hilbert space: the Lagrange identity method, *J.Differential Equations*, 24, 197, 1977.
148. Moravetz C.S., *Notes on time decay and scattering for some hyperbolic problems.* SIAM Regional Conf. Series in Applied Mathematics, Vol. 19, 1975.
149. Adams, R.A., *Sobolev Spaces*, New York, Academic Press, 1975.
150. Maz'ja V.G., *Sobolev Spaces*, Springer-Verlag Berlin Heidelberg New York Tokyo, 1985.
151. Straughan B., *Mathematical aspects of penetrative convection*, Pitman Research Notes in Math. Series, 288, 1993.
152. Murray A., Uniqueness and continuous dependence for the equations of elastodynamics without strain energy function, *Arch. Rat. Mech. Anal.*, 47, 195, 1972.
153. Lees M., Protter M.H., Unique continuation for parabolic differential equations and differential inequalities, *Duke Math. Journ.*, 28, 369, 1961.
154. Serrin J., *The initial value problem for the Navier-Stokes equations*, in "Nonlinear problems", Madison: Univ. Wiscounsin Press, 1963.
155. Rionero S., Galdi P.G., On the uniqueness of viscous fluid motions, *Arch. Rational Mech. Anal.*, 62, 295, 1975.
156. Rionero S., Continuous dependence and stability for nonlinear dispersive and dissipative Waves, *Rend. Sim. Mat. Univ. Padova*, 68, 269, 1982.
157. Rionero S., Salemi F., On some weighted Poincaré inequalities, *Fisica Matematica Suppl. U.M.I.* vol IV-5, n.1, 1985.
158. Capone F., De Angelis M., On the energy stability of fluid motions in exterior of a sphere under free boundary like conditions, *Rend. Acc. Sc. Fis. Mat.* serie IV, vol. LX, 1993.

159. Brezis H., *Analyse functionnelle-Theorie et applications*, Masson Ed., Paris, 1983.
160. Ladyzhenskaya O.A., *The mathematical theory of viscous incompressible flow*, Gordon and Breach, 1959.
161. Flavin J.N., Knops R.J., Some decay and other estimates in two-dimensional linear elastostatics, *Q.J. Mech. Appl. Math.*, 41, 223, 1988.
162. Galdi G.P., Rionero S., On the well-posedeness of the equilibrium problem for linear elasticity in unbounded regions, *Q.J. Mech. Appl. Math.*, 10, 333, 1980.
163. Galdi G.P., Knops R.J., Rionero S., Asymptotic behaviour in the nonlinear elastic beam, *Arch. Rational Mech. Anal.*, 87, 305, 1985.
164. Flavin J.N., Upper estimates for an elastostatic displacement problem, *Proc. R. Ir. Acad.*, 89, 175, 1989.
165. Sokolnikoff I.S., *Mathematical Theory of Elasticity* (Second Edition), McGraw - Hill, New York, 1956.
166. Flavin J.N., On Knowles' version of Saint-Venant's principle in two-dimensional elastostatics, *Arch. Rational Mech. Anal.*, 53, 266, 1974.
167. Knowles J.K., An energy estimate for the biharmonic equation and its application to Saint-Venant's principle in plane elastostatics, *Indian J. Pure Appl. Math.* 14, 791, 1983.
168. Vafeades P., Horgan C.O., Exponential decay estimates for the solutions of the von Kármán equations on a semi-infinite strip, *Arch. Rational Mech. Anal.* 104, 1, 1988.
169. Oleinik O.A., Yosifian G.A., On Saint-Venant's principle in plane elasticity theory, *Dokl. Akad. SSR*, 239, 530, 1978 [*Sov Math. Dokl.*, 19, 364].
170. Oleinik O.A., Yosifian G.A., The Saint-Venant's principle in the two-dimensional theory of elasticity and boundary value problems for a biharmonic equation in unbounded domains, *Sibirsk Mat. Z.*, 19, 1154, 1978 [Siberian Math. J., 19, 813].
171. Fichera G., Il principio di Saint-Venant: Intuizione dell'ingegnere e rigore del matematico, *Rend. di Matematica Roma*, V10, s.6, 181, 1977.
172. Oleinik O.A., Yosifian G.A., On the asymptotic behaviour at infinity of solutions in linear elasticity, *Arch. Rational Mech. Anal.*, 78, 29, 1982.
173. Oleinik O.A., Yosifian G.A., Saint-Venant's principle for the mixed boundary value problem of the theory of elasticity and its applications, *Dokl. Akad. Nauk SSSR*, 233, 824, 1977.
174. Oleinik O.A., Yosifian G.A., A priori estimates for solutions of the first boundary value problem for the theory of elasticity system of equations and their applications, *Uspehi Mat. Nauk*, 32, 197, 1977.
175. Berdichevski V.L., On the proof of the Saint-Venant's principle for bodies of arbitrary shape, *Prikl. Math. Mekn.*, 38, 851, 1974 [*J. Appl. Math. Mech.* 38, 799, 1975].

176. Knops R.J., Payne L.E., Rionero S., Saint-Venant's principle on unbounded regions, *Proc. R. Soc. Edinburgh A*, 115, 319, 1990.
177. Flavin J.N., Knops R.J., Payne L.E., Asymptotic and other estimates for a semi-linear elliptic equation in a cylinder, *Q.J. Mech. Appl. Math.*, 45, 617, 1992.
178. Berestycki H., Niremberg L., Some qualitative properties of solutions to semi-linear elliptic equations in cylindrical domains. Publications du Laboratoire d'Analyse Numerique (Université Pierre et Marie Curie, Paris 1988).
179. Breuer S., Roseman J.J., Decay theorems for nonlinear Dirichlet problems in semi-infinite cylinders, *Arch. Rati. Mech. Anal.*, 94, 363, 1986.
180. Kirchgassner K., Scheurle J., On the bounded solutions of a semilinear elliptic equation in a strip, *J. Diff. Equations*, 32, 119, 1979.
181. Horgan C.O., Payne L.E., Decay estimates for second-order quasilinear partial differential equations, *Advances in Appl. Math.*, 5, 309, 1984.
182. Horgan C.O., Payne L.E., On the asymptotic behaviour of inhomogeneous second-order quasi-linear partial differential equations, *Quart. Appl. Math.*, 47, 753, 1989.
183. Horgan C.O., Payne L.E., A Saint-Venant's principle for a theory of nonlinear plane elasticity, *Quart. Appl. Math.*, 50, 641, 1992.
184. Mora X., Semilinear parabolic problems define semiflows on C^k spaces, *Trans. Am. Math. Soc.*, 278, 21, 1983.
185. Markus L., Asymptotically autonomous differential system. Contribution to the theory of Nonlinear oscillations, *Ann. Math. Studies*, 3, 17, 1956.
186. Karciva P.M., Local movement in herbivorous insects: applying in passive diffusion model to mark-recapture field experiments, *Oecologica (Berline)*, 57, 322, 1983.
187. Aronson D.G., Density-dependent interaction-diffusion systems. In: *Dynamics and Modelling of Reactive Systems*. New York; Acad. Press, 161, 1980. (W.E.Stewart, W.H.Ray, C.C.Couley editors).
188. Newman W.I., Some exact solutions to a nonlinear diffusion problem in population denetics and combustion, *J. Theor. Biol.*, 85, 325, 1980.
189. Serrin J., On the stability of viscous fluid motions, *Arch. Rational Mech. Anal.*, 3, 1, 1959 B.
190. Galdi G.P., Padula M., A new approach to energy theory in the stability of fluid motion, *Arch. Rat. Mech. Anal.* 110, 187, 1990.
191. Berker R., Integration des equations du mouvement d'un fluide visqueux incompressible, *Handbuch der Physik* Bd Viii/2, 1, 1963.
192. Rionero S., Metodi variazionali per la stabilitá asintotica in media in magnetoidrodinamica, *Ann. Mat. Pura Appl.*, 78, 339, 1968.
193. Chandrasekhar S., *Hydrodynamic and Hydromagnetic stability*, Dover, New York, 1981.

194. Davis S.H., von Kerczek C., A reformulation of Energy stability theory, *Arch. Rat. Mech. Anal.*, 52, 112, 1973.
195. Rionero S.: Energy fluid motions stability for free boundary like problems in the exterior of starshaped domains, in *Proc. of the workshop: Energy Methods for free boundary problems in Continuum Mechanics*, Oviedo, Spain, March 21-23 1994, Kluwer Academic Publishers (to appear).
196. Mulone G., Rionero S., On the stability of the rotating Bènard problem, *Bull. Tech. Univ. Istanbul*, 47, 181, 1994.
197. Mulone G., Rionero S., On the non-linear stability of the rotating Bènard problem via the Lyapunov direct method, *J. Math. Anal. App.*, 144, 109, 1989.
198. Rionero S., On the choice of the Lyapunov function in the fluid motion stability, In *Energy stability and convection*, 168, Pitman Res. Notes in Math., 1988.
199. Rionero S., Mulone G., A nonlinear stability analysis of the magnetic Bènard problem through the Lyapunov direct method, *Arch. Rat. Mech. Anal.*, 103, 347, 1988.
200. Payne L., Weinberger H., *Non Linear Problems*, Ed. Langer R.E., Univ. of Wisconsin Press, 1963.
201. Gilbarg D., Trudinger N., *Elliptic Partial Differential Equations of Second Order*, Prentice-Hall, Englewood Cliffs, NJ. 1983.
202. Talenti G., Some estimates of solutions to Monge-Ampère equations in dimension two, *Ann. Scuola Norm. Pisa Cl. Sci.*, (4) 3, n.2, 183, 1981.
203. Efimov N.V., Generation of singularities of surfaces of negative curvature, *Mat. Sbornik*, 64, 286, 1964.
204. Rozendorn E., Approximate solution of the wind and pressure balance equation for anticyclones in tropical and subtropical latitudes, *Dokl. Akad. Nauk S.S.S.R.*, 3, 584, 1980.
205. Popivanov P., Kutev N., On the solvability of equations with prescribed Gauss curvature changing its sign, *Math. Nachr.*, 137, 159, 1988.
206. Caughey T.K., Ellison J., Existence, Uniqueness and Stability of a class of nonlinear P.D.Es., *Jour. Math. Anal. Appl.*, 51, 1, 1975.
207. Fitzgibbon W.E., Strongly Damped Quasi Linear Evolution Equation, *Jour. Math. Anal. Appl.*, 79, 1981.
208. Greenberg J., Mac Camy R.C., Mizel V., On the existence, uniqueness and stability of solutions of the equations $\sigma'(u_x)u_{xx} + \lambda u_{xxt} = \rho_0 u_{tyt}$, *J. Math. Mech.* 17, 707, 1968.
209. Levine H.A., Some uniqueness and growth theorems in the Cauchy problem for $Pu_{tt} + Mu_t + Nu = 0$ in Hilbert space, *Math. Z.* 126 (1972), 345-360.
210. Morro A., Payne L.E., Straughan B., Decay, growth, continuous dependence and uniqueness results of generalized heat conduction theories, *Appl. Anal.* 38, 1990.
211. Straughan B., Growth and instability for wave equations with dissipation,

with applications in contemporary continuum mechanics, *J. Math. Anal. Appl.* 61 (1977), 303-330.

212. Rionero S., Stability results for hyperbolic and parabolic equations, *Transport Theory and Statistical Physics* (to appear).
213. Logan J.D., *Applied Mathematics. A contemporary approach*, Wiley-Interscience Publication, 194, 1987.
214. Cattaneo C., *Atti Sem. Mat. Fis. Univ. Modena*, 3, 83, 1948.
215. Morro A., Ruggeri T., *Propagazione del calore ed equazioni costitutive.* Quaderno C.N.R. - G.N.F.M. Lessons given at the VIII Summer School of Math. Physics. Ravello, Villa Rufolo 12-30 Sept. 1984.
216. Arnold V.I., Conditions for nonlinear stability of the stationary plane curvilinear flows of an ideal fluid, *Doklady Mat. Nauk.* 162, 5, 773, 1965.
217. Arnold V.I., On a priori estimates in the theory of hydrodynamic stability, *English Transl.: A.M.S. Transl.* 19, 267-269, 1969.
218. Marchioro C., Pulvirenti M., *Mathematical theory of incompressible nonviscous fluids*, Appl. Math. Sciences vol. 93, Springer-Verlag, 1993.
219. Cercignani C., *Theory and application of the Boltzmann equation*, Scottish Ac.Press - Text in Mathematics 1975.
220. Waldmann L., *Transporterscheinangen in Gasen von mittlerem Druck*, Handbuch der Physik, vol.12, ed. S.Fluggen Springer-Verlag, Berlin, 1958.
221. Spiga G., Nonnenmacher T., Boffi V.C., Moment equations for the diffusion of the particles of a mixture via the scattering kernel formulation of the nonlinear Boltzmann equation, *Physica* 131A, 431, 1985.
222. Boffi V.C., Spiga G., Rigorous iterated solutions to a nonlinear integral evolution problem in particle transport theory, *J. Math. Phys.* 23, 2299, 1982.
223. Rionero S., Global solution to a nonlinear integral evolution problem in particle transport theory, *Atti Acc. Scienze Turin (Italy)* Suppl. vol. 120, 1986.
224. Kantarovic L.V., Akilov G.P., *Analisi funzionale*, Editori Riuniti, pp. 626, 1980.
225. Rionero S., Guerriero G., L_1-stability of the solutions to an integral evolution equation of the nonlinear particle transport theory, *Meccanica* 24, n.4, 1989.
226. Gutowski R., Investigation of the stability of transversal vibrations of a nonlinear string by the method of differential inequality, *Proc. V Int. Meeting on "Waves and stability in continuous media"*, Rionero S. editor, World Sc. Series on Adv. Math. Appl. Sc. 4, 168, 1989.
227. Knops R.J., Payne L.E., Some uniqueness and continuous dependence theorems for nonlinear elastodynamics in exterior domains, *Applicable Anal.* 15, 33, 1983.
228. Galdi G.P., Rionero S., Continuous data dependence in linear elastodynamics on unbounded domains without definiteness conditions on the elasticities, *Proc. Royal Soc. Edinburgh* 93A, 299, 1983.
229. Chirita S., Rionero S., Some continuous data dependence and uniqueness

results for nonlinear elastodynamics on an unbounded domain, *Ricerche di Matematica* 37, 2, 241, 1988.

230. Galdi G.P., Rionero S., Continuous dependence in linear elasticity on exterior domains, *Int. J. Engng. Sci.* 17, 521, 1979.
231. Knops R.J., Galdi G.P., Rionero S., Uniqueness and continuous dependence in the linear elastodynamic exterior and half-space problems, *Proc. Cambr. Philos. Soc.* 99, 357, 1986.
232. Hardy G.H., Littlewood J.E., Polya G., *Inequalities*, Cambridge University Press, 1959.
233. Gilbarg D., Trudinger N.S., *Elliptic partial differential equations of second order*, Springer-Verlag, Berlin, Heidelberg and New York, 1977.
234. Rektorys K., *Variational methods in mathematics, science and engineering*, D.Reidel Publ. Comp. Dordrecth-Holland/Boston U.S.A., 1975.
235. Sigillito V.G., *Explicit a priori inequalities with applications to boundary value problems*, Pitmann Publishing. London-San Francisco-Melbourne, 1977.

AUTHOR INDEX

SUBJECT INDEX